Negotiating in/visibility

Manchester University Press

Negotiating in/visibility

Women, science, engineering and medicine in the twentieth century

Edited by

Amelia Bonea and Irina Nastasă-Matei

MANCHESTER UNIVERSITY PRESS

Copyright © Manchester University Press 2025

While copyright in the volume as a whole is vested in Manchester University Press, copyright in individual chapters belongs to their respective authors, and no chapter may be reproduced wholly or in part without the express permission in writing of both author and publisher.

An electronic version of this book has been made freely available under a Creative Commons (CC BY-NC-ND) licence, thanks to the support of University of Manchester Library, which permits non-commercial use, distribution and reproduction provided the author(s) and Manchester University Press are fully cited and no modifications or adaptations are made. Details of the licence can be viewed at https://creativecommons.org/licenses/by-nc-nd/4.0/

Published by Manchester University Press
Oxford Road, Manchester, M13 9PL

www.manchesteruniversitypress.co.uk

British Library Cataloguing-in-Publication Data
A catalogue record for this book is available from the British Library

ISBN   978 1 5261 7838 1   hardback

First published 2025

The publisher has no responsibility for the persistence or accuracy of URLs for any external or third-party internet websites referred to in this book, and does not guarantee that any content on such websites is, or will remain, accurate or appropriate.

EU authorised representative for GPSR:
Easy Access System Europe, Mustamäe tee 50, 10621 Tallinn, Estonia
gpsr.requests@easproject.com

Typeset by Newgen Publishing UK

Copiilor noştri

Contents

List of figures — ix
List of tables — x
List of contributors — xi
Acknowledgements — xvii
List of abbreviations — xix
Foreword by Mariko Ogawa — xxi

Introduction: In/visible women, science, engineering and medicine in the twentieth century – Amelia Bonea and Irina Nastasă-Matei — 1

I Laboratory cultures: Visible scientific rebels, invisible innovators

1 Breaking down the barriers at Cambridge in the 1930s: Reinet Maasdorp's experience at Rutherford's Cavendish Laboratory – Kathryn Keeble — 39
2 'Your research is crap, do not bother to apply again': Female evolutionary biology theorists as scientific rebels and oppositional scientists – Nuala Proinnseas Caomhánach — 57

II In/visibilities across borders: Scientific collaborations and contestations

3 Inventing a career across borders in the early 1930s: The case of cytogeneticists Eileen W. Erlanson and E. K. Janaki Ammal – Savithri Preetha Nair — 79
4 Vlasta Kálalová Di-Lotti in Iraq: Medical practice and scientific research – Adéla Jůnová Macková — 103
5 Early years of the International Conference of Women Engineers and Scientists: Shaping transnational collaboration in the Cold War era, 1964–1975 – Emily Rees Koerner and Graeme Gooday — 125

III In/visibilities in medicine and care: Treating, teaching, reforming

6 'A model of devotion to the school': Female doctors in secondary schools in interwar Romania – Camelia Zavarache — 147
7 Women and the practice of Western medicine in late Republican China: Evidence from Sichuan – Jean Corbi — 172
8 Agency and coercion: Fighting 'women's illnesses' with grassroots science and medicine during the Great Famine in China, 1958–1962 – Kathryn Edgerton-Tarpley — 192

IV Intimate knowledge and in/visible domesticities: Science, medicine and the home

9 The curious case of Yashoda Devi, a woman Ayurvedic practitioner in colonial India – Saurav Kumar Rai — 217
10 Lady Irwin College: Domestic science post-secondary education for three women graduates in India – Anne Hardgrove — 236
11 Clara Park: A mother's intimate knowledge and child science – Marga Vicedo — 257

V Towards visible change? Publics, pedagogies and politics of science

12 The valuable 's': Publics and counterpublics of abortion and contraception in late-twentieth-century Greece – Evangelia Chordaki — 281
13 The power of autobiography: Documenting women scientists through a lecture series at the University of Illinois – Bethany G. Anderson and Kristen Allen Wilson — 301
14 How to do science as a woman and laugh? Insights and lessons from Hungary – Andrea Pető — 321

Select bibliography — 334
Index — 353

Figures

0.1	Support programmes for women in STEMM implemented by NSF (USA), EC (EU) and MEXT (Japan)	xxii
0.2	Nobel Prize women in science and international exchanges	xxix
1.1	Reinet Maasdorp and the Cavendish Research Group, Cambridge, 1935 (courtesy of Margaret Kettlewell)	40
3.1	Eileen Erlanson (later Macfarlane), dressed in Indian sari, undated (courtesy of the McCamic family)	95
4.1	Vlasta Kálalová Di-Lotti with nurse Marie Marianinová and son Radbor, Baghdad, 1928 (courtesy of the State Regional Archives in Trebon, Czech Republic)	115
9.1	List of 108 books published by Yashoda Devi (illustration from Yashoda Devi, *Samsar ka Nari Itihas (Bharat ka Nari Itihas)*, vol. 2, Allahabad, 1922)	223
9.2	Consequences of adultery (illustration from Yashoda Devi, *Vivah Vigyan Kamshastra: Anand Mandir*, Allahabad, n.d., p. 508)	228
11.1	Clara Park and her daughter Jessica, *c.*1960 (courtesy of Clara Park)	260
14.1	Geographer Jenő Cholnoky (1870–1950) with his students in a classroom at Pázmány Péter University on Museum Boulevard, Budapest, 1929 (photograph donated by Tamás Cholnoky, Fortepan 29841, open access, https://fortepan.hu/hu/photos/?id=29841)	322

Tables

0.1	Japanese female PhD holders prior to 1940, excluding PhDs in medicine (except Tada Urata)	xxvi
0.2	Japanese female PhD holders, 1905–45	xxvii
7.1	Physicians registered in Sichuan, 1937–47	176
7.2	Physicians practising in the hospitals and clinics surveyed in 1939	177
7.3	Physicians registered between 1929 and 1934	178
7.4	Male physicians more likely to graduate from a short-term medical course	180
7.5	Women as likely as men to graduate from a longer medical course	181
13.1	Count of how registrants learned about the lecture series for 2020–21 academic year	311
13.2	Number of participants who were staff, students, faculty, alumni and not affiliated with the University of Illinois for the 2020–21 academic year	312

Contributors

Kristen Allen Wilson is the Illinois Distributed Museum Coordinator. The museum documents innovations that have occurred on campus and leads visitors to important locations through a virtual platform. Kristen works with students, faculty and staff around campus to add new content to the website. Kristen is interested in creating interactive, self-guided digital museum experiences.

Bethany G. Anderson is the Natural and Applied Sciences Archivist at the University of Illinois Archives, and a PhD student in the Department of History at the University of Illinois Urbana-Champaign. In this role, she works with units across the University of Illinois campus to document the scientific enterprise. Bethany is especially interested in documenting women in science and technology, oral history and exploring computational approaches to digital archival materials.

Amelia Bonea is Lecturer in Global History of Science, Technology and Medicine at the Centre for the History of Science, Technology and Medicine, University of Manchester. Her first monograph, *The News of Empire: Telegraphy, Journalism, and the Politics of Reporting in Colonial India, c.1830–1900* (Oxford University Press, 2016), was awarded the 2017 Eugenia M. Palmegiano Prize by the American Historical Association. A second book, *Anxious Times: Medicine and Modernity in Nineteenth-Century Britain*, was co-authored with Melissa Dickson, Sally Shuttleworth and Jennifer Wallis, as part of the ERC-funded project 'Diseases of Modern Life' (2014–2019) at the University of Oxford. Her current work examines the global entanglements of Earth and environmental sciences in South and East Asia in the twentieth century and the role of women therein.

Nuala Proinnseas Caomhánach is a PhD candidate in the Department of History at New York University and Research Associate in the Invertebrate Department at the American Museum of Natural History. Her research examines the relationship between scientific knowledge, climate change and conservation law in Madagascar, illuminating how changes in botanical science have affected international conservation ideology, policy and practice. Nuala is a contributing editor at the *Journal of the History of Ideas* blog and Broadly Speaking series. She co-curated the *Black Botany: The Nature of Black Experience* exhibition as a Mellon Fellow at the New York Botanical Garden and co-produces the Not That Kind of Doctor podcast, which invites PhD students to discuss their research in an informal manner.

Evangelia Chordaki is a Postdoctoral Research Fellow at the Seeger Center for Hellenic Studies at Princeton University. She is a historian of science with a PhD in Science Communication and Gender Studies (2022), funded by the Hellenic Foundation for Research and Innovation. Her research interests include feminist epistemologies and the intertwinement of gender, technoscience and digitality, with an emphasis on cyberfeminism and xenofeminism. Her work also focuses on the communication of the Covid-19 pandemic, especially the impact of epistemic uncertainty, scientific expertise and boundary work on the performativities and practices of science communication. Since 2023, Evangelia is a research associate at the National Hellenic Research Foundation.

Jean Corbi is a PhD student in History at Sciences Po Paris, where he also teaches contemporary history. His research interests focus on the relationship between the state and the emerging medical profession in late Republican China, with an emphasis on Sichuan Province between 1928 and 1949.

Kathryn Edgerton-Tarpley is Professor of Late Imperial and Modern Chinese History at San Diego State University. She is the author of *Tears from Iron: Cultural Responses to Famine in Nineteenth-Century China* (University of California Press, 2008) and multiple articles on famines and floods. Her current book project analyses changes and continuities in Chinese responses to calamity by employing case studies of three major famines that struck North China under governments with markedly different ideological foundations.

Graeme Gooday is Professor of History of Science and Technology at the School of Philosophy, Religion and History of Science, University of Leeds. He has published widely on the social histories of measurement, electric lighting, patenting, energy consumption and hearing loss. More recently, he has been working collaboratively on the transnational history of women in engineering, building upon the AHRC-funded public engagement project 'Electrifying Women' (2019–2020).

Anne Hardgrove has a PhD from the Inter-Departmental Program in Anthropology and History from the University of Michigan, Ann Arbor. She is the author of *Community and Public Culture: The Marwaris in Calcutta, 1897–1997* (Columbia University Press, 2007). Her research focuses on gender, sexuality and colonialism in India and world history. Hardgrove is Associate Professor of History at the University of Texas at San Antonio.

Adéla Jůnová Macková studied economy and social history at the Faculty of Arts, Charles University, Prague. She works as a scientist at the Masaryk Institute and Archives of the Czech Academy of Sciences. Her research focuses on Czechoslovak relations with the Middle East, economic history, the history of science and historical correspondence.

Kathryn Keeble lectures in diverse areas of literary production, including literature and its philosophical contexts and how art and literature represent otherwise inexpressible conditions such as the experiences of trauma in wars. She also teaches study skills and critical thinking. Dr Keeble writes on the history of science, literature and the arts, biography, Australian literature, film and arts discourse. Published in the *Journal of Australian Studies*, *Historical Records of Australian Science*, *Antithesis* and various edited volumes for Lexington and Cambridge Scholars Press, she is also an academic peer reviewer for the *Quarterly Journal of Film and Television* and Bloomsbury and an editorial board member of the arts journal *Double Dialogues*. As an arts reviewer, she regularly writes on art forms as diverse as dance, opera, jazz, theatre, comedy and circus.

Savithri Preetha Nair received her PhD from the School of Oriental and African Studies (SOAS), University of London, for a dissertation on the museum and the shaping of the sciences in colonial India. Nair's research focuses on the history of science, modernity and enlightenment at the turn of the nineteenth century, the history and politics of collecting for science, sociology of knowledge, the public museum and women in science in colonial and post-colonial India. Her publications include *Science and the Changing Environment in India: A Guide to Sources in the India Office Records, 1780–1920* (co-authored with Richard Axelby, British Library, 2010) and *Raja Serfoji II: Science, Medicine and Enlightenment in Tanjore, 1786–1832* (Routledge, 2012) as well as several papers in peer-reviewed international journals and edited volumes. Nair's recently published biography of Indian cytogeneticist Edavaleth Kakkat Janaki Ammal (1897–1984) is the first archive-based study of the life and science of an Asian woman scientist. Nair is an independent scholar and divides her time between London and Kerala.

Irina Nastasă-Matei is a historian and political scientist working on transnational networks of education and science exchange, especially in the context of authoritarian regimes. Her publications include *Educație, politică și propagandă. Studenții români în Germania nazistă* [*Education, Politics and Propaganda: Romanian Students in Nazi Germany*] (Școala Ardeleană/Eikon, 2016) and, as co-author, *Cultură și propagandă. Institutul Român din Berlin (1940–1945)* [*Culture and Propaganda: The Romanian Institute in Berlin (1940–1945)*] (Editura Mega, 2018, also available in German). Her most recent projects investigate public health communication, especially surrounding the production and administration of vaccines and their instrumentalization as tools of public diplomacy and soft power by the Romanian communist regime, as well as Romania's academic and scientific exchanges with other socialist countries during the Cold War. Irina is Senior Lecturer in the Department of Political Sciences, University of Bucharest.

Mariko Ogawa is Professor Emerita of History and Philosophy of Science at Mie University and Executive Director of the Tokai Foundation for Gender Studies in Nagoya. She is the author of numerous publications, including *Feminizumu to kagaku/gijutsu* [*Feminism and Science/Technology*] (Iwanami Shoten, 2001), *Yomigaeru Darwin* [*Darwin Redux*] (Iwanami Shoten, 2003) and *Byōgenkin to kokka* [*Germs and the State*] (Nagoya University Press, 2016). Co-translations include five of Londa Schiebinger's major works and the European Commission's *Women in Industrial Research: A Wake Up Call for European Industry* (European Communities, 2003). Mariko has received the Sawayanagi Masataro Memorial Tohoku University Gender Equality Award (2017) and the Prime Minister's Commendation for Gender Equality (2022).

Andrea Pető is a historian and Professor in the Department of Gender Studies at the Central European University, Vienna, a Research Affiliate at the CEU Democracy Institute, Budapest, and a Doctor of Science of the Hungarian Academy of Sciences. Her works on gender, politics, the Holocaust and war have been translated into twenty-three languages. She was awarded the 2018 All European Academies (ALLEA) Madame de Staël Prize for Cultural Values and the 2022 University of Oslo Human Rights Award. She is Doctor Honoris Causa of Södertörn University, Stockholm. Recent publications include *The Women of the Arrow Cross Party: Invisible Hungarian Perpetrators in the Second World War* (Palgrave Macmillan, 2020) and

Forgotten Massacre: Budapest 1944 (De Gruyter, 2021). She writes op-ed pieces on academic freedom and illiberal higher education for many international and national media outlets.

Saurav Kumar Rai completed his MPhil and PhD in the History Department, University of Delhi, on the social history of health and medicine, under the supervision of the late Dr Biswamoy Pati. His monograph, *Ayurveda, Nation and Society: United Provinces, c.1890–1950*, was published with Orient BlackSwan in 2024, as part of the York University Series 'New Perspectives in South Asian History'. He has several articles to his credit in reputed journals, including *Indian Economic and Social History Review*, *History and Sociology of South Asia* and *Social Scientist*. Previously a senior research assistant at the Nehru Memorial Museum and Library, Saurav is currently Research Officer at the Gandhi Smriti and Darshan Samiti, New Delhi.

Emily Rees Koerner is a historian specializing in gender and technology and Visiting Fellow at the School of Philosophy, Religion and History of Science, University of Leeds. Her doctoral research, undertaken at the University of Nottingham, was on the domestication of television in Britain, with a particular focus on the gendered aspects of this process. She was the postdoctoral research assistant on the AHRC-funded public engagement project 'Electrifying Women' (2019–2020) on the history of women in engineering. Her most recent research continues this theme within a transnational context.

Marga Vicedo is Professor of the History of Science in the Institute for the History and Philosophy of Science and Technology (IHPST) at the University of Toronto. She has published extensively on the history of genetics, animal behaviour, attachment theory and autism in journals like *Isis*, *British Journal for the History of Science*, *History of Psychiatry*, *Journal of the History of Biology* and *Journal of the Behavioral Sciences*. She is the author of *The Nature and Nurture of Love: From Imprinting to Attachment in Cold War America* (Chicago University Press, 2013). Her most recent book, *Intelligent Love: The Story of Clara Park, Her Autistic Daughter and the Myth of the Refrigerator Mother* (Beacon Press, 2022), was awarded the History of Science Society 2022 Watson Davis and Helen Miles Davis Prize for a history of science book that effectively appeals to general readers.

Camelia Zavarache's research interests focus on public education, gender studies, childhood studies and family history. She received her PhD in 2016 for a dissertation on 'Moral, Health and Marriage Education of the Young Generations in Romania, 1918–1939' and has since been Research

Assistant at the Nicolae Iorga Institute of History in Bucharest. After a fellowship at the École des hautes études en sciences sociales, Paris, she joined the 'Fragmented Modernities' project, financed through a grant from the Romanian National Authority for Scientific Research and Innovation, CNCS-UEFISCDI. Since 2020, Camelia has been a member of the COST Action transdisciplinary network 'Women On the Move', Working Group 3: 'Labour, Belonging and Economy'.

Acknowledgements

We would like to thank all the participants in the 'Hidden Histories: Women and Science in the Twentieth Century' virtual conference convened with the institutional support of the Karl Jaspers Centre for Transcultural Studies, University of Heidelberg, and the Department of Political Sciences, University of Bucharest, in May 2021. We also extend our gratitude to the Deutsche Forschungsgemeinschaft (German Research Foundation), for funding the conference through Amelia Bonea's 'Archives of the Earth: Fossils, Science and Historical Imaginaries in Twentieth-Century India' project (Grant Agreement No. 423157196). The University of Manchester Library generously agreed to cover publication costs through its Open Access Monograph Competition scheme. Our heartfelt thanks to everyone at Manchester University Press who believed in the idea of this edited volume, especially Meredith Carroll, Humairaa Dudhwala and the anonymous reviewers for their detailed and constructive feedback. We are also grateful to Lizzie Evans and Dan Shutt, who oversaw the production of the book and copy-edited the manuscript, respectively.

The volume also owes its current shape to several colleagues and friends who generously offered support at various stages, including reading drafts, suggesting useful references, double-checking transliterations and engaging in extensive conversations about women in science: Özlem, Temitope and Pelumi Deniz Akinbosoye-Zaimoğlu, Kathryn Edgerton-Tarpley, Graeme Gooday, Aya Homei, Vladimir Janković, Luciana Jinga, Jaehwan Hyun, Savithri Preetha Nair, Mariko Ogawa, Andrea Pető, Zoltán Rostás, Ashok Sahni, Sally Shuttleworth and Meng Zhang. We are also indebted to Emily Rees Koerner for bringing to our attention the wonderful photograph taken at the Third International Conference of Women Engineers and Scientists held in Turin, Italy, in 1971, and to the Institution of Engineering and Technology Archives/Women's

Engineering Society for granting permission to feature it on this volume's cover.

We dedicate this volume to our children, Armina, Petra and Vlad, in the hope that by the time they reach adulthood, the inequities discussed here will have become obsolete and irrelevant.

Abbreviations

ACWF	All-China Women's Federation
AIWC	All India Women's Conference
AKPR	Archiv kanceláře prezidenta republiky (Archive of the Office of the President of the Republic), Prague
ANIC	Arhivele Naţionale Istorice Centrale (National Archives of Romania, Central Branch), Bucharest
A.Sc.W.	Association of Scientific Workers
BCP	British Communist Party
BHL, UoM	Bentley Historical Library, University of Michigan
BMJ	*British Medical Journal*
CCP	Chinese Communist Party
CSAWG	Cambridge Scientists' Anti-War Group
DAC	Delfis Archival Center, Athens
EPMEWSE	The Japan Inter-Society Liaison Association Committee for Promoting Equal Participation of Men and Women in Science and Engineering
GLF	Great Leap Forward, 1958–62
ICWES	International Conference of Women Engineers and Scientists
IDM	Illinois Distributed Museum
IFUW	International Federation of University Women
LAPNP	Památníku národního písemnictví (Museum of Czech Literature), Prague
MEXT	Ministry of Education, Culture, Sports, Science and Technology, Japan
MTA	Magyar Tudományos Akadémia (Hungarian Academy of Sciences)
MÚA, AÚTGM	Masarykův ústav a Archiv AV ČR, Archiv Ústavu Tomáše Garrigua Masaryka (Masaryk Institute and Archives of the Czech Academy of Sciences, Archive of the Institute of Tomáš Garrigue Masaryk), Prague

MWIA	Medical Women's International Association
NA	Národní Archiv (National Archives of the Czech Republic)
NSAC	National Society for Autistic Children
NSF	National Science Foundation, United States
PUMC	Peking Union Medical College
SJWS	Society of Japanese Women Scientists (Nihon josei kagakusha no kai)
SMA	Shanghai Municipal Archives
SOkA	State Regional Archives, Písek, Czech Republic
SPA	Sichuan Provincial Archives
STEMM	Science, Technology, Engineering, Mathematics and Medicine
SWE	Society of Women Engineers, United States
SWG GRI	Standing Working Group on Gender in Research and Innovation, European Research Area and Innovation Committee
UBL	University Botanical Laboratory, University of Madras, Chennai
U of I Archives	University of Illinois Archives
TCM	Traditional Chinese Medicine
WCA	Wuwei County Archives
WCCU	West China Union University (Huaxi xiehe daxue)
WES	Women's Engineering Society, United Kingdom

Foreword

Mariko Ogawa

What is the significance of the twentieth century for women and science? This question was at the heart of the joint conference convened in spring 2021 by scholars at the Universities of Heidelberg and Bucharest. Held virtually at the height of the Covid-19 pandemic, the conference brought together researchers from different parts of the world to discuss hidden histories of women in science. The present volume is the outcome of that conversation. In this foreword, I would like to build upon my keynote speech to address the above question not necessarily by discussing individual scientists, but rather the changing activities of women scientists in the twentieth century. In this way, I hope to set the ground for the rich discussion of case studies the volume showcases.

Strong programmes to support women researchers in science, technology, engineering, mathematics and medicine (STEMM) have been implemented in the United States since 1980 and in the European Union since around 2000 (Figure 0.1). Until the end of the twentieth century, support programmes for women researchers consisted primarily of grants and other types of assistance for individual researchers. However, by the end of twentieth century, it was recognized that changes in scientific culture and organizations were necessary to increase the number of women researchers.[1] Therefore, in the US and the EU, additional funds were provided to universities and research institutions with the expectation of organizational change. In 2006, Japan also launched a model development project to increase the number of women researchers in universities and research institutions.

Before turning our attention to a few important episodes relating to the history of women in STEMM fields during the twentieth century, it may be useful to reflect on the meaning of this century for women researchers.[2] If the twenty-first century is often regarded as the century of women in STEMM, the twentieth century may be said to have been 'a long runway' towards many dramatic changes. I suggest that four topics are particularly important when discussing women researchers in the twentieth century: the impact of

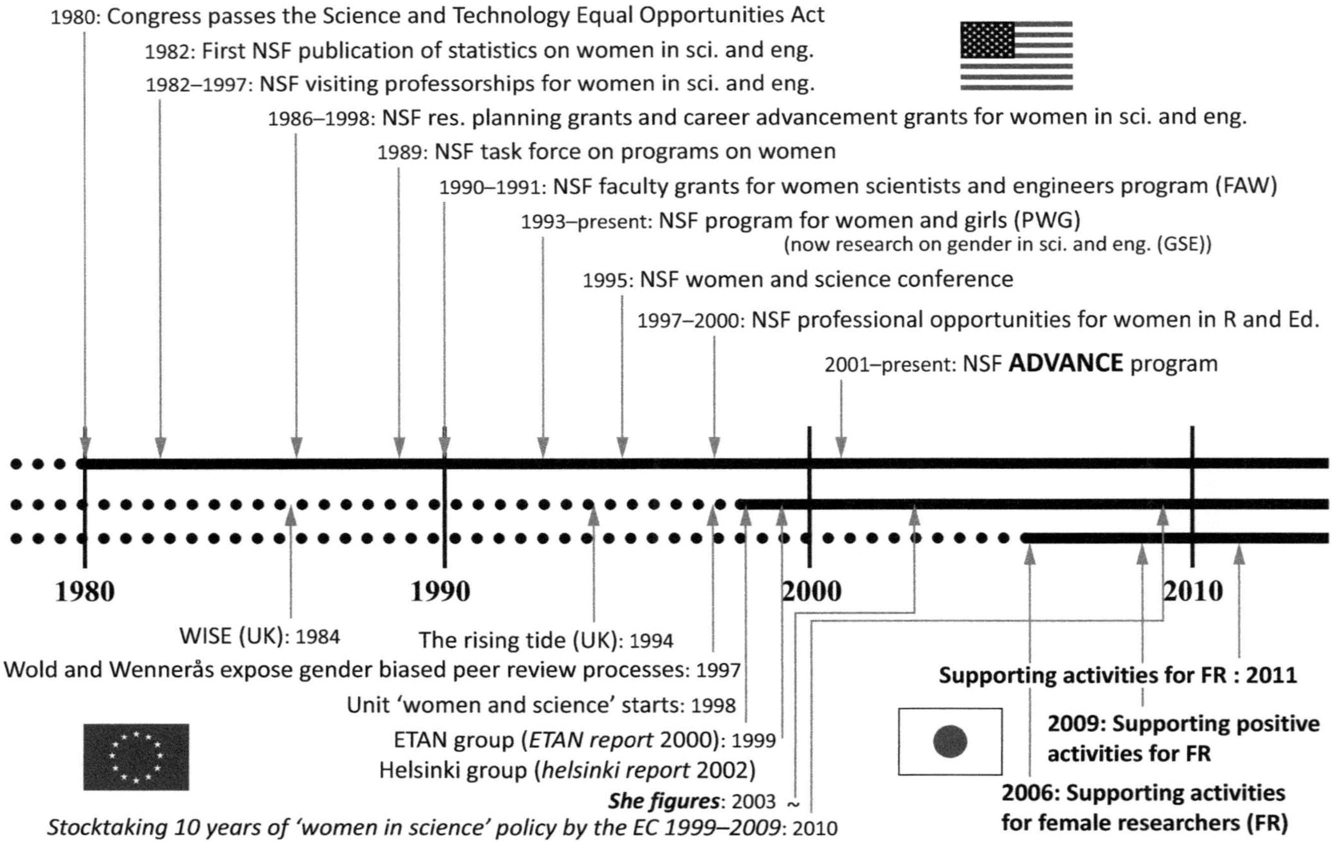

1980: Congress passes the Science and Technology Equal Opportunities Act

1982: First NSF publication of statistics on women in sci. and eng.

1982–1997: NSF visiting professorships for women in sci. and eng.

1986–1998: NSF res. planning grants and career advancement grants for women in sci. and eng.

1989: NSF task force on programs on women

1990–1991: NSF faculty grants for women scientists and engineers program (FAW)

1993–present: NSF program for women and girls (PWG)
(now research on gender in sci. and eng. (GSE))

1995: NSF women and science conference

1997–2000: NSF professional opportunities for women in R and Ed.

2001–present: NSF **ADVANCE** program

WISE (UK): 1984

Wold and Wennerås expose gender biased peer review processes: 1997

The rising tide (UK): 1994

Unit 'women and science' starts: 1998

ETAN group (*ETAN report* 2000): 1999
Helsinki group (*helsinki report* 2002)

She figures: 2003 ~

Stocktaking 10 years of 'women in science' policy by the EC 1999–2009: 2010

Supporting activities for FR : 2011

2009: Supporting positive activities for FR

2006: Supporting activities for female researchers (FR)

1980　　1990　　2000　　2010

Figure 0.1 Support programmes for women in STEMM implemented by NSF (USA), EC (EU) and MEXT (Japan)

the two world wars, the introduction of PhD (Doctor of Philosophy) programmes, the Nobel Prize and the creation of international organizations.

The two world wars

The two world wars were undoubtedly the most significant events of the twentieth century. These tragic events ironically provided opportunities for women to advance in STEMM fields previously dominated by men. Studies show that in the UK and the US, the two world wars promoted women's participation in the rapidly developing STEMM fields to make up for shortages in the male workforce.

As Sally Horrocks discusses, during and after World War I, women scientists gained footholds in academia, industry and government research, despite facing prejudice and many other barriers.[3] Furthermore, women's scientific career opportunities received a boost during World War I because of the realignment of science and military activities. Patricia Fara has demonstrated that British women scientists gained a foundation in academia as well as in industry and government research activities after World War I.[4] Similarly, Jordynn Jack has documented the opportunities that existed for women scientists in the US during World War II.[5] Examining gender in relation to war and STEMM is an extremely important and wide-ranging area that will require further research. It is encouraging to see that some of the chapters in this volume engage with the topic in contexts that go beyond Western Europe and the US, such as Hungary, offering fresh insights into the relationship between war, science and gender.

Higher education, especially PhD programmes

Higher education in the US and Europe

In Europe and the US, women gained access to undergraduate higher education in the last quarter of the nineteenth century. At that time, there were no opportunities for women to continue their studies after graduating from university. Thus, the next goal was to gain entrance to graduate schools. In the early 1890s, some American graduate schools decided to admit women and award them PhD degrees. Based on women's access to PhD programmes, we can identify three distinct university groups in the US.[6] The first group, which included Yale University and the University of Pennsylvania, was so conservative that it continued to refuse women's entry into its undergraduate colleges. The second group, which included

Columbia and Brown Universities, based its decision to admit women into its graduate schools on the formation of a coordinated college for women undergraduates. The third group, which included Stanford University and the University of Chicago, was the most liberal of all and, in 1891, instituted full coeducation. As Margaret Rossiter concludes, by the early twentieth century, academic feminists could rejoice that their efforts had played some role in opening the highest degrees for women.[7] However, postdoctoral opportunities, which were necessary for successful careers in science, remained far from equal among men and women scientists. The next goal, therefore, was to increase equal opportunities for employment and research after completing doctoral studies.

The situation in British universities was similar in some regards.[8] The PhD degree came to Britain in the 1920s. The motivation behind introducing a PhD award system stemmed from the desire to imitate the German higher-education system and the need to attract students from foreign and colonial countries. Oddly enough, the role of PhD degrees in the careers of women scientists has not been a topic of much interest among historians.[9] The Sex Disqualification (Removal) Act, issued on 23 December 1919, marked an important moment. This was demonstrated by the fact that the very next day, on Christmas Eve, Helena Normanton successfully became the first woman to be admitted to the Middle Temple, a development that immediately highlighted the impact of this Act. Moreover, the Act brought about great changes for women researchers.

In 1920, two male students received a PhD degree at the University of Oxford for the first time, and in 1921, twelve universities, including the Universities of Cambridge, London and Edinburgh, followed Oxford in awarding PhD degrees.[10] Lillian M. Penson was the first woman to be awarded a PhD in history as one of the nine PhD awardees at the University of London in 1921. In 1922, Evelyn Simpson became the first woman to be awarded a PhD in literature at Oxford; three years later, Sylva Thurlow became the first woman to be awarded a PhD at Cambridge (in chemistry).[11] After earning her degree, she moved to the US and appears to have worked for the Women's Medical College in Pennsylvania.

The University of London went on to produce more male and female PhD holders than any other university in Britain. Before 1920, the University of London awarded a specific degree called a London Doctorate, which was an advanced research degree that was almost equivalent to the later PhD. Martha Annie Whiteley and Ida Smedley obtained their London Doctorates in 1902 and 1905, respectively. Apart from university degrees, which began at Oxford in 1920 and Cambridge in 1948,[12] the most prestigious universities in the UK were also open to women who wanted to pursue doctoral studies. Overall, it is fair to conclude that the twentieth century was a time of struggle for women's postgraduate education and the right to study for a PhD degree.

Higher education in Japan

In Japan, Tohoku Imperial University admitted its first three women students, two in chemistry and one in mathematics, in 1913; even before 1913, some women had travelled to Europe and the US in search of higher education.[13] Table 0.1 shows that Japanese women earned PhDs in science, agriculture and pharmacy before 1940. As several contributors to this volume also discuss (e.g., Corbi, Edgerton-Tarpley and Zavarache), medicine was an important field of activity for women. Japan was no exception to this, but MDs (other than Tada Urata) have been excluded from Table 0.1 because of their large number. The figure also records women who obtained PhDs abroad in fields other than the sciences. Japan's Imperial Universities were not enthusiastic about accepting women in humanities and social sciences.

Japanese universities first awarded a PhD in science to Kono Yasui in 1927 and a PhD in medicine to Kanaeko Miyagawa in 1931. As Table 0.1 shows, except for Urata (1905), Ōhashi (1926) and Tange (1927), who obtained PhDs abroad, almost all women in the sciences received their doctoral degrees (in science, agriculture, pharmacy and medicine) from Japanese universities after 1927, although universities were hardly open to women. The medical faculties of all imperial universities did not admit women but did examine them for doctoral degrees. Although missing from Table 0.1, Toshiko Yuasa received her PhD in physics in France in 1943.

Tada Urata and Hiro Ōhashi's lives in the early twentieth century bear testimony to the tremendous parental pressure on women to get married. Urata had no option but to disappear on her wedding day at the age of eighteen. Ōhashi, at the request of her parents, was forced to interrupt her studies and was unhappily married off at the age of sixteen. Many Japanese women with strong academic aspirations were forced to choose between marriage and education early in their careers; many PhD holders were single.

Table 0.2 presents statistics on all Japanese female PhD holders between 1905 and 1945.[14] More than 80 per cent of the total number of PhD awardees were in medicine, but there was limited access to undergraduate education in medicine for women during that time. Before World War II, Japanese universities awarded women doctorates in science only in the fields of chemistry and biology. However, Toshiko Yuasa earned her PhD in physics in France in 1943. Doctoral degrees in mathematics, physics and astronomy were awarded to women in Japan only after World War II. In 1959, Sawako Gohara became the first woman to obtain a doctorate in engineering.

Table 0.1 Japanese female PhD holders prior to 1940, excluding PhDs in medicine (except Tada Urata)

Name	Information on doctorate and career	Marital status
Urata, Tada	1905 Marburg Univ., PhD, medicine (later, clinical doctor)	married
Haraguchi (Arai), Tsuru	1912 Columbia Univ., PhD, psychology (deceased in 1915, aged twenty-nine)	married
Kōra (Wada), Tomi	1922 Columbia Univ., PhD, psychology (later, politician)	married
Ōhashi, Hiro	1926 Chicago Univ., PhD, botany (fifth president of JWU)	divorced
Kan (Inoue), Shina	1927 Yale Univ., PhD, philosophy (professor, JWU)	married
Yasui, Kono	1927 Tokyo Imperial Univ., botany (professor, TWHNS)	single
Tange, Ume	1927 Johns Hopkins Univ., PhD, chemistry (professor, JWU) 1940 Tokyo Imperial Univ., PhD, agriculture	single
Kuroda, Chika	1929 Tohoku Imperial Univ., chemistry (professor, TWHNS)	single
Kasuya, Yoshi	1930 Colombia Univ., PhD, education (professor, TC)	single
Takizawa, Matsuyo	c.1930 Columbia Univ., PhD, economics and sociology	married
Katō, Sechi	1931 Kyoto Imperial Univ., chemistry (researcher, RIKEN)	married
Tsujimura, Michiyo	1932 Tokyo Imperial Univ., agriculture (professor, TWHNS)	single
Honma, Yasu	1936 Hokkaido Imperial Univ., agriculture (assistant, HIU)	single
Hatagoshi, Yasu	1937 Kyoto Imperial Univ., PhD, agriculture (professor, NWHNS)	married
Suzuki, Hideru	1937 Tokyo Imperial Univ., pharmacy (professor, JWU)	single
Ogawa, Fumiyo	1938 Tohoku Imperial Univ., zoology (professor, Kyoritsu Women's Univ.)	married
Matsumoto, Shizuko	1940 Kyoto Imperial Univ., chemistry (professor, Osaka Women's College)	single

JWU: Japan Women's University; TC: Tsuda College, later Tsuda University; TWHNS: Tokyo Women's Higher Normal University, later Ochanomizu Women's University; RIKEN: Rikagaku Kenkyūsho (Institute of Physical and Chemical Research)

Table 0.2 Japanese female PhD holders, 1905–45

Discipline	Japan	Germany, US or France	Total	Percentage (%)
Medicine	95*	1 (Urata)	96	81
Science	9	3 (Ōhashi, Tange and Yuasa**)	12	10
Agriculture	4	0	4	3
Pharmacy	1	0	1	1
Psychology	0	2 (Haraguchi and Kōra)	2	2
Philosophy	0	1 (Kan)	1	1
Education	0	1 (Kasuya)	1	1
Economics	0	1 (Takizawa)	1	1
Total	109	9	118	100

* Several women earned doctoral degrees from the sixth and seventh Imperial Universities created in Korea (Keijō Imperial University) and Taiwan (Taihoku Imperial University) under Japanese rule.
** Toshiko Yuasa (PhD in physics, France, 1943).

In contemporary Japan, women scientists continue to face many hurdles. The findings of a 2007 online survey conducted by The Japan Inter-Society Liaison Association Committee for Promoting Equal Participation of Men and Women in Science and Engineering (EPMEWSE) showed that Japan had extremely few academic couples.[15] Japanese academia was comprised mainly of male researchers, about 56 per cent of whom were married to unemployed housewives. The remainder was represented by single female researchers. By contrast, in the US, many famous couples in science had already existed before 1940. Rossiter has shown that there were four times more famous couples in the period 1950–72. Sharon McGrayne has also observed that husband-and-wife research teams were exceedingly popular in the US. However, being a research couple is not always a good thing. From the point of view of mobility, it is difficult to move around as a couple.[16] If the husband's mobility is prioritized, the wife is forced to become a trailing spouse, as was the case of several female Nobel Prize laureates.

The Nobel Prize in the twentieth century

The Nobel Prize was first awarded in 1901. There were ten female Nobel Prize laureates in science in the twentieth century, representing only 2.3 per cent of the total number of awardees. Some studies demonstrate that many outstanding female scientists have been overlooked in the past. The pertinent question to ask here is whether double standards existed in awarding the Nobel Prize.

The low number of female laureates was long believed to be due to the oversight of the Nobel Prize selection committee, but in 2017 the American Chemical Society made a positive statement that outstanding female researchers had not received sufficient recognition.[17] Although information on the Nobel Prize selection process remained undisclosed for the first fifty years, by that time the status of nominations of Nobel Prize winners up to the mid-twentieth century had become clear. The number of female candidate nominations, which was almost zero for the first twenty years, was counted at around four since the 1930s.[18] The first woman who received many nominations but did not go on to win a prize was Lise Meitner in nuclear physics. She was nominated every year between 1924 and 1965 in chemistry or physics or both, with a total of forty-eight nominations. Incidentally, this is a much larger number than Maria Goeppert-Mayer's twenty-seven nominations for the physics prize by 1963 and Dorothy C. Hodgkin's thirty-two nominations for the chemistry prize by 1964. One reason for the historically low number of women nominees in science fields was that the institution was male-dominated; the low ratio of women among nominators was another problem. Rossiter has described this disadvantageous situation for women in science as the 'Matthew Matilda effect'.[19]

Figure 0.2 (top of the timeline) divides the first nine female Nobel laureates in the twentieth century into three generations, each comprising three women. The first generation of female Nobel laureates conducted their research alongside their husbands, achieved significant results and even shared the Nobel Prize with their spouses.[20] The next generation conducted their research independently of their husbands. The third generation consisted of single women who received their Nobel Prizes in the 1980s.

The first and second generations were all married and bore children. Marie Curie and her daughter were awarded the Nobel Prize in their thirties, with the subsequent four women following in their fifties. This does not preclude the possibility that decisions to award Nobel Prizes to single women may have been delayed in their earlier careers. Feminism had more than a little to do with the fact that single women were finally allowed to be honoured in the 1980s. Their ages, in order, were eighty-one, seventy-seven, and seventy-one. Lise Meitner was unmarried and died at the age of ninety in 1968, just before the second wave of the feminist movement.[21] In Figure 0.2, the third generation is indicated by the oval located after the second wave of feminism on the timeline; it is difficult to deny the existence of a double standard in the twentieth century, especially before the second

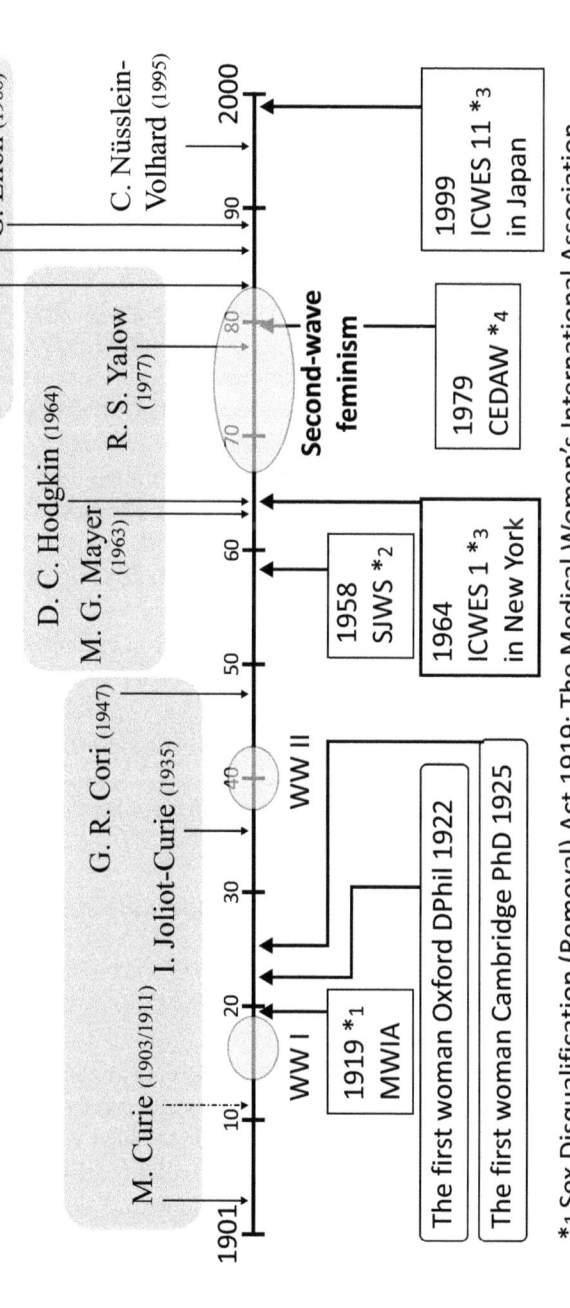

Figure 0.2 Nobel Prize women in science and international exchanges

wave of feminism, when even top scientists were assumed to be 'good wives' and 'wise mothers'.

Gerty and Carl Cori received the Nobel Prize in 1947. For thirty-five years, they formed a close scientific partnership. They started their careers at a medical school in Prague, in what is now the Czech Republic, and moved to the US in 1922 to avoid antisemitism.[22] However, for a long time, they could not find jobs that allowed them to work together because of anti-nepotism rules. Maria Goeppert Mayer was another typical example of a trailing spouse. She married Joseph Mayer in Göttingen in Germany. Eventually, Joseph was hired by the Johns Hopkins University, which had strict anti-nepotism rules, prohibiting the employment of relatives. Maria had neither a job nor a salary, while Joseph gained a post at Columbia, following his term at Hopkins. In 1960, she was finally appointed a full professor of physics at the University of California. Three years later, she won the Nobel Prize. Even Nobel Prize-winning women suffered from the campaign against nepotism. The second generation of Nobel Prize women pursued their own research in different fields from those of their husbands. Universities may have found it easier to hire couples specializing in different fields.

In conclusion, the twentieth century was a challenging time for academic couples. The circumstances and position of married women in the history of science have been deeply related to their education. Higher education promised women researchers more opportunities for activities either as a spouse in an equal relationship or as an independent researcher, but in practice there were many difficulties to overcome. In fact, although universities have recently begun to actively support academic couples, finding jobs in proximity to each other remains notoriously difficult to achieve.[23]

International exchanges encourage women scientists

The empowerment of female researchers in STEMM fields developed alongside the international cooperation that would unite them. The Medical Women's International Association (MWIA) founded in 1919 was one of the earliest organizations that connected women in the medical field.[24] In Japan, Sonoko Maeda, the thirteenth female doctor to receive a Japanese national medical licence, established the Japan Medical Women's Association (JMWA) in 1902 to promote research and collaboration and elevate their social position. The MWIA created a female medical doctors' international network, which remains active to this day.

As Graeme Gooday and Emily Rees Koerner also discuss in this volume, in 1964, almost twenty years after World War II, the first International Conference of Women Engineers and Scientists (ICWES) was held in

New York. The Society of Japanese Women Scientists (Nihon josei kagakusha no kai, hereafter SJWS), established six years earlier, planned to send a delegation to ICWES.[25] Notwithstanding the difficult circumstances of the Cold War, the rise of international exchanges among female researchers had a significant impact on the world academic community, including the Japanese one.[26] Leading members of the SJWS participated in the ICWES in New York, which proved to be an eye-opening experience. Maria Goeppert-Mayer, Nobel laureate physicist, Grace Hopper, American computer scientist, and Dorothy Hodgkin, Nobel laureate chemist, visited Japan in 1965, becoming a source of inspiration for many Japanese women scientists. Since the second ICWES, held in Cambridge in 1967,[27] international exchanges among women scientists have become increasingly popular. The feminist movement that followed further encouraged many Japanese female researchers.

The second wave of feminism led to the establishment of the Convention on Elimination of All Forms of Discrimination Against Women in 1979. In Japan, the ratification of the Convention in 1985, including the Equal Employment Opportunity Law, resulted in a slight increase in the number of students in science and engineering. Japan has produced more than two-dozen male Nobel Prize winners, yet there have been no female laureates to date. Japan's ranking on the gender gap index is notoriously poor, standing at 120 on a list of 156 countries. The gender gap in Japan is worse in politics and the STEMM fields. As women scientists remain a minority in their own countries, such international exchanges are especially beneficial.

In conclusion, I propose we keep in mind the following four messages, to foreground the detailed analysis of women's experiences of in/visibility in science in the twentieth century.

- The two tragic world wars provided women scientists with the opportunity to realize their potential in ways previously unimaginable. However, the work was sometimes dangerous and the working conditions were not comparable to those of men.
- The twentieth century also brought graduate education and PhD programmes to women, enabling them to acquire an education on the same footing as their male peers. This, however, did not necessarily bring equal opportunities for employment in research institutions.
- The feminist movements for suffrage and equal opportunities promoted the emancipation of women scientists from the old gender roles of 'good wives' and 'wise mothers', while also drawing attention to the many unseen forms of scientific labour women performed in the confines of their homes.
- The twentieth century accelerated international exchanges among women scientists, helping them to forge collaborations across regional and disciplinary boundaries and support each other in their attempts

to overcome the many barriers they faced in their pursuit of scientific careers.

Taken together, these developments became the runway for the flight of the twenty-first century, with its equal – in principle, if not necessarily in practice – opportunities for women to work in various scientific fields. To what extent the developments of the previous century made women more visible in science and how they experienced those instances of in/visibility is a topic the following chapters explore in more detail.

Notes

1 L. Schiebinger, *Has Feminism Changed Science?* (Cambridge, MA: Harvard University Press, 1999).
2 For a global overview of women researchers in the twentieth century, see Organization for Economic Co-operation and Development, *Women in Scientific Careers: Unleashing the Potential* (Paris: OECD Publishing, 2006); N. Kumar (ed.), *Gender and Science: Studies across Cultures* (New Delhi: Cambridge University Press, 2012).
3 S. Horrocks, 'The women who cracked science's glass ceiling', *Nature*, 575: 7781 (November 2019), 243–46; S. Horrocks, 'Promising pioneer profession? Women in industrial chemistry in inter-war Britain', *British Journal of the History of Science*, 33: 3 (2000), 351–67; S. Horrocks, 'World War II, post-war reconstruction and British women chemists', *Ambix*, 58: 2 (2011), 150–70.
4 P. Fara, *A Lab of One's Own: Science and Suffrage in the First World War* (Oxford: Oxford University Press, 2018). For medicine, see L. Leneman, 'Medical women at war, 1914–1918', *Medical History*, 38 (1994), 160–77. For mathematics, see J. Barrow-Gree and T. Royle, 'The work of British women mathematicians during the First World War', in C. G. Jones, A. E. Martin and A. Wolf (eds), *The Palgrave Handbook of Women and Science since 1660* (Cham: Palgrave Macmillan, 2022), pp. 549–72.
5 J. Jack, *Science on the Home Front: American Women Scientists in World War II* (Chicago: University of Illinois Press, 2009). As Jack discusses, many women researchers worked on the Manhattan Project. See also R. H. Howes and C. L. Herzenberg, *Their Day in the Sun: Women of the Manhattan Project* (Philadelphia, PA: Temple University Press, 1999); M. W. Rossiter, *Women Scientists in America: Before Affirmative Action 1940–1972* (Baltimore, MD: Johns Hopkins University Press, 1995), especially chapters 1 and 2.
6 M. W. Rossiter, *Women Scientists in America: Struggles and Strategies to 1940* (Baltimore, MD: Johns Hopkins University Press, 1982), pp. 33–34.
7 Rossiter, *Women Scientists in America: Struggles and Strategies to 1940*, p. 50.
8 R. Simpson, *How the PhD Came to Britain: A Century of Struggle for Postgraduate Education* (Guildford: Society for Research into Higher Education,

1983). For undergraduate education, see C. Dyhouse, *No Distinction of Sex? Women in British Universities, 1870–1939* (London: Routledge, 2003).
9 Joan Mason has published extensively on women's admission to the Royal Society. E.g., J. Mason, 'The admission of the first women to the Royal Society of London', *Notes and Records of the Royal Society of London*, 46: 2 (1992), 279–300.
10 Simpson, *How the PhD Came to Britain*, p. 164.
11 Sylva Thurlow published several single-authored and co-authored papers in *Biochemical Laboratory* in 1924–25. Nobel Laureate D. M. Crowfoot (later Hodgkin) earned her PhD in 1937 from the University of Cambridge. The award of a bachelor's degree and the award of a PhD degree are two different things.
12 'Timeline: 100 Years of Women's History at Oxford', University of Oxford, www.ox.ac.uk/about/oxford-people/women-at-oxford/centenary-womens-timeline (accessed 1 June 2024); R. M. Tullberg, *Women at Cambridge* (Cambridge: Cambridge University Press, 1975), especially chapter 11.
13 E.g., 'Japanese Alumnae Digital Museum', Wellesley College, www.wellesley.edu/ealc/alum-corner/japan-alum (accessed 20 December 2022).
14 M. Ogawa, 'History of women's participation in STEM fields in Japan', *Asian Women*, 33: 3 (2017), 65–85; N. Kodate and K. Kodate, *Japanese Women in Science and Engineering: History and Policy Change* (New York: Routledge, 2016); Y. Arikawa (ed.), *Joshi rigaku kyōiku o rīdo shita josei kagakushatachi: Rekimeiki, Meijiki kōhan kara no kiseki* [Women Scientists Leading Science Education among Women: From the Dawn of Women's Participation, or the Late Meiji Period] (Tokyo: Akashi-shoten, 2013).
15 EPMEWSE: The Japan Inter-Society Liaison Association Committee for Promoting Equal Participation of Men and Women in Science and Engineering, https://djrenrakukai.org/en/index.html (accessed 20 December 2022). Especially, EPMEWSE, 'Large-scale survey of actual conditions of gender equality in scientific and technological professions', July 2008, fig. 1.60: Spouse's job, p. 13, https://djrenrakukai.org/doc_pdf/h19enquete_report_en.pdf (accessed 20 December 2022). EPMEWSE has no data about the jobs of spouses in the twentieth century because it was established in 2002.
16 S. B. McGrayne, *Nobel Prize Women in Science: Their Lives, Struggles, and Momentous Discoveries* (New York: A Birch Lane Press Book, 1993).
17 V. V. Mainz and E. T. Strom (eds), *The Posthumous Nobel Prize in Chemistry: Ladies in Waiting for the Nobel Prize*, vol. 2 (Washington, DC: American Chemical Society Symposium, 2019); 'Nomination archive', The Nobel Prize, www.nobelprize.org/nomination/archive/ (accessed 20 December 2022); H. Rose, *Love, Power and Knowledge: Toward a Feminist Transformation of the Science* (Cambridge: Polity Press, 1994); S. Modgil et al., 'Nobel nominations in science: Constraints of the fairer sex', *Annals of Neurosciences*, 25 (2018), 63–78.
18 Modgil et al., 'Nobel nominations in science', tables 1–3.

19 'The "Matilda effect" is an implicit bias against acknowledging the achievements of women scientists, whose work is often attributed to their male colleagues.' Quoted in Mainz and Strom (eds), *The Posthumous Nobel Prize in Chemistry*, p. ix; M. W. Rossiter, 'The Matthew Matilda effect in science', *Social Studies of Science*, 23 (1993), 325–41; A. E. Lincoln, S. Pincus, J. B. Koster and P. S. Leboy, 'The Matilda effect in science: Awards and prizes in the US, 1990s and 2000s', *Social Studies of Science*, 42 (2012), 307–20.

20 I have highlighted three couples who were married and both partners won Nobel Prizes. Needless to say, there are several examples of married couples who worked together in the course of their careers but only the husband won a Nobel Prize. Further research is needed on such couples. See J. Harvey, 'The mystery of the Nobel Laureate and his vanishing wife', in A. Lykknes, D. L. Opitz and B. Van Triggelen (eds), *For Better or For Worse? Collaborative Couples in the Sciences* (Birkhäuser: Springer Verlag, 2012), pp. 57–77.

21 Many reasons have been identified for the failure to recognize Meitner's achievements. In particular, the malicious actions of Manne Siegbahn and Otto Hahn have been corroborated by recent documents. R. L. Sime, *Lise Meitner: A Life in Physics* (Berkeley: University of California Press, 1996), especially chapter 14.

22 McGrayne, *Nobel Prize Women in Science*, chapter 5; H. M. Pycior, N. G. Slack and P. G. Abir-Am (eds), *Creative Couples in the Sciences* (New Brunswick: Rutgers University Press, 1996), chapter 3. These references were consulted for more information on Nobel Prize-winning women.

23 L. Schiebinger, A. D. Henderson and S. K. Gilmartin, *Dual-Career Academic Couples: What Universities Need to Know* (Stanford, CA: M. R. Clayman Institute for Gender Research, Stanford University, 2008).

24 E. Pohl Lovejoy, *Women Physicians and Surgeons: National and International Organizations* (Livingston, NY: The Livingston Press, 1939), pp. 15–24.

25 *SJWS Newsletter*, No. 1 (1958) special issue reporting on the opening ceremony (in Japanese); No. 9 (1964) special issue for the first ICWES. The following five members were sent to New York: Mizoguchi, Saruhashi, Kagawa, Yamada and Kumagai.

26 See Rees Koerner and Gooday's chapter in this volume.

27 The SJWS sent eight members to Cambridge, whose reports on the ICWES were published in *SJWS Newsletter*, No. 12 (1968).

Introduction: In/visible women, science, engineering and medicine in the twentieth century

Amelia Bonea and Irina Nastasă-Matei

Genesis of the volume

This is a book about women and their experiences in science, engineering and medicine in the twentieth century. Some, like the American evolutionary biologist Lynn Margulis, were fairly visible actors in the academic and public arenas of professional science. Others, like the doctors working in secondary schools in interwar Romania or those who struggled to alleviate 'women's illnesses' in famine-stricken rural areas during China's Great Leap Forward have been largely invisible – as medical practitioners, creators of knowledge, educators and subjects of historical inquiry. This volume probes the nature and extent of women's in/visibility in the making, pedagogy, institutionalization and communication of science, broadly conceived, in the twentieth century, seeking to document the factors that circumscribed it and understand how women navigated their circumstances.

The project grew out of our own experiences of and scholarly preoccupations with gender inequality in academia and beyond. As fellow Romanians, we have long wondered why women should be so conspicuously absent from history (text)books and institutions of education whose narratives of science continue to be dominated by the achievements of a handful of white male scientists. The alienation we experienced as students in such institutions – indeed, that many students continue to experience – is fresh in our minds. This is a paradoxical situation, if we consider that the communist regimes in Eastern and Central Europe actively encouraged the participation of women in science. As a result of that legacy, countries in the former communist bloc like Romania continue to have a higher proportion of female scientists than counterparts with stronger economic indicators, such as Germany or Japan.[1] Of course, numbers only tell an incomplete story of what it means to be a woman in STEMM, not only because reliable statistics are notoriously difficult to come by, but also because communist regimes produced their own forms of gender inequality.[2]

The process of recognizing and naming such absences for what they are – instances of gender-based discrimination, rooted in a long history of structural inequality and marginalization of women – is rarely straightforward.³ In our case, the crystallization of a research agenda around this topic happened gradually, while pursuing projects that hadn't initially set out to investigate the role of women in science. For Nastasă-Matei, this was a study of Romanian students in Nazi Germany, undertaken as part of her doctoral research, followed by a postdoctoral project investigating Romanian Humboldt fellows in West Germany during the Cold War. Women were under-represented in both cases and were targeted through mechanisms of exclusion that can be clearly documented in the archives but have seldom been recognized in historical literature. For Bonea, the impetus came while working on a project about the global entanglements of palaeontology in South Asia funded by the German Research Foundation (Deutsche Forschungsgemeinschaft, Project No. 423157196, 'Archives of the Earth: Fossils, Science and Historical Imaginaries in Twentieth Century India'). The project investigated how the fossil record of the Indian subcontinent was incorporated into institutions of research, education and public edification, such as museums, in India and abroad. Although there was little expectation of finding female voices in a field of science that has been notoriously dominated by men, it soon became clear that women had engaged with palaeontology in colonial and post-colonial South Asia in a variety of guises, as scientists, fieldwork companions, popularisers of science and institution builders.

The experience of working on these projects reminded us of Anne Firor Scott's pertinent observation that 'people see most easily things they are prepared to see and overlook those they do not expect to encounter'.⁴ Indeed, women have always engaged with science, but the 'canon' of history-writing has been heavily skewed towards rendering them invisible, even when traces of their lives and work did survive in the archives.⁵ In her Foreword to this volume, Mariko Ogawa reminds us that the twentieth century was a period when women started gaining access to science education and careers in unprecedented numbers. However, as we discuss below, this growing presence in the scientific establishment has not prevented them from being persistently absent from the 'collective memory of science', as Pnina Abir-Am puts it.⁶ This shows how enduring perceptions of women as insignificant historical and scientific actors have been. Our volume seeks to investigate the politics of invisibility that helped relegate women to such peripheral positions in the annals of twentieth-century science, engineering and medicine, and understand how they negotiated their circumstances in regional contexts that transcend the usual focus on the US and Western Europe.

The invisibility of women in twentieth-century science, engineering and medicine was context-specific and layered. Put differently, it was shaped by a host of intersecting factors like gender, race, ethnicity, class, caste, religion, marital status, sexuality, age, urban–rural divides and so on[7] that played out differently in different geographical and temporal contexts. To borrow Anna Menyhért's beautiful metaphor in her discussion of women writers in twentieth-century Hungary, women's engagement with science in the twentieth century appeared to be 'both hidden and [hiding] itself away in the manner of a sinking stream which comes to the surface only to vanish underground again'.[8] As scholars, we became increasingly preoccupied to capture the nature of this 'stream' to understand the circumstances that made women and their lives in science surface or 'vanish underground'. When, why and how did women become invisible? When, why and how did they seek visibility? Was invisibility always a form of discrimination, exclusion and misrecognition or could it also become a strategy of resistance and survival?

Insights and questions like these prompted us to convene an international conference on 'Hidden Histories: Women and Science in the Twentieth Century' in May 2021. Held online at the height of the Covid-19 pandemic, the event was organized in connection with Bonea's 'Archives of the Earth' project based at the Heidelberg Centre for Transcultural Studies, funded by the German Research Foundation and co-hosted by the Faculty of Political Science, University of Bucharest. The initial call for papers attracted almost 100 submissions, leading to fifty-four presentations. More than half introduced case studies from the UK, France, Germany, the US and Russia, but a substantial number discussed histories of women in science, engineering and medicine in South Asian, East Asian, as well as Southern, Eastern and Central European contexts. A paper each was also presented on Trinidad and Tobago and Israel, with the remaining two focusing on the colonial contexts of Portuguese Africa and the French Pacific. Mariko Ogawa and Andrea Pető delivered keynote lectures.

In a volume concerned with absences and invisibility, it is essential to acknowledge the gaps in our own research. Despite efforts to circulate the call for papers in relevant academic forums, we received only one submission each pertaining to South America and Africa for the conference and no submissions for the present volume. As a result, African and South American case studies are absent, although Rees Koerner and Gooday's chapter discusses the participation of engineers from Morocco, Uganda, Nigeria and Ghana in the International Conference of Women Engineers and Scientists (ICWES). Such limitations notwithstanding, the volume prides itself on its geographical and linguistic diversity, featuring three chapters on India, two on China, one each on Romania, Hungary, the Czech Republic and Greece, two on the UK and three on the US. Mariko Ogawa's foreword discusses the changing contours of women's engagement with professional science in Japan against broader international developments in the field. Equally importantly, many of the chapters engage with the history of women in science, engineering and medicine from border-defying, trans-regional perspectives, demonstrating that the 'science' they document was itself a transregional enterprise.

The volume combines individual and collective portraits of women in science, engineering and medicine with discussions of institutional structures, work and associational cultures, medical practice and education, science and domesticity, science communication and activism, and science policy. By bringing together case studies that are not usually discussed alongside each other, it seeks to expand the geography of research and destabilize the conventional focus on North America and Western Europe characteristic of much scholarly research on the topic. What do we learn about the experiences of women in STEMM in the twentieth century when we ponder examples from East Asia and South Asia alongside those from Eastern, Central and Southern Europe and the Anglo-American world? How can we rethink gender and science in the twentieth century when we move away from a monolingual archive dominated by the English language? To answer these questions, the chapters bring together a wide range of material – official archives, personal collections, oral history interviews, correspondence, press articles, memoirs, statistics, legislation, lecture series – in languages as diverse as Chinese, Czech, English, Greek, Hindi, Hungarian, Japanese and Romanian. This multilingual archive is examined against the background of changing local and global contexts that shaped the extent and ways in which women became in/visible in STEMM. The essays thus make important qualifications to the widespread narrative that the twentieth century was a 'century for women' in science by showing what exactly that meant in different geographical and socio-political contexts and identifying, as J. Devika aptly proposed, 'points of contact' between this kaleidoscope of experiences, rather than trying to fit them into 'a single unified history'.[9]

Numbers and the politics of in/visibility

In a volume concerned with the experiences of women in STEMM, it seems appropriate to discuss, albeit briefly, the paradoxical relationship between numbers and in/visibility. As many of us can attest, it is not uncommon for scholars investigating the gendered histories of science to be asked to explain the relevance of their research by reference to the historically low number of women who have pursued professional careers in STEMM. In her book on African American women mathematicians at NASA, Margot Lee Shetterly reveals that the 'magnitude of the story' she wanted to write was questioned, especially in the early stages of research: 'How many women are we talking about? Five or six?'[10] Since the proportion of women was often limited to begin with, the argument usually goes, their contributions to STEMM and the topic itself must also be of little historical consequence. 'Real' science, we are told, happened elsewhere, in the laboratory of an Ernest Rutherford, Albert Einstein and, occasionally, Marie Curie. Low numbers, real or imagined, are used to justify the exclusion of marginalized groups from historical accounts and the public imagination of science because low numbers are equated with insignificant, unrepresentative contributions. Documenting diversity in science – and the lack thereof – thus becomes a dangerous fad rather than a crucial step towards understanding how knowledge was produced and circulated.

This volume suggests that arguments about the low number of women in science are particularly problematic with regard to the twentieth century. As we discussed above, the last century brought increased opportunities for women's participation in science, but those developments did not necessarily translate into higher visibility for them as historical actors of science-making, education, institutionalization and communication. On the contrary, with the possible exception of Russia, regions of the world with the highest proportions of women in STEMM – such as Central Asia, Latin America and the Caribbean, the Arab States, and Central and Eastern Europe – remain poorly documented in the history and historiography of science. At the other end of the spectrum, countries like Japan and India, with notoriously low numbers of women in STEMM, often feature in relevant literature in tokenistic ways that remind us of Abha Sur's cogent point that 'The twinning [sic] of exclusion and exoticism continues to be a staple of the recent historiography of science.'[11]

Biographical dictionaries are emblematic of these trends. The *Biographical Dictionary of Women in Science* features primarily scientists from North America, the UK and Western Europe, and only seventeen (out of a total of 2,500) from India, China and Japan. The first volume contains one entry each for Romania and Bulgaria and ten for Hungary, which also points to significant regional differences in representation. A similar trend can be observed in *The Palgrave Handbook of Women and Science since 1660*, which includes one chapter each on Japan and India (out of twenty-nine) and no contributions on Central and Eastern Europe. *International Women in Science: A Biographical Dictionary to 1950* features short biographies of more than 350 women in science, a field defined broadly to include artistic endeavours connected to science. All except eight of the scientists hail from the US and Western Europe, while the Japanese and Chinese scientists included were educated or worked there.

Magdolna Hargittai's collection of portraits of women scientists makes for an interesting contrast, aiming as it does for a more global approach, albeit one heavily dominated by the Global North (about three-quarters of the case studies), with separate sections dedicated to scientists from Russia, Turkey and India. Japanese chemist Reiko Kuroda and Chinese American physicist Chien-Shiung Wu are also included.[12] Since the collection is based primarily on interviews, recollections and photographs from Hargittai and her husband Istvan's personal archives, we can assume it reflects the scientific networks forged by this academic couple over the years. At the same time, their collection underscores the significance of international collaborations to the making and memorialization of science and of the different importance attached to the act of orienting oneself towards the wider world by scientists based in academic establishments in the 'Western' world and those outside of it. While for scientists in Eastern and Central Europe, the 'West' has been an important point of reference – more so after the collapse of the communist regimes in the region – the opposite has not necessarily been the case (the former Soviet Union represents perhaps an interesting exception, as some of the chapters in this volume also discuss).

Research on women, gender and science in Japan and South Korea documents a similar scenario in which comparisons with the US and Western Europe feature prominently in a scholarly agenda intimately connected to efforts to shape government policies around the dismal representation of women in science.[13] However, as Jaehwan Hyun points out, in the 1980s, female scientists in South Korea also drew on Eastern European models of women's participation in science to demand more inclusive science policies and increased government support for women in science in the context of the Cold War. This history is yet to be told.[14] The point to remember is one that scholars of science and European imperialism and, to some extent, scholars of Cold War science have long been making, namely that networks of science-making and communication have often been 'polycentric'. Attending to other types of relationships than those between the 'West' and its 'Others' – for example, intra-Asian scientific exchanges, or those between Eastern European and East Asian and African countries – is essential if we are to move beyond an understanding of 'modern' science as something that was exported from a metropolitan centre to its peripheries.[15]

Biographical dictionaries are, of course, only one genre of writing about women in science. As the next section discusses in more detail, other types of publications, especially those that have engaged with the topic from feminist and post-colonial perspectives, have done a better job of attempting to correct this imbalance.[16] However, dictionaries are a good index of 'dominant trends in the historiography' which, as Abha Sur cogently reminds us, 'are but refractions of the power relations in the world'.[17] The point remains: the visibility of women in the historiography of science has not necessarily been a function of their numbers, but of politics and power asymmetries, many of which stem from lingering imperial legacies about regimes of knowledge-making and scientific rationality. We have come some way from the days of publications of 'scientific types' and encyclopaedias of scientists that listed no to very few women,[18] but we are still a long way from a time when women scientists, engineers and doctors, in their various guises, will have entered the historiographical mainstream.

Indeed, the proportion of women in science often appears lower than it was because much of the scientific work women have engaged in historically – within or without the consecrated sites of professional science – has not been recognized and counted as such. One is reminded here of earlier feminist critiques, especially Rossiter's famous 'Matthew Matilda effect' in science, used to explain how women have been systematically denied credit for their scientific work and have been erased from histories of science, their achievements credited to (white) male colleagues who could leverage more scientific prestige and important institutional positions.[19] 'When it [was] no longer possible ... to render the women's share of the work invisible', Abir-Am reminds us, 'society's solution has been to regard [their] contribution[s] as merely derivative'.[20]

A study of the journal *Theoretical Population Biology* provides a striking illustration of the mechanisms by which female scientists' work has been rendered invisible. Only 7 per cent of the authors published in the 1970s were women, but the situation changed dramatically when researchers turned their attention to footnotes, where they represented more than half (59 per cent) of the programmers acknowledged in the pages of the journal. This example demonstrates that women's computing work was inseparable from knowledge-making in population genetics, but it was relegated to the footnotes of professional science and, until very recently at least, the footnotes of history-writing as well.[21] As we go forward, we should strive to mainstream such 'footnotes' into the history of science through research, writing, teaching and public engagement – until that is achieved, our understanding of processes of knowledge-making in STEMM can only be incomplete at best.

Indeed, as Maria Bucur emphasizes in a related context, 'writing histories that highlight the specific qualities, achievements, failures, and overall contributions' of women to science is a 'core element of how we understand our past'.[22] Studying the lives, collective and individual, of women in STEMM is an opportunity to reflect on how we have defined 'science', what has counted as 'scientific labour' and who has counted as a 'scientist' in different geographical and historical contexts. As the next section discusses, there is much to learn in this respect by bringing different geographies and strands of scholarly literature in conversation with each other, especially gender and feminist scholarship, histories of science and imperialism, and investigations of gender and power in (post-)socialist societies.

Multisited histories of gender, science and power

Any assessment of the present volume's contribution to the extant literature on women, gender and science must begin with the recognition that there are significant variations in how scholars have engaged with this topic in different regional and academic contexts. Since the early 1980s, a growing body of literature has been published in various parts of the world examining the lives and career trajectories of women scientists as well as the sociopolitical, economic and cultural circumstances that have circumscribed their access to STEMM.[23] Some of these publications follow a compensatory approach, seeking to uncover forgotten figures of science and searching for historical proof of women's participation in it.[24] Others move beyond recovering lost voices to examining the complex interplay of factors that contributed to the structural exclusion and marginalization of women, illuminating, for example, how gendered identities were constructed and how they came to structure interactions between men and women in science, as well as processes of knowledge-making.[25]

Compensatory writing might go some way towards making women in STEMM more visible, but it is not, in and by itself, sufficient to challenge the epistemological assumptions underlying historical analyses that have long treated them as lesser actors of knowledge-making. In fact, as Devika cautions in a related context, this type of writing can end up reinforcing hegemonic conceptions of science and gender roles, for example, when 'the biography does the job of filling a gap rather too literally and faithfully, by upholding the figure of the transgressive woman author who was "as good as" or "better than" her male counterparts'.[26] As other scholars have also noted, understanding the historical trajectory of 'processes of gendering' and, indeed, the power relations that circumscribed them is more pressing and radical a research agenda than simply filling a gap.[27]

The compensatory approach has been influential in many areas of the world. In South-Eastern and Central Europe, writing in this vein has sometimes acquired a celebratory or exceptionalist tone, likely an outcome of the fact that the study of gender and science is a relatively new area of research. Until 1989, there was only a small and mostly underground feminist movement in these regions, although sometimes with relevant impact on the global women's movement.²⁸ The authoritarian regimes in the region imposed a cultural and political agenda that encouraged the entrance of women into scientific fields, but their activities became subsumed under a broader and, for the purposes of perpetuating the regimes in power, politically more significant identity as 'working people'. In the post-1989 period, investigations of science under communism, including the role of women therein, have often been incorporated into a broader repertoire of anti-communist writing.²⁹

Scholarly engagement with women's history and gender studies varies across countries of the former communist bloc, a point valid for Europe as a whole, as Andrea Pető has argued: 'In each country the development of writing gender history is deeply connected to national historiography and the characteristics of historians as an institutionalized profession. The bias towards gender or towards women's history depends on the local intellectual tradition regarding difference.'³⁰ In Romania, there has been a gradual increase in gender studies publications since the 1990s,³¹ but there is a noticeable lack of qualitative and quantitative studies that engage with science, a situation mirrored in the older and academically more established field of women's history. As we have emphasized above, the omission is significant, considering the relatively high proportion of women in STEMM. It points not only to dominant perceptions of science as a masculine pursuit – statistics notwithstanding – but also the fact that writing about women and gender is a political act with tangible consequences. Recent attempts to ban the teaching of 'gender ideology' in schools and universities are a chilling reminder of this,³² although overall the push against gender studies has been less successful politically in Romania than in Hungary or Poland. The move is ironic, considering that neither gender studies nor women's history have garnered much scholarly attention to date, certainly not in relation to women's participation in science.

As in other parts of the world, much of the writing on women and science in Eastern and Central Europe has originated with women who have themselves juggled careers as professional scientists. This further illustrates how the 'location' or position from which one speaks comes to circumscribe the 'locution', or what one can afford to say.³³ In her paper on women in technological higher education in twentieth-century Hungary, Mária Palasik hints at such dilemmas when she asks readers not to construe her research as an attack on the patriarchal structures responsible for the marginalization of women in technoscience:

> The aim of this paper is ... not to attack the male scientific community for its closed and exclusive structure, but rather to present women's achievements in selected areas, achievements that may have contributed to the destruction of

overly narrow disciplinary boundaries. My primary purpose is therefore to provide a realistic picture of the Hungarian scientific community in the area of technology.[34]

Notwithstanding the reluctance to challenge openly gender-based structural inequalities and offend the powers that be, the above passage does suggest a way forward: women's engagement with science is best examined across multiple sites of activity and 'narrow disciplinary boundaries'. Grażyna Kubica's research on Polish female anthropologists who worked in the shadow of the famous Bronisław Malinoswki is a relevant example which documents the intersections of gender and the politics of knowledge production at the beginning of the twentieth century. In particular, her biography of Maria Czaplicka, which has been translated into English, examines Czaplicka's literary and anthropological output on gender, shamanism and race and moves beyond the study of a single individual towards a documentation of an 'anthropological collective' well anchored in the intellectual and political milieus of its time.[35]

Documenting communities of science-making, as some of the essays in this volume also attempt to do, adds crucial context to the history of women and gender in science, enabling us to move away from the traditional heroine model of history-writing, with its emphasis on individual, elite, 'star' scientists. Similarly, conceiving of women's engagement with science as a multisited activity that straddled the laboratory, the school, the clinic, the home, the media and so on anchors science firmly within society, illuminating how invisibility operated in science, for example, by exposing those facets of women's work that were most likely to be ignored. Yasu Furukawa's recent biography of Umeko Tsuda is exemplary in this respect, in that it not only recovers Tsuda the well-known educator, but also Tsuda the forgotten biologist – indeed, as he points out, many in Japan do not even know that Tsuda was a trained biologist.[36]

For these reasons, the contributions in this volume seek to bring together under one umbrella women's engagements with science, engineering and medicine rather than treating them separately. As one of the first scientific professions to be considered acceptable and, therefore, open to women in many countries, medicine has attracted sustained scholarly attention to date, often at its intersections with missionary work, education and imperialism, since many (Western) European and American women doctors joined missionary societies in India, China, Korea or Japan.[37] Examining science, technology and medicine alongside each other also enables us to understand how medical practice and education were intertwined with research, even though women's identities as scientists were usually trumped by their more visible lives as mothers, wives, educators or doctors, as Tsuda's example above demonstrates (see also Macková's chapter in this volume).

Similarly, investigating women's activities across multiple sites of science that straddled institutional and professional divides is an important step towards documenting larger communities of knowledge-making and practice and capturing their trans-regional and trans-disciplinary dimensions. As several scholars have noted, mobility was central to many women's lives in science. In her ground-breaking biography of Indian botanist and cytogeneticist E. K. Janaki Ammal, Savithri Preetha Nair likens Janaki's 'nomadic' life in pursuit of science to a pilgrimage: 'Janaki was a curious pilgrim; not only was mobility a way of being for her, but also eminently a way of making knowledge.'[38] Like Janaki, many women travelled abroad in search of education, knowledge and career opportunities in the twentieth century. This was the case with the European and American medical doctors who engaged in missionary work in East and South Asia. Research on colonial Korea and India demonstrates that women's mobility was often circumscribed by imperial circumstances and highlights the different ways in which colonizing and colonized women were able to access education and careers in science.[39] At the same time, as Grace Shen argues with regard to Chinese women scientists' transnational mobility in the early twentieth century, the decision to travel abroad was framed not only by strategic career considerations and politically infused agendas like 'national salvation or modernization', but also by the complicated personal circumstances women tried to navigate.[40] Mariko Ogawa makes a similar argument in her discussion of Japanese women scientists in the Foreword to this volume.

Mobility was made possible by formal and informal networks of exchange and collaboration. Georgeta Nazarska has used prosopography as well as historical and social network analysis to investigate associational cultures in hitherto-little-explored settings like Bulgaria, Romania, Yugoslavia, Greece and Turkey (e.g., in relation to the activity of the International Federation of University Women in the Balkans in the 1920s–1950s). She has also used this approach to study professional subgroups among Bulgarian female doctors in the nineteenth century, demonstrating that they were not a monolithic entity.[41] As the following section discusses in more detail, several chapters in this volume pursue similarly fruitful lines of investigation, exploring formal and informal networks of support, collaboration and, sometimes, contestation between women in STEMM (e.g., Nair, Macková, Rees Koerner and Gooday). All point to the importance of gender solidarity in the pursuit of scientific careers by women and the validation that participation in such networks afforded them, difficult to find in other types of professional, male-dominated contexts.

Broadening the conversation to include different academic cultures and strands of writing also reveals the context-specific ways in which power asymmetries have worked to render women in science, engineering and medicine invisible in the twentieth century. Earlier North American feminist scholarship connected women's invisibility in science to the ideal of scientific objectivity, which helped devalue certain forms of knowledge and scientific labour associated with women by cementing the notion of the scientist as a rational (white) man engaged in the pursuit of knowledge and unhindered by his own subjectivity. As Donna Haraway argued, knowledge could only be fragmentary and 'feminist objectivity mean[t] quite simply *situated knowledges*'.[42] But understanding how knowledge is 'situated' requires us to document the socio-economic, political and material contexts of its production and circulation; among other things, this should also mean investigating histories of gender and women in science and theorizing them based on polycentric historical archives and scholarship that are not always dominated by the English language. In fact, as Naomi Oreskes has pointed out, even in the US context, 'The types of work that women have done in many branches of American science [in the twentieth century] fit conventional notions of objectivity', rather than the opposite. Scientific objectivity, Oreskes countered, was not the most appropriate value for understanding women's invisibility in science; the ideology of scientific heroism was, especially in field sciences like geology and geodesy.[43]

At the heart of it, women's invisibility in science is the outcome of power asymmetries, power being a concept which has long engaged feminist scholarship itself. As the chapters in this volume also discuss, across different times and geographies, some historical actors have managed to acquire sufficient political, economic, social and symbolic capital to be in a position to define what 'science' is, what problems are worthy of investigation, who counts as a 'scientist', what counts as 'scientific' labour, who should be rewarded for it and who, indeed, should be preserved in the archives of science. A substantial body of scholarship on science in imperial settings has added flesh to this argument, demonstrating that many of these asymmetries can be traced back to colonial regimes of knowledge and power. In colonial and quasi-colonial contexts like those of India and China, science-making was inseparable from the business of empire-making; heroes – and scientific rationality – were defined by insidious considerations of race, not only gender, sexuality, class, caste and so on.[44] In fact, this is also the case in the American context discussed above; it is one of the reasons Lee Shetterly was asked to justify the significance of her research on Black women mathematicians at NASA. Scholarship on science and imperialism has also been instrumental in recovering the contributions to science of many invisible colonized actors, by moving away from the scientific establishment into the field, the bazaar, the entrepôt and so on, and reconceptualizing science as practice, not only knowledge.[45] It is clear that not all marginalized actors were marginalized in the same way; in/visibility in science was intersectional.

If we are to capture their contributions to science, we need to broaden our understanding of what constitutes a 'legitimate' site of knowledge-making and 'legitimate' scientific work.

Seen from such comparative perspectives, the near silence of post-1989 historiography on the topic of women and science in the former communist bloc is even more striking, given that many women were not active on the peripheries of the scientific establishment, but at its very heart. Equally striking is the fact that the shadow of empire(s) lingers on – often unacknowledged – in discussions of science in South-Eastern and Central Europe, where it is not uncommon for public debates and even academic scholarship to take for granted the notion that 'modern' equals 'Western' and that 'Western' science is the yardstick by which all science is to be measured.[46] Although recent scholarship engages with the 'interimperial' legacies that have shaped Eastern European regions like Transylvania,[47] the domain of science has only occasionally been interrogated in the manner that scholarship on South Asia and other former colonial contexts has done, by moving away from George Basalla's influential diffusionist model of 'Western science' towards more sophisticated understandings of 'modern' science and empire as co-constituted, or a questioning of the moral and political underpinnings of 'Western science'.[48] Recent work that investigates how the implementation of a doctrine of 'self-management' – of scientific workers and their labour – by Yugoslavian socialist elites was linked to the creation of certain forms of scientific 'independence' from the 'West' represents a similar attempt in this direction.[49]

In short, approaching the topic holistically, with a mind prepared to question well-established orthodoxies of geography, science, discipline, gender, race, labour and so on, is essential to considering the task at hand: piecing together histories of women in science from fragmentary, often silent archives. One is reminded here of Saidiya Hartman's wonderfully instructive work on Atlantic slavery: 'I had been looking for relatives whose only proof of existence was fragments of stories and names that repeated themselves across generations.'[50] Piecing together a story from such fragmentary archives should perhaps begin with the recognition of a salient, albeit rarely acknowledged truth – one amply demonstrated by Hartman's work – namely, that writing histories of erased, invisible or marginalized groups is difficult, time-consuming work.[51] Not infrequently, such work is rendered even more complicated by the entrenched power dynamics of academic research and translation in(to) the English language, which ensure, for example, that vibrant intellectual debates, often published in other languages, are only occasionally available beyond their country of publication.

Building on such scholarly precedents but also attempting to overcome some of their limitations, the chapters in this volume examine the in/visibility of women across several specialized and non-specialized sites of science-making, practice and circulation: the laboratory, the university, the clinic, the hospital, the home, the school and the media. By zooming in on various sites of science, but treating them in an interconnected manner, we also hope to probe the ways in which the construction of gendered identities in science, engineering and medicine in the twentieth century was not simply a matter of discourse but also of practice. This line of investigation allows us to address a related conundrum, namely the common conflation between being present in science and exercising agency. Understanding how women negotiated their positions and the gendered norms of science, engineering and medicine both at the level of discourse and practice will hopefully take us a step closer towards documenting their agency, not only their presence, in STEMM.

In/visibilities across sites of science

Laboratory cultures: Invisible innovators, visible scientific rebels

As several scholars have discussed, the laboratory was one of the most exclusionary sites of professional science for women in the twentieth century, where social and institutional hierarchies were most likely to be enforced and replicated.[52] Although the number of women working in laboratories around the world increased gradually during the twentieth century, many remained invisible inside and outside them, as the contributions in the first section also illuminate. The authors reveal how women from the white colonial elite were able to access higher education in the metropole in ways that were closed to many colonized women, but often faced insurmountable challenges in securing degrees and building a career in science in metropolitan scientific establishments. They also show how female scientists were more likely to acquire visibility in the academia and beyond as masculinised 'rebels', rather than theorists and innovators in their own.

Kathryn Keeble's chapter engages with scientific mobility in the context of the British Empire, examining specifically the case of Johannesburg-born scientist and teacher Reinet Maasdorp, who travelled to Britain in 1935 to pursue postgraduate studies in physics, upon securing a Beit Railway Trust Rhodesian Fellowship and a grant from Cape Town University. Maasdorp joined Ernest Rutherford's Cavendish Laboratory at Cambridge, the only woman of fifty postgraduate students to do so, and worked alongside her male peers, sometimes performing the most important parts of experiments. Notwithstanding Rutherford's support for women's rights and their full membership of the university, Maasdorp endured systematic gender-based discrimination in Cambridge and found it difficult to continue her work despite having published two co-authored papers in the *Proceedings of the Royal Society*. She ultimately failed to secure a degree and an academic career in science. As Keeble points out, Maasdorp's professional trajectory as a secondary-school science teacher was typical of many female Oxbridge graduates of her generation, who were discriminated against in institutions of higher education until the anti-discrimination and equal pay legislation of the 1970s.

The fact that Maasdorp became involved in political activism together with her husband John Fremlin reinforces the point that women's scientific careers are best examined through the lenses of inter-related domains of activity. Maasdorp understood all too well that scientific and political invisibility went hand in hand: there was little hope of changing one without trying to change the other (in this connection, see also Pető's chapter in this volume). Political activism – in Maasdorp's case, joining the Cambridge Socialist Society and the Cambridge Scientists' Anti-War Group – was not only a way to compensate for her invisibility and lack of recognition in professional science, but also an attempt to shape the political foundations of science. As Keeble explains, 'lateral thinking', that is, the practice of investigating alternative scientific roles ascribed to or assumed by women, as science educators, members of scientific organizations, and voluntary, unpaid scientific workers, is essential to recovering hidden historical figures like Maasdorp. This must, however, be coupled with a recognition of the fact that, as a member of the white colonial elite in South Africa and Rhodesia, Maasdorp was nevertheless able to acquire a certain degree of visibility that was systematically denied to Black women.

Nuala Proinnseas Caomhánach's discussion of Lynn Margulis' scientific career as a 'rebel' evolutionary biologist further illuminates the struggles women faced inside and outside the laboratory and testifies to their willingness to challenge disciplinary and patriarchal boundaries. Combining insights from cell biology, ecology and geoscience, Margulis' much-maligned theory of evolution was as synthetic in its origin as the idea it propagated, namely that evolution had begun with microbes, not animals, and that symbiosis, not competition, had been its driving force. Although Margulis did manage to forge a relatively successful career in academia, her visibility was predicated on her ability to leverage the power of media and navigate all the risks such a process of exposure involved. As Caomhánach cogently points out, 'without the making of a persona, the scientist remains obscure and invisible'. In the case of women scientists – Rosalind Franklin being another example – that invisibility was not only 'ascribed' by men, but it was also sometimes 'curated' by them.

The construction of a scientific persona in Margulis' case must also be seen against the background of American media's growing concern with science reporting in the 1960s–1970s. As Caomhánach discusses, the process of becoming visible involved several layers of 'de-gendering' and 're-gendering'. Media reporting on Margulis replicated entrenched gender biases by attaching non-scientific attributes to her. Dismissed as a theorist because she was a woman, she eventually 'gain[ed] acceptance and ... inclusion in the field' because of her image as a masculinized scientific rebel. The point then is that Margulis was able to control her image as an 'outsider' and 'scientific rebel' only to a certain extent; the visibility afforded her did not shelter her from insidious forms of sexism and discrimination.

In/visibilities across borders: Scientific collaborations and contestations

Keeble's chapter also draws attention to the importance of mobility in women's engagement with science in the twentieth century. The three chapters in this section provide further proof of the vital role mobility as well as personal and professional networks played in facilitating women's access to scientific careers and recognition. While men enjoyed easier access paths to scientific positions, for women the prestige and capital provided by travelling abroad was sometimes the only avenue available to improve their chances of securing positions in science. Forging international collaborations acquired particular urgency in situations of political and economic crisis, as Christine von Oertzen's study of the International Federation of University Women also shows. The Federation played an important role in the support of academic women, especially in the context of the Nazi dictatorship and the Holocaust.[53]

The chapters in this section also reveal how in/visibility operated in relation to women's scientific networks. Strategic friendships between women scientists from the Global North and Global South, such as that discussed in Savithri Preetha Nair's chapter, have been rather invisible. The same can be said of the quasi-imperial networks some Eastern and Central European states attempted to build after 1918, as they sought to expand their cultural, economic and scientific influence in the Middle East and the Balkans. In this context, some female medical doctors attempted to forge careers abroad, combining medical work with research on 'tropical diseases', as Adéla Jůnová Macková's chapter shows. By contrast, Emily Rees Koerner and Graeme Gooday's chapter on the International Conference of Women Engineers and Scientists addresses a much more visible and formalized type of scientific network.

Nair's chapter makes a case for using biographical exploration and studies of friendships between women scientists to uncover 'invisible academic networks' and understand women's experiences and strategies of survival in science. The discussion focuses on the life-long friendship and scientific collaboration between the British American botanist and cytogeneticist Eileen Jessie Whitehead Erlanson and Indian botanist and cytogeneticist Edavalath Kakkat Janaki Ammal. While Janaki Ammal's life is slowly becoming better known in India, thanks also to Nair's scholarship, Eileen Erlanson has been largely forgotten, despite being one of the first women in the world (if not, indeed, the first) to be awarded a PhD and DSc in genetics. Both women were high achievers, Eileen as a three-time National Research Council Fellow and Janaki as the first Indian woman to obtain a doctorate in botany and the first to secure a teaching position in an all-male institution of higher education. Both were also very committed to being acknowledged not only in traditional roles afforded to women – as educators – but also as researchers or 'recognized scientists'. As Nair points out, Eileen believed she 'could do more good for science in Research than in pure teaching', a position Janaki heartily endorsed. The analysis reveals the importance of female solidarity as a strategy of survival for women in science: Eileen left the US for India because she was unable to secure a job amid the worsening circumstances of the Great Depression, while colonial scientists like Janaki developed a 'creative counterculture' by forging transnational connexions with other women scientists outside South Asia to sustain scientific endeavours and advance their careers. Although different in temperament and family background, they used their friendship to support their careers and goals. This case study also illustrates women scientists' different perceptions of marriage and family life: Janaki, like some of the Japanese scientists discussed in Ogawa's foreword, regarded marriage as an impediment to a career in science, while for Eileen the two were not necessarily incompatible. Quite the contrary, she believed that finding a spouse could contribute to a woman's financial independence and her mental and social well-being, ultimately benefiting her career.

Macková's chapter discusses Czech physician Vlasta Kálalová Di-Lotti's efforts to establish a medical and scientific career for herself in the Middle East, revealing the alliances women forged with (powerful) men, in this case the first Czechoslovak president, Tomáš Garrigue Masaryk, in their quest to acquire visibility as scientists. Kálalová Di-Lotti, a graduate of the Faculty of Medicine in Prague, belonged to the second generation of university-educated women, who were generally expected to become physicians in their hometowns upon graduation. Like Erlanson and Janaki Ammal, she harboured the ambition of being recognized as a researcher, not only as a medical practitioner. Keen to pursue research on the transmission and prevention of tropical diseases like leishmaniasis, Kálalová Di-Lotti first moved to Istanbul and later to Baghdad, with the intention of establishing an institute of research on tropical diseases there, to be linked to the National Institute of Health in Prague.

In the event, she was unable to escape the patriarchal structures of science in her home country, despite having distanced herself physically from it. Her ambition to establish an institute and herself as a researcher of tropical diseases failed to materialize. Although she shared her research on leishmaniasis with scientists from the State Institute of Health in Prague, as Macková points out, 'scientific cooperation with Czechoslovak scientists, which was to be the main focus of her future work in Iraq, failed'. When the gates of professional science were almost impenetrable, as was the case with Kálalová Di-Lotti, it is perhaps no surprise that many women researchers did not leave a substantial scientific output behind (in this case, a paper published in *The Journal of Czech Physicians*). By contrast, Kálalová Di-Lotti's impact as a medical practitioner is more visible and somehow easier to document through her small clinic in Baghdad and her work as a physician who helped deliver babies and performed complicated surgical procedures, including hymenoplasties.

As Macková points out, President Masaryk's financial and institutional support for Kálalová Di-Lotti's project must be seen in light of post-World War I Czech efforts to expand the country's influence in the Balkan Peninsula and the Middle East and secure new economic markets for its industrial products. The institute she wanted to establish was to function not only as a research centre on tropical diseases, but also as an outpost that advanced Czech cultural and commercial interests in the region. Although unsuccessful, Kálalová Di-Lotti's involvement in this project highlights the ways in which career opportunities for women were intertwined with the economic and geopolitical (quasi-imperial) agendas of the new states that emerged in East-Central Europe after World War I, seldom examined in previous scholarship.

By contrast, Rees Koerner and Gooday's chapter on the International Conference of Women Engineers and Scientists discusses a more visible type of scientific network which created opportunities for women to discuss technoscientific ideas and practices, as well as the challenges they encountered in their respective fields. The paper focuses on the first four meetings, held in the US, UK, Italy and Poland between 1964 and 1975, to show how participants challenged the dominant scientific agendas of the Cold War world, such as nuclear war race and Space Race, and how ICWES provided a platform for collaboration for women scientists beyond the East–West divide. Although Rees Koerner and Gooday caution against 'glib[ly] describ[ing] their geographical scope as global', the available data show that ICWES was regularly attended by women delegates from around thirty countries, with variations in regional visibility between different conferences: 'South American nations were most visible in ICWES 1 (USA); African nations were first apparent at ICWES 2 (UK); and only at ICWES 4 (Poland) was there broad participation from Eastern European nations, with the USA constituency (44 attendees) for the first time overshadowed by a large majority of native (Polish) participants (423 attendees).'

As the two authors emphasize, the ICWES archive provides an opportunity to place engineering firmly on the agenda of future historiographies of women in 'science', not least because engineering was less exclusionary a field than the fundamental or 'pure' sciences, as the participation of delegates from African, Asian and South American countries also demonstrate. This is not to say that certain countries did not feature more prominently in the organization of these conferences than others. Indeed, as the authors point out, 'the planning and programming of successive meetings relied on the national-level organisations for women in science and engineering most prominently in eight countries: Brazil, France, Italy, Japan, the Philippines, Poland, UK and USA'. Rees Koerner and Gooday's chapter also renders visible the intersections of collective and individual agencies of women in STEMM, for example, by showing that transnational collaboration between national bodies and the untiring effort of individual engineers like Ira Rischowski were essential to the success of these events.

The trans-regional approach they employ is particularly effective at exposing the politics and anxieties over the low number of women in science in the US and Western Europe, especially in the charged geopolitical climate of the Cold War, when comparison with the countries of the communist bloc became topical. As the authors point out, 'it was left ... to the communist countries participating in ICWES meetings to highlight their more developed capacity to educate, equip and enrol women for roles in engineering and applied science'. The political undertones of this mission cannot be ignored and delegates from communist countries likely regarded the promotion of women's participation in engineering and applied science as a tool of science diplomacy. Of course, the link between science, gender and politics was not emblematic of that context alone. All the contributions in this section show that the examination of personal and professional networks in science and their role in rendering women more visible cannot be separated from an examination of the broader political context in which their lives and careers unfolded.

In/visibilities in medicine and care: Treating, teaching, reforming

As in the case of science and engineering, some of the chapters in this volume also provide an opportunity for comparative study between different areas of medicine. Medicine and care have been among the most examined sites of science in relation to gender, education and women's work, but the papers in this section suggest new avenues of investigation. They focus in particular on neglected arenas like women's work as doctors and educators in secondary schools in Romania, women physicians' role in training midwives in Republican China and attempts made by female cadres, researchers, doctors and medical students to alleviate 'women's illnesses' and promote 'grassroot science' in Maoist China. These chapters highlight often-overlooked institutional contexts, ranging from secondary schools to rural health stations, and examine how political events, such as World War I and the Great Leap Famine,[54] both expanded and constrained the range of opportunities available to female physicians and scientists.

Camelia Zavarache's contribution addresses the importance of political change, border reconfiguration and state-building in the appointment of women as doctors in secondary-school education in interwar Romania, a time when the country significantly expanded its territory and minority population.[55] Zavarache argues that the state's agenda of providing hygiene education to secondary-school students was intertwined with the aim of producing a healthy 'body' for the Romanian nation. This played a crucial role in allowing women access to positions as school doctors, especially after the passing of the 1928 Law on Secondary Education which stipulated that female doctors should be employed to teach hygiene courses and conduct medical examinations in girls' schools. Although the socio-political and economic circumstances that followed World War I created a situation that helped render women more visible both as health professionals and 'modernising agents' of the new, enlarged Romanian state, Zavarache demonstrates that women often had to fight for the school positions they were legally entitled to.

Using archival documents which have been largely neglected in Romanian historiography, especially correspondence between women doctors and local authorities – as opposed to the more visible documents created by the central authorities which have usually attracted scholarly attention – she examines the complicated intersections of gender, age, language, ethnicity and rural–urban divides that enabled or hindered women's access to secondary-school positions. The chapter captures women's determination to fight for their careers, for example, through their constant, sometimes successful, petitioning of the Ministry of Public Instruction to be appointed to positions already occupied by male doctors. At the same time, however, it shows that these jobs were less financially appealing, on account of the low salaries they provided – indeed, male doctors usually juggled several positions, including as secondary-school doctors, in an attempt to secure a living. Put differently, the women doctors discussed here were not fighting for positions at the heart of the scientific establishment, which remained male-dominated, but for less attractive, gendered jobs which were considered more appropriate for them.

Jean Corbi's and Kathryn Edgerton-Tarpley's chapters also demonstrate how exceptional circumstances such as war and human-made disasters like famines provided opportunities for women to become active and exercise a considerable degree of agency in science. However, like Pető's concluding chapter on Hungary, they also qualify that narrative by showing that the opportunities created by situations of crisis did not necessarily translate into long-term structural change or increased visibility for women in the public and professional arenas of science (in fact, in Pető's case, women even ended up being excluded from universities again after the introduction of new legislation in 1921).

Corbi's chapter shows that the number of women doctors in Sichuan at the end of the Republic was significant both nationally and internationally. As Corbi puts it, 'almost one in ten doctors was a woman, bringing Sichuan closer to the national average in 1936 and placing the province on a par with American cities at the same time'. Women practitioners of Western medicine in Sichuan gained prominence and visibility between 1937 and 1945 because of the political context of the time – China's partial occupation by Japan, which saw Sichuan become the new centre of the country and resulted in extensive documentation produced by the public health administration. This extensive documentation generated by a situation of crisis ironically made women practitioners of Western medicine in Sichuan more visible in historical archives but, as Corbi shows in his analysis of the tensions between Western and Chinese medicine, it did not make them more visible in historiography and public memory. By focusing on the feminization of the medical profession in the interior, rather than the eastern parts of China, as previous scholarship has done, Corbi's chapter is also an attempt to render women medical practitioners in Sichuan more visible. The chapter shows that women's invisibility in historiography and public memory stemmed from their conceptualization as citizens in need of education and improvement, rather than educators in their own right, who 'fully participated in the propagation of "science" in China'. In a discussion that mirrors other contexts like India, Japan and Romania, Corbi points out that, 'Science for women generally derived from the gendered role they were assigned to, as mothers or homemakers, focusing on health, hygiene, and domestic science.' One of the implications for the history of medicine has been that women have become visible mostly as nurses or midwives in need of training,[56] rather than physicians who played an important role in training others, such as midwives.

In line with the more general focus of the volume, Corbi's contribution also documents women's transnational mobility in science, through its engagement with American missionary work in China, which enabled growing numbers of Chinese women to train and practise as physicians. In particular, the chapter provides fresh insights into the West China Union University, one of the main institutions where Chinese women physicians were trained. This institution of education was organized by a branch of the Women's Missionary Society, which had been active in Sichuan since the end of the nineteenth century.

The negotiation between Traditional Chinese Medicine (TCM) and 'modern' medicine is also captured in Edgerton-Tarpley's chapter, which discusses women as practitioners and subjects of medical care during the Great Leap Famine of the Mao era. The period witnessed a dramatic increase in 'women's illnesses', especially uterine prolapse and amenorrhea, an outcome of deteriorating socio-economic conditions, including malnutrition and overwork. As Edgerton-Tarpley discusses, women doctors became particularly visible in a political context that emphasized women's productive and reproductive roles and made acknowledgement of famine conditions taboo. The tragic circumstances of the famine made Chinese women doctors important and visible as a group on account of the medical, social and political functions they fulfilled, but they continued to remain invisible at an individual level.

This situation is perhaps best illustrated by the fact that most of the women Edgerton-Tarpley discusses remain anonymous in the archives of science. Invisible but not absent, 'women health workers and researchers' nevertheless 'played a key role in fostering Maoist "grassroots science" during China's famine-era campaign to treat "women's illnesses" (*funü bing*).' In a remarkable display of agency under exceptionally difficult circumstances, they not only tried to alleviate women's suffering, but also devised TCM therapies and published important texts on them, challenging official policies around women and medicine on the ground.

Intimate knowledge and in/visible domesticities: Science, medicine and the home

As a growing body of literature discusses, the private lives of women (and men) are as important to understanding their contributions to science as their professional, more public, scientific personas.[57] This also means that women did not become involved in the making, pedagogy and communication of science only in heavily institutionalized and professionalized settings like universities, laboratories or professional associations, but also in the confines of their homes. Their contributions to science were frequently intertwined with specific concerns of domesticity and motherhood that shaped their lives in many other ways. Focusing on the stereotype of the scientist and the institutionalized, professionalized sites of 'modern' science favours the 'accumulators' of scientific prestige and credits, rendering invisible the myriad contributions of women who rarely fit the profile of the model scientist. By including domestic spaces and domesticity in the equation, we can offer new insights into processes of knowledge-making and bring to life portraits of women involved in both small scientific innovations and large scientific breakouts.

Indeed, all chapters in this volume engage with women's personal lives at their intersection with science, underscoring a point made earlier in the introduction, that the five thematical sections into which the volume is divided should be treated as overlapping and interconnected. They also show how diverse women's experiences of domesticity and science were. For example, Keeble's chapter discusses Reinet Maasdorp's relationship with her husband John Fremlin and the ways in which marriage circumscribed her career options, even though Fremlin was supportive of her scientific activities. Two other scientific couples, William and Nora Wooster and Norman and Antoinette Pirie, are also discussed in that context. Somewhat similarly, Lynn Margulis' scientific achievements were overshadowed by her scientist superstar first husband, Carl Sagan, to whom she was sometimes compared. Vlasta Kálalová Di-Lotti's activity in Iraq was also profoundly shaped by her status as a wife and mother, as Macková discusses, eventually prompting her return to Czechoslovakia. Women also had diverging views of marriage and its impact on their careers: some, like Janaki Ammal or the Japanese scientists discussed in Ogawa's Foreword, chose celibacy, while others, like Eileen Erlanson, contracted several marriages in their quest for a fulfilled personal and professional life. When celibacy was not an option, as in the case of physician Tada Urata and botanist Hiro Ōhashi, women either disappeared on their wedding day or were trapped in unhappy marriages.

The three chapters in this section examine the intersections of domesticity and science especially in relation to medicine and education. Saurav Rai's contribution focuses on Yashoda Devi, a woman Ayurvedic practitioner in colonial and post-colonial India, who navigated the social and gender biases of this field to promote a reformist agenda of healthcare and domestic education for women, eventually becoming a prominent and influential health practitioner. Devi's remarkable public visibility as an Ayurvedic practitioner was made possible by media of communication like the post and the exploding print market in India in the first half of the twentieth century. Her medical practice was so successful that correspondence addressed to her only by name, by women seeking medical advice, usually found its way to the addressee. Her success as a best-selling author is even more remarkable in a context of low literacy levels, especially among female readers.

Public popularity notwithstanding, Devi was and continues to be regarded as a marginal figure in the Ayurvedic movement. As Rai explains, Ayurvedic practice reinforced patriarchy and regarded women as preservers of the family's health with responsibility for the 'scientific' management of the household. This was a view that Devi herself propagated, even as she simultaneously criticized gender hierarchies and violence against women, for example, in her writings on women's sexual consent or sexual pleasure. Devi's case demonstrates how women struggled to have their voices heard in an Ayurvedic practice dominated by a patriarchal discourse. It also shows that, even within this patriarchal framework, her status as a woman was instrumental in challenging ideas about women's social roles and making Indian women in general more visible, for example, by editing an encyclopaedia of 100 Indian female historical figures.

Anne Hardgrove uses historical analysis and interviews with three graduates of Lady Irwin College in New Delhi – Shyamala Gopalan, 'Ritu' and Flower Siliman – to document the establishment of domestic science for women in India, a development which was crucial to opening scientific opportunities for them, but also played an important political and ideological role. As Hardgrove puts it, 'domestic science became the perfect "back door" to women seeking careers, and an innovative way of bringing [them] into both higher education as well as scientific fields'. Studying domestic science offered a certain degree of invisibility to women, allowing them to escape the male gaze, if temporarily, and in some cases becoming a stepping stone towards a career in science. Hardgrove argues that this was because the 'scientific curriculum specially designed and carved out for women ... did not interrupt the male prerogative in science, and [was] seen as in support of, and not at counterpoint to, women's roles as good wives and mothers'. The chapter also moves beyond the usual concern with the British Empire to investigate how American home economics shaped the development of domestic science in India, for example, during the 1950s–1960s, when collaborations were initiated with institutions like the University of Tennessee-Knoxville, leading to exchanges of experts in this field. As Hardgrove shows, considerations of class and caste circumscribed the teaching of home science in India, a subject that paradoxically rendered invisible domestic servants, a ubiquitous presence in most middle- and upper-class Indian households. In this particular context, the visibility of middle- and upper-class women came at the cost of the visibility of their lower-class (and -caste) servants.

Finally, Marga Vicedo's study of the American college educator Clara Parks' contributions to the development of child medicine and the science of autism draws attention to mothers as a social group whose scientific contributions have been neglected even in scholarship that has examined the intersections of domesticity and science. As she points out, this neglect is largely the outcome of the 'modern conception of scientific objectivity as the absence of subjective emotional attachment'. Drawing on personal and institutional archives, she documents Parks' challenging journey as the mother of an autistic child, Jessica, and parent-expert in autism who lacked professional scientific credentials and was even diagnosed as a 'refrigerator mother' – a mother whose cold and intellectual nature was responsible for her daughter's 'retreat into autism'.

Vicedo places the discussion of Clara Park's research on autism within a longer twentieth-century trend of devaluing women's observational skills, a devaluation which was intimately connected with their inability to access formal education and enter commensurate scientific careers. As she explains, 'not only were women kept out of science, but science increasingly disregarded skills that were associated with women'. Park's encounter with the US Navy psychologist Bernard Rimland, himself the father of an autistic son and published critic of psychoanalytic accounts that blamed mothers for their children's autism, marked a turning point in her life and career as a parent-expert and, eventually, published author and public speaker on autism. Here, again, we are reminded of the importance of personal and scientific solidarities to navigating difficult circumstances and building visibility that can help open a conversation about, if not always legitimize, forms of knowledge that have been rejected by the professional scientific establishment.

Towards visible change? Publics, pedagogies and politics of science

The first four sections of the volume focus primarily on identifying and discussing historical patterns of exclusion and obscuration of women, as well as the strategies they employed to forge careers in science and receive recognition for their scientific work. The final section is oriented towards the future, as attention shifts to the audiences of science, epistemologically just practices of science communication and a more inclusive politics of science-making and scientific writing. Indeed, audiences have long been ignored in research on the history of science, not least because their historical engagement with science is notoriously difficult to document. This has been particularly the case with segments of the population such as women and minority groups that have been most affected by policies and practices which have contributed to making women invisible.

Evangelia Chordaki's chapter discusses the social and gender power relations that shape science communication around a topic that is central to women's lives – gynaecology, birth control and contraception – but in which they have been paradoxically marginalized as producers of scientific knowledge and actors of scientific debate. She argues for an understanding of the public sphere that underscores its political dimensions and its interconnections with 'social and epistemic justice', pointing out that traditional conceptualizations have promoted the invisibility of marginalized social groups. Her analysis of public debates around birth control in Greece after the end of the far-right military dictatorship known as 'the Greek junta' (1967–74) reveals the complicated links between scientific discourse, social pressure, activist practices and political agendas. Debates about women's health, she argues, offer a compelling illustration of how the public spheres and the audiences of science are constructed, how scientific authorities and politicians are controlling the production and circulation of knowledge, how they gender communication roles and how they rank audiences, rendering suitable audiences relevant and marginalizing other groups.

In other words, both in the Greek case discussed here and in other contexts – for example, the recent re-evaluation of the legislation on abortion rights in the US – those who are most affected by these policies are largely excluded from the production and communication of relevant scientific messages and are marginalized as audiences. The scientific message is not addressed to women but is transformed into a political message that targets more politicized and more simplified audiences. Chordaki's chapter moves beyond a simple analysis of these exclusionary practices of science communication to become an exercise in rendering both marginalized science communicators and audiences more visible. She argues for the incorporation of feminist perspectives into the study of women in science, by focusing on feminist actors' scientific communications on birth control in Greece, disseminated through feminist media and targeted at women.

Bethany G. Anderson and Kristen Allen Wilson continue this focus on audiences, this time at their intersection with public engagement and the pedagogy associated with the history of science. Attention shifts from the mass media of science communication to archival repositories and museums to explore the ways in which curatorial practices have rendered women and ethnic and racial minorities largely invisible in the archives of science, but also to consider concrete strategies of elevating the visibility of these scientific actors among the collections. The chapter discusses a joint initiative by the University of Illinois Archives and the Illinois Distributed Museum to create a more inclusive archive of science through the organization of a Women in Science Lecture Series that advocates new ways of engaging with archives and museums. Acknowledging the fact that women scientists usually conduct interdisciplinary research as members of several campus units and that their under-representation across disciplines takes different forms, the project provides a platform for women from different scientific fields to discuss their careers and lives in science. Their lectures, which are widely publicized, are recorded and used as 'biographical "vertical file[s]"' to create a different archive of science for future generations. The initiative not only provides a voice to women scientists, but also enables audiences to become involved in this process of knowledge- and archive-making by providing feedback via questionnaires.

Andrea Pető's contribution concludes the volume with a discussion of women in science in twentieth-century Hungary, against the backdrop of alternating authoritarian and more democratic political regimes. Like the other papers in this section, she looks both to the past and the future, in this case to answer a question with bittersweet undertones: how to be a woman in science and laugh? Turning to the past, Pető argues, is a necessary step in identifying historical patterns of exclusion and discrimination and attempting to make predictions for the future. But it is also a warning that the fight for equality and equity is an ongoing project. For instance, she points out that in the past women often preferred to remain in less-prominent, less-visible positions to survive the patriarchal structures of the Hungarian scientific and academic establishment. The situation is similar to the contemporary one, where male project directors have a greater chance of receiving research funding than women. To have their research funded, women often write grant applications, but prefer to appoint male colleagues as project directors, working as researchers under their leadership. As Pető pertinently concludes, 'women researchers have a better chance of long-term scientific success if they choose the invisible status of "servant" rather than seeking individual, i.e., institutionally recognized, laurels'.

The past, Pető seems to remind us, can be a disheartening country, but the future need not necessarily be so; indeed, the present offers opportunities for change. The chapter bridges historical research with contemporary politics and policies around women's participation in science to propose a three-pronged agenda for a different future. First, Pető argues, we need to change the numbers of women in science, which requires more inclusive policies of recruitment and affirmative action. Second, the institutions of science must change. To achieve gender equity, a strong top-down, political intervention is crucial, and it must include equal treatment, equal opportunities and coherent and intensive efforts at gender mainstreaming. Finally, the knowledge field must also change. Investigating women's role in the production and communication of knowledge and changing the historiographical paradigm are essential to rescuing the perspectives and achievements of women from the peripheries of science-making, policies and practice, and bringing them into the political and historiographical mainstream.

It is difficult to write the concluding paragraphs for a volume whose main agenda has been to show how diverse women's experiences of in/visibility in twentieth-century science, engineering and medicine have been. One point that stands out is that the relationship between the percentage of women in these fields (to the extent that official statistics can be trusted) and their visibility in STEMM and the historiography of science is rarely straightforward, since even in those settings where women were well represented in science, such as Central and Eastern Europe, they have been largely invisible as scientific and historical actors. The reasons for this have to do with the asymmetrical power relations that have informed not only the making of science and scientists, but also the writing of their histories. Women's in/visibility in twentieth-century STEMM, this volume argues, was intersectional, with factors like gender, class, caste, race, sexuality, marital status, rural-urban divides and so on playing out differently in different geographical and historical settings. Although invisibility was more often than not a mark of structural inequality, discrimination and lack of recognition, in some cases it ironically offered opportunities for women to pursue science and a strategy to circumvent some of the obstacles they faced within the scientific establishment. Indeed, moving away from an archive that focuses only on professional science towards a multisited exploration of women's engagement with science is an important strategy for recovering their experiences as producers of knowledge, educators, medical practitioners and science communicators in the twentieth century. The volume not only shows that women were present in STEMM, but also seeks to document how they exercised agency by negotiating their circumstances and in/visibility in these fields.

The chapters are not only an exercise in rendering women more visible but also discuss practical solutions to challenging gender inequalities in STEMM, showing that change must be concerted and it must happen at an institutional, political and intellectual level. Changing the field of knowledge, for example, requires mainstreaming women's contributions to STEMM into the archives, histories and teaching of science, as well as its public engagement agendas. Instead of being dismissed as insignificant and unrepresentative, women's participation in STEMM should be regarded as essential to understanding process of knowledge-making in these fields. Such an approach should also seek to expand the geography of research and destabilize some of the power asymmetries that have long circumscribed scholarship on gender and women in STEMM, for example, by moving away from the conventional focus on the 'Western' world and archives dominated by the English language.

One point on which all the papers in this volume seem to agree is that much difficult work still lies ahead of us. The trick, especially in the face of challenging political circumstances, is to find ways to deliver some of this work and not collapse under the magnitude of the task. Perhaps laughter is a useful antidote to some of the adversities we are facing (if not, indeed, the absurdity of the situations we often find ourselves in). In this particular case, it can also act as a powerful reminder of the importance of leaning on each other, of forging networks of solidarity and collaboration and of the passion that drove many women to science in the first place, as the photograph on the cover of this volume also attests.

Notes

1 In 2019, UNESCO reported that 45.8 per cent of science researchers in Romania were female, as opposed to 28 per cent for Germany and 16.2 per cent for Japan. Region-wise, the following percentages were reported, based on data from 2016: 48.2 per cent for Central Asia, 45.1 per cent for Latin America and the Caribbean, 41.5 per cent for the Arab States, 39.3 per cent for Central and Eastern Europe, 32.7 per cent for North America and Western Europe, 31.8 per cent for Sub-Saharan Africa, 23.9 per cent for East Asia and the Pacific, 18.5 per cent for South and West Asia. Institute for Statistics, 'Women in science', Fact Sheet No. 55, June 2019, FS/2019/SCI/55, http://uis.unesco.org/sites/defa ult/files/documents/fs55-women-in-science-2019-en.pdf (accessed 17 August 2023).
2 E.g., A. H. Koblitz, 'Life in the fast lane: Arab women in science and technology', *Bulletin of Science, Technology & Society*, 36: 2 (2016), 107–17. On gender inequality in communist Romania, see L. M. Jinga, *Gen și reprezentare în România comunistă, 1944–1989: Femeile în cadrul Partidului Communist Român* [*Gender and Representation in Communist Romania, 1944–1989: Women and the Romanian Communist Party*] (Bucharest: Polirom, 2015).
3 Rossiter's comment about the situation of women scientists in post-World War II US could well be applied to post-communist societies like Romania: 'For many years there was little consciousness that these attitudes and practices might constitute something as ugly as discrimination. It was just the way it was.' M. Rossiter, *Women Scientists in America: Before Affirmative Action, 1940–1972* (Baltimore, MD: Johns Hopkins University Press, 1995), p. xvi.
4 A. F. Scott, 'On seeing and not seeing: A case of historical invisibility', *The Journal of American History*, 71: 1 (1984), 405.
5 E.g., M. B. Ogilvie and J. D. Harvey (eds), *The Biographical Dictionary of Women in Science: Pioneering Lives from Ancient Times to the Mid-Twentieth Century*, 2 vols. (New York and London: Routledge, 2000); C. G. Jones, A. E. Martin and A. Wolf (eds), *The Palgrave Handbook of Women and Science since 1660* (Cham: Palgrave Macmillan, 2021). On gender and the historical 'canon', see A. Pető, 'From visibility to analysis: Gender and history', in C. Salvaterra and B. Waaldijk (eds), *Paths to Gender: European Historical Perspectives on Women and Men* (Pisa: PLUS Pisa University Press, 2009), pp. 1–9.

6 P. G. Abir-Am, 'Series foreword', in H. M. Pycior, N. G. Slack and P. G. Abir-Am (eds), *Creative Couples in the Sciences* (New Brunswick, NJ: Rutgers University Press, 1996), p. ix.
7 On intersectionality, see K. W. Crenshaw, 'Demarginalizing the intersection of race and gender: A Black feminist critique of antidiscrimination doctrine, feminist theory and antiracist politics', *University of Chicago Legal Forum* (1989), 139–68. On intersectionality and in/visibility in STEMM, see M. Nash and R. Moore, 'In/visible: The intersectional experiences of women of color in science, technology, engineering, mathematics, and medicine in Australia', *Gender, Work & Organization*, 31: 3 (2024), 693–709.
8 A. Menyhért, *Women's Literary Tradition and Twentieth-Century Hungarian Writers: Renée Erdős, Ágnes Nemes Nagy, Minka Czóbel, Ilona Harmos Kosztolányi, Anna Lesznai*, trans. A. Bentley (Leiden: Brill NV, 2020), p. 4.
9 J. Devika, *En-gendering Individuals: The Language of Re-forming in Early Twentieth-Century Keralam* (New Delhi: Orient Longman, 2007), p. 32.
10 M. L. Shetterly, *Hidden Figures: The Untold Story of the African American Women Who Helped Win the Space Race* (London: William Collins, 2016), p. xv.
11 A. Sur, *Dispersed Radiance: Caste, Gender, and Modern Science in India* (New Delhi: Navayana Publications, 2011), p. 16.
12 Ogilvie and Harvey (eds), *The Biographical Dictionary of Women in Science*; Jones, Martin and Wolf (eds), *The Palgrave Handbook of Women and Science since 1660*; C. M. C. Haines and H. M. Stevens, *International Women in Science: A Biographical Dictionary to 1950* (Santa Barbara, CA and Oxford: ABC-Clio, 2001); M. Hargittai, *Women Scientists: Reflections, Challenges, and Breaking Boundaries* (Oxford: Oxford University Press, 2015).
13 E.g., J.-R. Yoon, 'Korean women in science and technology', *Asian Women*, 10 (2000), 33–42; G. Kawano and M. Ogawa (eds), *Josei kenkyūsha shienseisaku no kokusaihikaku: Nihon no genjō to kadai [International Comparison of Support Policies for Women Researchers: Current Scenario and Issues in Japan]* (Tokyo: Akashi Shoten, 2021). We are grateful to Jaehwan Hyun for drawing our attention to Eunkyoung Lee's Korean-language publications on the history of science and women's policy and her work as the 'architect of the women in science policy in South Korea'. Jaehwan Hyun, electronic communication, 8 May 2024.
14 Jaehwan Hyun, electronic communication, 8 May 2024.
15 E.g., F.-T. Fan, 'Science in cultural borderlands: Methodological reflections on the study of science, European imperialism, and cultural encounter', *East Asian Science, Technology and Society: An International Journal*, 1: 2 (2007), 213–31; D. Arnold, *Everyday Technology: Machines and the Making of India's Modernity* (Chicago, IL: University of Chicago Press, 2013); A. Ghosh, *Making It Count: Statistics and Statecraft in the Early People's Republic of China* (Princeton, NJ: Princeton University Press, 2020); B. C. Iacob, 'Malariology and decolonization: Eastern European experts from the League of Nations to the World Health Organization', *Journal of Global History*, 17: 2 (2022), 233–53.

On intra-Asian knowledge exchanges, see N. Green, *How Asia Found Herself: A Story of Intercultural Understanding* (New Haven, CT: Yale University Press, 2022).
16 E.g., S. G. Harding, *Sciences from Below: Feminisms, Postcolonialisms, and Modernities* (Durham, NC: Duke University Press, 2008); L. Schiebinger, 'Feminist history of colonial science', *Hypatia*, 19: 1 (2004), 233–54; M. Mayberry and B. Subramaniam (eds), *Feminist Science Studies: A New Generation* (New York and London: Routledge, 2001).
17 Sur, *Dispersed Radiance*, p. 17.
18 E.g., J. G. Crowther, *Scientific Types* (London: Barrie & Rockliff, 1968); J. Daintith, S. Mitchell, E. Tootill and D. Gjertsen, *Biographical Encyclopedia of Scientists*, 2 vols. (Bristol: Institute of Physics Publishing, 1994).
19 M. W. Rossiter, 'The Matthew Matilda effect in science', *Social Studies of Science*, 23: 2 (1993), 325–41.
20 Abir-Am, 'Series foreword', p. ix.
21 E. Yong, 'The women who contributed to science but were buried in footnotes', *The Atlantic* (11 February 2019), www.theatlantic.com/science/archive/2019/02/womens-history-in-science-hidden-footnotes/582472/ (accessed 11 September 2023).
22 M. Bucur, 'From invisibility to marginality: Women's history in Romania', *Women's History Review*, 27: 1 (2018), 49.
23 E.g., L. Jordanova, 'Gender and the historiography of science', *The British Journal for the History of Science*, 26: 4 (1993), 469–83; S. G. Kohlstedt, 'Women in the history of science: An ambiguous place', *Osiris*, 10 (1995), 39–58; L. Schiebinger, *The Mind Has No Sex? Women in the Origins of Modern Science* (Cambridge, MA: Harvard University Press, 1989); N. Kumar (ed.), *Women and Science in India: A Reader* (New Delhi: Oxford University Press, 2009); Sur, *Dispersed Radiance*; N. Kodate and K. Kodate, *Japanese Women in Science and Engineering: History and Policy Change* (London: Routlege, 2017); M. Ogawa, 'Nihon no STEMM bun'ya ni okeru josei jinzai no rekishi' ['A history of women's resources in STEMM fields in Japan'], *Kagaku gijutsu shakairon kenkyū*, 19 (2021), 43–52.
24 See, e.g., a recent volume on women in Earth sciences in India, a field long neglected by historians of science: V. Rai, *Leading Ladies in the Earth Sciences in India* (Lucknow: Book Rivers, 2022).
25 E.g., Jordanova, 'Gender and the historiography of science'; Schiebinger, *The Mind Has No Sex?*; Sur, *Dispersed Radiance*; A. Shteir and B. Lightman (eds), *Figuring It Out: Science, Gender and Visual Culture* (Hanover, NH: University Press of New England, 2006).
26 Devika, *En-gendering Individuals*, p. 14. E.g., a recent volume reflects on the position of trailblazing Hungarian women in fields like astronomy, botany, mineralogy, mathematics, medicine and biochemistry, but ends up reinforcing gender stereotypes instead of critically interrogating them: 'As Margaret Alic writes, the cultivation of science takes intelligence, creativity, proper upbringing and dedication. These four prerequisites certainly apply to astronomy,

particularly if augmented by endurance, vigilance, outstanding mathematical skills (obviously) and proficiency in the use of various devices and techniques. Although a survey of the career of women engaged in astronomy through history cannot pretend to account for the role of women across the board in the civilization of a given era, it can clearly supply important details for a better understanding of that role.' R. M. Cristian and A. Kérchy (eds), *Pioneering Hungarian Women in Science and Education* (Budapest: Akadémiai Kiadó, 2022), p. 50.

27 Devika, *En-gendering Individuals*, p. 14; Pető, 'From visibility to analysis', p. 3.

28 National differences can be observed here. While Romania had no grassroots feminist movement, women's organizations with transnational connections emerged in Bulgaria and Yugoslavia during the Cold War.

29 E.g., Jinga, *Gen și reprezentare*; K. Ghodsee, *Second World, Second Sex: Socialist Women's Activism and Global Solidarity During the Cold War* (Durham, NC: Duke University Press, 2019); S. Penn and J. Massino (eds), *Gender Politics and Everyday Life in State Socialist Eastern and Central Europe* (New York: Palgrave MacMillan, 2009); C. Donert and C. Moll-Murata (eds), 'Women's rights and global socialism', special issue, *International Review of Social History*, 67, 30 (2022): 1–262; Z. Lóránd, *The Feminist Challenge to the Socialist State in Yugoslavia* (London: Palgrave Macmillan, 2018); A. Pető, 'A history of the Hungarian Women's Movement and feminism', *Feminizmas, visuomenė, kultūra* (January 2001), 151–60; C. Bonfiglioli and S. Žerić, 'Working class women's activism in socialist Yugoslavia: An exploration of archives from Varaždin, Croatia', *Comparative Southeast European Studies*, 70: 1 (2022), 80–102; C. Bonfiglioli, 'Women's internationalism and Yugoslav-Indian Connections: From the Non-Aligned Movement to the UN Decade for Women', *Nationalities Papers*, 49: 3 (2020), 446–61.

30 Pető, 'From visibility to analysis', p. 6. See also N. Funk and M. Mueller (eds), *Gender Politics and Post-communism: Reflections from Eastern Europe and the Former Soviet Union* (New York and London: Routledge, 1993).

31 E.g., Jinga, *Gen și reprezentare*; O. Băluță, *Gen și putere: Partea leului în politica românească* [*Gender and Power: The Lion's Share in Romanian Politics*] (Iași: Polirom, 2006); G. Kligman, *Politicile de gen în perioada postsocialistă* [*Gender Policies in the Post-Socialist Period*] (Iași: Polirom, 2006); Ș. Mihăilescu, *Din istoria feminismului românesc: Studiu și antologie de texte (1929–1948)* [*From the History of Romanian Feminism: Study and Anthology of Texts (1929–1948)*] (Iași: Polirom, 2006); M. Bucur, *Gendering Modernism: A Historical Reappraisal of the Canon* (New York: Bloomsbury, 2017); T. Văcărescu, *Personajele acestea de a doua mână: Din publicațiile membrelor Școlii Sociologice de la București* [*These Second-Hand Characters: From the Publications of the Members of the Bucharest Sociological School*] (Bucharest: EIKON, 2018); J. Massino, *Ambigous Transitions: Gender, the State, and Everyday Life in Socialist and Postsocialist Romania* (New York and Oxford: Berghahn, 2019); O. Zamfirache (ed.), *Ea: Perspective feministe*

asupra societății românești [*She: Feminist Perspectives on Romanian Society*] (București: Curtea Veche, 2019).

32 In June 2020, the Romanian Parliament passed an amendment to Article 7 of the Law of National Education which prohibited, in all institutions of education, activities that 'spread the theory or opinion of gender identity, understood as the theory or opinion that gender is a different concept from biological sex and that the two are not always the same'. The amendment was met with considerable public backlash, especially from universities and researchers, who signed a memorandum opposing it. The Romanian president Klaus Iohannis notified the Constitutional Court which ruled, in December 2020, that the amendment was unconstitutional because it violated freedom of thought, opinion and expression.

33 D. N. Livingstone, *Putting Science in Its Place: Geographies of Scientific Knowledge* (Chicago, IL and London: University of Chicago Press, 2003), p. 7.

34 M. Palasik, 'Women in technological higher education and in the sciences in 20th century Hungary', *Hungarian Studies Review*, 29: 1–2 (2002), 25.

35 G. Kubica, trans. B. Koschalka, *Maria Czaplicka: Gender, Shamanism, Race: An Anthropological Biography* (Lincoln: University of Nebraska Press, 2020).

36 Y. Furukawa, *Tsuda Umeko: Kagaku he no michi, daigaku no yume* [*Umeko Tsuda: The Road to Science and the Dream of Founding a University*] (Tokyo: Tokyo University Press, 2022).

37 E.g., three of the four papers featured in the 'Historical context' section in Kumar (ed.), *Women and Science in India* are on medicine. See also G. Forbes, *Women in Colonial India: Essays on Politics, Medicine, and Historiography* (New Delhi: Chronicle Books, 2005); S. Sehrawat, *Colonial Medical Care in North India: Gender, State, and Society, c.1830–1920* (Oxford: Oxford University Press, 2013); H. Fujimoto, 'Women, missionaries, and medical professions: The history of overseas female students in Meiji Japan', *Japan Forum*, 32: 2 (2020), 185–208. While gender, medicine and European imperialism have attracted a considerable degree of attention, more recent research has also started engaging with the topic in the context of the Japanese Empire. E.g., S. M. Kim, *Imperatives of Care: Women and Medicine in Colonial Korea* (Honolulu: University of Hawaii Press, 2019); H. Fujimoto, A. Homei and E. Nakamura (eds), *Medical Women in the Japanese Empire: Sources and Critique* (London: Routledge, forth. 2025).

38 S. P. Nair, *Chromosome Woman, Nomad Scientist: E. K. Janaki Ammal, A Life 1897–1984* (Abingdon and New York: Routledge, 2023), p. 537.

39 We are grateful to Jaehwan Hyun for emphasizing this point in relation to the scientific education of women in colonial Korea. See, e.g., A. Burton, 'Contesting the zenana: The Mission to "Make Lady Doctors" for India, 1874–1885', *Journal of British Studies*, 35: 3 (1996), 368–97; Y.-J. Sun, 'The emergence of a pioneering female scientist in Korea: Biographical research on Sam Soon Kim', *Asian Women*, 35 (2019), 69–89.

40 G. Y. Shen, 'Women and the transnational dynamics of science education in early twentieth century China: A quiet revolution', *Chinese Annals of History of Science and Technology*, 3: 2 (2019), 89.

41 G. Nazarska, 'An (un)established academic and scientific network: Branches of the International Federation of University Women on the Balkans (1920–1950s)', *Balkanistic Forum*, 31: 1 (2022), 32–58; G. Nazarska, 'Opportunities for an academic career of women scientists at the Bulgarian Academy of Sciences (mid-1940s–1980s)', *Balkanistic Forum*, 30: 1 (2021), 120–37.

42 D. Haraway, 'Situated knowledges: The science question in feminism and the privilege of partial perspective', *Feminist Studies*, 14: 3 (1988), 575–99. Original emphasis. In this connection, see also Livingstone's argument that 'What passes as science is contingent on time and place'. D. N. Livingstone, *Putting Science in Its Place: Geographies of Scientific Knowledge* (Chicago, IL: University of Chicago Press, 2003), p. 13.

43 N. Oreskes, 'Objectivity or heroism? On the invisibility of women in science', *Osiris*, 11 (1996), 89.

44 E.g., A. Digby, W. Ernst and P. B. Mukharji (eds), *Crossing Colonial Historiographies: Histories of Colonial and Indigenous Medicines in Transnational Perspective* (Newcastle upon Tyne: Cambridge Scholars Publishing, 2010); M. Elshakry and S. Sivasundaram (eds), *Science, Race and Imperialism*, vol. 6 of *Victorian Science and Literature* (London: Chatto and Pickering, 2012); Fan, 'Science in cultural borderlands'; Special Issue: 'Colonial science in former Japanese Imperial Universities', *East Asian Science, Technology and Society: An International Journal*, 1: 2 (2007).

45 Fan, 'Science in cultural borderlands'; J. A. Secord, 'Knowledge in transit', *Isis* 95: 4 (2004), 654–72; K. Raj, *Relocating Modern Science: Circulation and the Construction of Knowledge in South Asia and Europe, 1650–1900* (Basingstoke: Palgrave Macmillan, 2007); L. Fleetwood, *Science on the Roof of the World: Empire and the Remaking of the Himalaya* (Cambridge: Cambridge University Press, 2022); A. Bonea, 'Owning the (deep) past: Paleontological knowledge and the political afterlives of fossils', *History of Knowledge* (25 July 2023), https://historyofknowledge.net/2023/07/25/paleontological-knowledge/. The discourse/practice duality has also been discussed in feminist history, e.g., J. W. Scott, *Gender and the Politics of History* (New York: Columbia University Press, 1988); L. Jordanova, *History in Practice* (London: Bloomsbury Academic, 2019).

46 The literature on this topic in relation to the history of science in this region is sparse, but some works discuss the 'Orientalization' of Eastern Europe and its struggle to construct a 'Western' identity for itself. E.g., M. Buchowski, 'The specter of Orientalism in Europe: From exotic other to stigmatized brother', *Anthropological Quarterly*, 79: 3 (2006), 463–82; M. Todorova, *Imagining the Balkans* (Oxford: Oxford University Press, 1997). See also Katherine Verdery's nuanced interrogation of the Cold War dichotomies which replaced the 'self' versus 'other' and 'the metropole' versus 'the colony' with 'West' versus 'East': K. Verdery, 'Nationalism, postsocialism, and space in Eastern Europe', *Social*

Research, 63: 1 (1996), 77–95; K. Verdery, 'Whither postsocialism?' in C. Hann (ed.), *Postsocialism: Ideals, Ideologies and Practices in Eurasia* (London: Routledge, 2002), pp. 15–22.
47 A. Pârvulescu and M. Boatcă, *Creolizing the Modern: Transylvania across Empires* (Ithaca, NY and London: Cornell University Press, 2022).
48 G. Basalla, 'The spread of Western science', *Science*, 156: 3775 (1967), 611–22. For critiques of this model, see D. Raina, *Images and Contexts: The Historiography of Science and Modernity in India* (Oxford: Oxford University Press, 2003); Raj, *Relocating Modern Science*; F.-T. Fan, *British Naturalists in Qing China: Science, Empire and Cultural Encounter* (Cambridge, MA: Harvard University Press, 2004); H. Tilley, 'Global histories, vernacular science, and African genealogies; or, is the history of science ready for the world?' *Isis*, 101: 1 (2010), 110–19.
49 We are grateful to Vladimir Janković for bringing this scholarship to our attention.
50 S. Hartman, *Lose Your Mother: A Journey Along the Atlantic Slave Route* (New York: Farrar, Straus and Giroux, 2007), p. 13.
51 Joanna Behrman's description of the painstaking research involved in identifying the protagonists of a 1913 photograph taken in Pierre Weiss' laboratory at ETH Zurich, three of whom were women, is a good illustration of this. J. Behrman, 'A history mystery: Adventures identifying people in a photograph', *Ex Libris Universum*, Niels Bohr Library & Archives and Center for History of Physics of the American Institute of Physics, 16 February 2023, www.aip.org/history-programs/niels-bohr-library/ex-libris-universum/history-mystery-adventures-identifying?fbclid=IwAR0y923WJ3vJUwjNvRv7W-EByIlYZMAT5vAv-0-lFtBQMVOZffw63AYCONk> (accessed 21 September 2023).
52 Rossiter, 'The Matthew Matilda effect in science'; A. Sur, 'Dispersed radiance: Women scientists in C. V. Raman's laboratory', *Meridians*, 1: 2 (2001), 95–127.
53 C. von Oertzen, *Science, Gender, and Internationalism: Women's Academic Networks, 1917–1955* (New York: Palgrave Macmillan, 2014).
54 See P. Fara, *A Lab of One's Own: Science and Suffrage in the First World War* (Oxford: Oxford University Press, 2018) and Ogawa's Foreword in this volume.
55 On post-World War I state-building in Eastern Europe, see B. Trencsényi, *The Politics of 'National Character': A Study in Interwar East European Thought* (London and New York: Routledge, 2012); B. Olschowsky, P. Juszkiewicz and J. Rydel (eds), *Central and Eastern Europe after the First World War* (Oldenbourg: De Gruyter, 2021); I. Livezeanu, *Cultural Politics in Greater Romania: Regionalism, Nation Building, and Ethnic Struggle, 1918–1930* (Ithaca, NY and London: Cornell University Press, 1995).
56 E.g., N. E. Barnes, *Intimate Communities: Wartime Healthcare and the Birth of Modern China, 1937–1945* (Oakland: University of California Press, 2018).
57 E.g., Abir-Am and Outram (eds), *Uneasy Careers and Intimate Lives*; C. von Oertzen, M. Rentetzi and E. S. Watkins (eds), 'Beyond the academy: Histories of gender and knowledge', Special Issue of *Centaurus*, 55: 2 (2013); D. R.

Coen, 'The common world: Histories of science and domestic intimacy', *Modern Intellectual History*, 11: 2 (2014), 417–38; D. L. Opitz, S. Bergwik and B. Van Tiggelen (eds), *Domesticity in the Making of Modern Science* (New York: Palgrave Macmillan, 2015).

I

Laboratory cultures: Visible scientific rebels, invisible innovators

1

Breaking down the barriers at Cambridge in the 1930s: Reinet Maasdorp's experience at Rutherford's Cavendish Laboratory

Kathryn Keeble

The acceptance of women as full members of the academy was a long, drawn-out process. Tracing the biographical steps of Reinet Maasdorp indicates the severity of the gender bias and the discrimination perpetrated by scientific institutions like the Royal Society. This chapter examines the work Maasdorp did initially in the Cavendish Laboratory and the significance of her overall contribution to science. Maasdorp's story reminds one of the absurdity of excluding one half of the population from potential engagement in scientific projects that aim to improve the human condition. It is also a reminder that gender bias must be continuously obliterated; one of the ways we can guard against such injustice is to document past treatments handed out to exceptional women scientists. The sheer male hysteria that marked the occasions women sought equality in academia indicates the pernicious historical reality of discrimination against women.

Born in Johannesburg, South Africa, in 1912, Maasdorp moved with her family to Rhodesia, now Zimbabwe, when she was very young. Bright and studious, she graduated her secondary schooling with 'outstanding' results and secured a scholarship to university.[1] Maasdorp began her degree in 1930 and graduated with a master of science, which enabled a scholarship and a job as a demonstrator in first-year physics. This led to a recommendation that she should apply to Cambridge University. A senior lecturer visiting Cambridge from Cape Town obtained permission for Maasdorp to apply to the Cavendish Laboratory as a postgraduate student, the norm at the time.[2] Maasdorp arrived in Cambridge as a recipient of a Beit Railway Trust Rhodesian Fellowship and a grant from Cape Town University in 1935. She was attached to one of the two women's colleges, Newnham. The day after she arrived, she reported to the Cavendish Laboratory to meet with Deputy Director James Chadwick. At this somewhat tense meeting, Maasdorp asked Chadwick why women could not be members of the university. In the resulting discussion, Chadwick became agitated as, according to him, 'women were bad at administration, and couldn't manage their own affairs let alone the affairs of a university'.[3] After this

Figure 1.1 Reinet Maasdorp and the Cavendish Research Group, Cambridge, 1935 (courtesy of Margaret Kettlewell)

first encounter with the deputy director of the Cavendish Laboratory, Maasdorp could have been forgiven for thinking this was not an auspicious start to her studies.

Highlighting the changing attitude, or lack thereof, towards women's participation in science during the 1930s and beyond, Maasdorp's experience uncovers the barriers that discouraged women from pursuing a career in science in the twentieth century. Based on a single case study, I do not want to give the impression that Maasdorp is representative or prototypical of an entire group, 'women'. An understanding of intersectionality, as first articulated by Kimberlé Crenshaw in her landmark paper 'Demarginalizing the intersection of race and sex', is implicit in the analysis. Crenshaw argues that an analysis of racism and sexism can be distorted if attention is paid to members of privileged groups because these ideas are founded on experiences that only apply to some, not all.[4] In this context, the gap between Maasdorp's experience and those of Black South African women at the time is immense. Maasdorp was a white woman living in a country that perpetuated intense discrimination against the non-white population. The privilege afforded to Maasdorp in gaining entry to Cambridge University was something undreamed of for non-white women at that time. The situation for Black researchers is today still highly disappointing, with only 2 per cent of Black researchers, both women and men, participating in STEMM doctoral programmes in the UK.[5] What Maasdorp's experience at Cambridge in the 1930s can show is how white women from across the British Empire entered the workplace and navigated social and professional networks in a male-dominated profession such as academic science. Recovering women's experience and history, Guglielmo suggests, is a way of 're-collecting women's stories ... [often] written out of or written differently within public memory'.[6] There is a benefit in looking at a single case study to determine what barriers were put in place to prevent a successful outcome.

This chapter begins with an examination of women's situation before Maasdorp arrived at Cambridge. As Mariko Ogawa also discusses in the Foreword to this volume, the acceptance of women to full membership of the university was a drawn-out and often fraught process. It then outlines Maasdorp's experience as a postgraduate student working in the laboratory and underlines the discrimination against women. Next, it looks at gender bias and how the discrimination against women was perpetrated by institutions such as the Royal Society. The last section of this chapter looks at the paid and unpaid contributions of Maasdorp to science.

Women at Cambridge

When Maasdorp arrived at Cambridge University in 1935, the Cavendish Laboratory was at the forefront of experimental voyages into the atomic structure. Among many other ground-breaking discoveries, Cavendish researchers discovered the neutron and photographed isotopes of chemical elements. They produced the first controlled nuclear disintegrations induced by accelerating high-energy particles, thus demonstrating Einstein's 1905 special theory of relativity, $E = mc^2$. Between 1904 and 1935, the year Maasdorp arrived at Cambridge University, ten Cavendish researchers won a Nobel Prize.[7]

The Cavendish Laboratory first officially accepted women students in 1882.[8] Despite this, women graduates at the University of Cambridge could cite their qualifications but could not receive their degree in the ceremonies at Senate House. Unlike Deputy Director James Chadwick, Director of Cambridge's Cavendish Laboratory Ernest Rutherford promoted women's rights and supported extending full membership of women at university, including the conferring of degrees. Originally from New Zealand, Rutherford studied for his doctorate at the Cavendish Laboratory under J. J. Thomson. In 1898, he took up a professorship at McGill University, Montreal, Canada. His first graduate student was a woman, the physicist Harriet Brooks. In 1907, Rutherford returned to Britain, accepting the post of director of physics at Victoria University of Manchester. He was vice-president of the Manchester Society for Women's Suffrage and the Manchester Branch of the Men's League for Women's Suffrage.[9] In 1919, Rutherford returned to Cambridge's Cavendish Laboratory as director.

The University of London became the first British university to concede degrees to women in 1878. Other universities followed suit so that, by the turn of the century, only Oxford and Cambridge excluded women from full eligibility for degrees.[10] The push for accepting women into degrees at Cambridge produced a bitter contest. The two women's colleges had been established by the late 1800s, Girton in 1869 and Newnham in 1871. Women were able to attend lectures at the discretion of lecturers. From 1881, women could sit examinations, but were not afforded full university membership and, therefore, were not awarded degrees. In 1897, the university proposed a motion to allow women formal recognition of degrees. The motion was rejected, prompting a riot, and an angry mob of triumphant male students marched on Newnham College, 'revealing a bitter misogyny'.[11] Ramming the bronze Newnham gates with a handcart, students chanted, 'We won't have women'.[12]

Despite the outcome, those in favour of including women as part of the university were unwilling to accept the decision and the debate about/for granting women membership at Cambridge gained momentum. The arguments about accepting women as university members played out in the press, with both sides arguing their position through vitriolic letters published in the daily newspapers. For example, during the campaign, before a further vote on granting women full membership to the university in 1921, one such letter in *The Times* contended that:

> The higher education of men is far more important for the community than that of women ... women who take up valuable space in laboratory and the time of university teachers marry soon after they go down, and have no longer time for a professional career or research ... Psychologists and nerve-specialists

in this country ... hold that the strain of University examinations tells much more in after life on women than men.

If so far his letter had not convinced his readers, the author then tried another well-worn argument, that women were morally unrestrained and that young men would be unable to resist their feminine wiles: '[T]he tendency of women students to distract the men from their proper pursuits and to waste their time and their parent's money will be greatly increased, as also the risk of ill-advised or improvident marriages.'[13]

Soon after being appointed director, Rutherford lent his name to the campaign in favour of women. In a letter to *The Times* on 8 December 1920, Rutherford and Professor of Chemistry William Pope implored their fellow academics to extend full rights to women at Cambridge:

> [W]e welcome the presence of women in our laboratories on the grounds that residence in this University is intended to fit the rising generation to take its proper place in the outside world ... For better or worse, women are often endowed with such a degree of intelligence as enables them to contribute substantially to progress in the various branches of learning; at the present stage in the world's affairs we can afford less than ever before to neglect the training and cultivation of all the young intelligence available.[14]

The University of Oxford granted women full membership in 1920, when fifty women graduates were admitted to degrees, including several women tutorial staff members. The historic occasion was marked in newspaper editorials: 'The leaders of women's education in Oxford may be congratulated on this triumphal recognition of the cause for which they have worked so long and so arduously, and on a victory won by courtesy, patience and merit alone.'[15]

A similar vote in Cambridge was once again defeated. In the aftermath, women students experienced hostility. In science lectures, they were greeted with a cacophony of foot-stamping from male students, had derogatory notes passed to them and were victims of other antisocial behaviours.[16] Letters in the press illustrate the hostile climate to which women students were subjected. For example, a letter-writer in 1922 suggested that 'the English University woman ... is making a grave mistake in aping man. Her interests are no longer in the home or in the training of children ... but rather in a fight for recognition in the professions ... in the struggle she is becoming unsexed. She has lost her feminine charm, and is virtually a neuter.'[17] Drawing on a familiar rhetoric about ideal womanhood, which could be found, in various iterations, as far as Japan and China,[18] Cambridge remained adamant about not conferring degrees on women by making them full members of the university.

A consequence of Cambridge's reluctance to confer membership and degree status on women was that this university fell behind the University of Oxford, whose student body recorded 18-per-cent women students in 1934–35. In contrast, the share of women full-time students stagnated in Cambridge from 10 per cent in 1900–01 and dropped to 9 per cent in 1934–35, when Maasdorp entered the Cavendish as a doctoral student.[19] Cambridge would not grant women equal status until 1948. Despite the university's adamantine position, women academics had begun to be appointed. In 1926, the University of Cambridge recruited ten women lecturers, increasing the percentage of women academics to 7 per cent.[20] Ann Davies was the first tenured woman lecturer in physics at the Cavendish Laboratory. Davies completed her Bachelor of Science in 1915, majoring in physics at Royal Holloway College. She then completed her PhD in 1922, researching radiation and ionization potentials of the rare gases.[21] In 1935, Davies was appointed a fellow and lecturer in physics at Newnham College and a lecturer in physics at the Cavendish Laboratory. Davies remained the only woman lecturer in physics at the Cavendish for the next twenty years.[22] Dyhouse suggests that '[t]here is indeed a sense in which any woman who achieved public recognition as a scholar in universities before 1939 could only have managed this through formidable persistence and application'.[23]

Life at the lab

When Maasdorp arrived at the Cavendish Laboratory in 1935, there were around fifty postgraduate students. Maasdorp was the only woman.[24] In 1934–35, there were 507 full-time women students at Cambridge; male students outnumbered women by ten to one.[25] Proportionally, most women students studied the liberal arts or social sciences; very few women pursued the sciences, much less the so-called hard sciences such as physics. It would also be a year before the appointment of Davies as the first woman lecturer.

Despite Rutherford's advocacy, female students did not experience equal privileges with their male counterparts. Female physics students were required to sit on the front benches of lecture theatres 'for fear their attention might be distracted by too much male proximity'.[26] Maasdorp needed to carry a card stating that she was 'a fit and proper person', which was necessary to present before entering the library.[27] Male students did not have to carry a similar card. Even when some of these restrictions were lifted, and female students could participate in university events and activities, they often experienced considerable male opposition. For example, for the first two years of her time at Cambridge, Maasdorp was not allowed to attend the annual Cavendish Laboratory Christmas Dinner, which she describes as 'a bang-up evening dress affair' with entertainment, such as comedy skits by staff members and students. By her third year, there were two other women postgraduate research students and the rule was relaxed. There was still some hostility, and Maasdorp recalled that upon seeing females at the dinner, an eminent guest remarked: 'Good God! Women!'[28] In her history of women at Cambridge, McWilliams Tullberg states that when the women heads of Girton and Newnham and the women staff of these colleges attended university functions, ceremonials and social gatherings, it was by courtesy only, and they were counted as honorary 'wives'.[29] Despite being a member of the Cavendish faculty, this would have been the status of Ann Davies attending the annual dinner and other university functions.

Women students found ingenious ways to work around the impediments imposed upon them. Two undergraduate physics students at the Cavendish in 1932, Marie Sparshott and Helen McGaw, outraged that the Cavendish 'kindergarten' training course in machine tool technology was out of bounds to them, got around this by enrolling in a short course offered at the local Cambridge technical college instead.[30]

Maasdorp began working with her supervisor, the Australian physicist Mark Oliphant. A grant of £250,000 made possible the construction of a High-Tension Laboratory designed by Oliphant.[31] He organized his students into a team; Maasdorp worked with two other postgraduate students, Kempton and Browne, using an old Cockroft-and-Walton particle accelerator. Maasdorp recalled that 'the apparatus itself was a bit temperamental, [the experimenters] always seemed to be repairing something, [or] replacing filaments which meant taking everything down, [and] making it vacuum-tight, all very time-consuming'. This type of early work in experimental physics was also dangerous. Maasdorp remembered two occasions when she received electric shocks. Once, when working on equipment for the High-Tension Laboratory, she was 'up on a ladder trying to adjust part of the circuit … 50,000 volts travelled through [her] coming out just above [her] knee which was supporting [her] against a shelf. [The current] drilled a neat little hole' into her leg. On another occasion, she came into contact with two pieces of wire connected to the terminals of a large battery of accumulators. The resulting flash, Maasdorp reported in her memoir, 'burned and copper-plated the backs of my fingers, painful for some days afterwards'.[32]

In one incident, Maasdorp records a degree of sexist behaviour where her fellow male students took advantage of her conscientiousness. Maasdorp stated that while working on a project for Oliphant, the design of a new Van der Graaf electrostatic generator, an incident occurred whereby increasing the voltage to test the machine was immediately followed by 'a big spark, a loud crack, and glass and oil all over the floor'. Maasdorp and her fellow male researchers decided to take a lunch break before tackling the lab-cleaning. Maasdorp returned and began the clean-up; however, her colleagues did not return until the following morning, when the lab was clean.[33] This behaviour reflects cultural attitudes towards what was considered women's work.

The Royal Society

The first issue of the Royal Society's journal, *Philosophical Transactions*, the world's longest-running scientific journal, was published in 1665. Røstvik and Fyfe argue that from its inception, the journal was tightly linked with the masculine culture of the Royal Society, where to be elected a Fellow was 'seen as a significant accolade for scientists since the mid-19th century'.[34] While women could submit papers, 'they were excluded from all editorial and evaluation roles: such gate-keeping roles were reserved for Fellows of the Society'.[35] The first woman to publish a paper in the journal was astronomer Caroline Herschel in 1787. From the 1900s onwards, women authors' names were marked by 'Miss'.[36] During the 1930s, approximately 4 per cent of all papers submitted to the Royal Society's journals had a woman scientist as an author or co-author.[37] Hertha Ayrton was the first woman nominated as a Fellow. In 1904, Ayrton was also the first woman to read her paper before the Royal Society; two years later, she even received the Royal Society's Hughes Medal. Although nominated and with such success recognized by the Society, Ayrton was ultimately denied the Fellowship.

Maasdorp and her colleagues submitted two research papers to the Royal Society journal, *Proceedings of the Royal Society*.[38] Discrimination against women persisted and was still the norm when Maasdorp and her colleagues' papers were published in November 1936. Until 1990, authors could only submit papers to the Society with the support of a Fellow, known as a 'communicator'. Rutherford acted as a communicator for papers submitted by the Cavendish, and he did not distinguish women researchers by marking them as 'Miss', so Maasdorp's gender was not evident in the published papers. Subsequently, Maasdorp and her fellow researchers received an invitation to the Society's annual party for published authors. Addressed to 'Mr Maasdorp', disappointingly, two days later, Maasdorp received word that her invitation had been withdrawn because the Royal Society had discovered that she was a woman.[39]

The Royal Society operated like a private men's club. As in Cambridge, women were only welcomed as guests to its formal social events, such as the annual dinner for authors. Guests were typically 'wives', and permission for guests' dispensation to attend social events was requested annually. Despite the appointment of the first women Fellows in 1945, crystallographer

Kathleen Lonsdale and biochemist Marjorie Stephenson, the Royal Society dining clubs remained male-only until the mid-1970s.[40]

Maasdorp did not submit her doctoral thesis. In her memoir, she states that she did not feel she had done enough original work.[41] A recent study by Clark et al. demonstrates that a feeling of 'impostorism was associated with lower self-efficacy and sense of belonging, both of which correlated with a higher likelihood of considering dropping out of one's graduate program'.[42] This was particularly the case where there was a perception of sexism in an academic department. Clark et al. conclude that in academia, 'individual-level competence beliefs, confidence, and belonging ... converge to ultimately undermine women's persistence in STEM'.[43] Given that Maasdorp had already experienced sexism in the laboratory, at the Royal Society and in the larger Cambridge environment due to exclusion from official functions, it is not surprising that this culture may have influenced her perception of herself as an 'imposter'. Yet, it is clear that she had considerable ability. The fact that a student with two papers published in a renowned journal such as *Proceedings of the Royal Society* was allowed to discontinue their doctoral study speaks of the many challenges and missed opportunities women scientists had to contend with.

There are several possible explanations for Maasdorp's feelings of impostorism. First, there was a lack of women as role models at the Cavendish Laboratory during her time there. The absence of women role models at Cambridge during the 1930s was acute. While Ann Davies was appointed in Maasdorp's second year, she was not Maasdorp's supervisor. In their study on the influence of contemporary women role models, Young et al. conclude that 'female STEM professors not only provide positive role models for women, but they also help to reduce the implicit stereotype that science is masculine in the culture-at-large'.[44] Another possible explanation is that Maasdorp may have felt more of an 'outsider' than her peers. Pickles suggests that international students like Maasdorp, who had come to the Cavendish from other countries such as South Africa, were considered 'outsiders' on account of 'their sex and the patriarchal attitudes of the time'.[45] The fact that Maasdorp was the only woman postgraduate student during her first year may have contributed to her outsider status. Research students at the Cavendish would often spend their afternoon break in the library with tea and buns provided. Maasdorp records that for the first few weeks, 'I stood in one corner and for ages no one addressed a single remark to me – shyness I was told later, but it was rather lonely.'[46] This ostracism, whether deliberate or not, may have contributed to Maasdorp finding other groups with which to collaborate.

In 1937, Maasdorp married fellow Cavendish postgraduate student John Fremlin. After World War II, Fremlin began working with Mark Oliphant on constructing the world's first proton synchrotron at Birmingham University. The Birmingham machine was completed in 1953. Maasdorp performed mathematical calculations as a human computer for the Birmingham machine. Once, while in hospital, she completed calculations surrounded by 'formulae, paper, pencils and a book of log tables ... performing ... calculations and tabulating results'.[47] This was another typical position relegated to women before the widespread advent of electronic computers. Whether paid or unpaid, women computers have been mostly invisible in the history of technology, such as in the case of the construction of the Birmingham synchrotron.[48] University-educated women made up the majority of human computers. Ruth Howes argues that women computers 'played a major part' in the development of the atomic bomb. At Los Alamos, Hanford, Columbia and Berkeley, the many wives of scientists employed on the Manhattan Project worked as computers performing a 'vast number' of calculations, often unpaid and largely forgotten in historical records.[49] After World War II, in the United States, a job classification of 'engineering computer' was created for women undertaking computing tasks to distinguish them from male positions such as 'junior engineer'.[50] In one example of recovering women's history, Harvard University's Project Phaedra initiative aims to uncover not only data from 2,500 astronomical logbooks and notebooks, but also the women who created them. More than 140 names of women working as astronomical computers have been recovered from notebooks spanning from 1881 to the 1950s.[51]

Activism

Maasdorp immersed herself in student political activism at Cambridge. She joined the Cambridge Socialist Society, which had grown from 200 members in 1933 to over 1,000 in 1938. Of the approximately 5,000 undergraduates at Cambridge in 1938, the year that Maasdorp joined the Society, 20 per cent were members.[52] Maasdorp also joined the Cambridge Scientists' Anti-War Group (CSAWG). The CSAWG was founded in 1932 by Cavendish scientist J. D. Bernal as a progressive force for social reform, a 'grassroots' organization concerned with the social responsibilities of the scientist.[53] Bernal, his biographer Andrew Brown attests, argued that 'scientists could exert a powerful influence in modern states only by organizing into cohesive groups. He saw the opportunity to display their implacable opposition to war and fascism as the rallying cry around which scientists could unite.'[54] Membership of the CSAWG was widespread and included faculty members, researchers and students. Unlike academic organizations like the Royal Society, the CSAWG encouraged women as members. The CSAWG met weekly in the basement of a King's Parade café.

Brown suggests that sympathizers to the causes highlighted by the group represented approximately 40 per cent of Cambridge's Cavendish and Dunn laboratories and 10 per cent of scientists from other laboratories.[55] The CSAWG 'kept up a constant stream of meetings, demonstrations and marches', and Maasdorp was 'one of its staunchest supporters'.[56] Physicist Maurice Wilkins, another of Oliphant's students, stated in his autobiography that he and his Cavendish peers, including Maasdorp, would spend 'much time drawing attention to the Nazi threat, the Spanish Civil War and the acute problems of Indian Independence'.[57] Davis contends that, for many involved in the British peace movements during the 1930s, 'the start of Franco's attempt to overthrow the Republican government in Spain was a decisive turning-point and led many of them to conclude that the need to defend democracy and socialism overrode their belief in pacifism'.[58] In 1937, the CSAWG published leaflets supporting the Spanish Republican Popular Front 'standing up to the fascist tanks and aeroplanes'.[59] Under the increasing threat from Nazi Germany, the group then turned its attention to an experimental programme testing government-issued gas masks.

The government issued a gas mask to every citizen in the face of a possible poison gas attack. John Fremlin lent the group his room at Trinity College to experiment with the government-issued plans for gas-proofing a room. The room was carefully sealed, with ten members inside. Each member was given specific instructions, including measuring hourly carbon dioxide concentration, temperature, humidity, and breathing and pulse rates.[60] Despite the suggested gas-proofing measures in place, the experiment demonstrated that, alarmingly, 'air passed readily from the outside into a gas-proofed room under normal atmospheric conditions'.[61] The experiments and tests on survival rates from high explosives and incendiary bombs culminated in the publication of a book, *The Protection of the Public from Aerial Attack* (1937). The publication caused a stir, with a review in *Nature* stating that the book represented:

> a slashing attack on the recommendations put forward by the Air Raid Precautions Department of the Home Office ... for the protection of the public ... The authors of this book, in their destructive criticism, seem to ignore the fact that some degree of protection is better than none ... This book can do nothing but harm ... it is calculated to destroy confidence in them and to create panic.[62]

The group responded stating that while it had 'no desire to create panic ... We would be lacking in our duty as scientists and citizens if we were to accept, without question, assurances of the validity of which we have not been convinced.'[63]

It is critical to situate Maasdorp in the wider context of the times with respect to the development of British science during the 1930s and in postwar international affairs. The 'British scientific Left', as Gary Werskey terms them, argued that socialism was 'a rational approach to society' that would enable 'the relief of human misery, transcendence of nation and class, progressivism, internationalism'.[64] While Bernal remained committed to the British Communist Party (BCP), with the signing of the Treaty of Non-Aggression between Germany and the USSR in August 1939, Left Book Club publisher Victor Gollancz called on members to resign from the BCP in protest. Many, including Reinet Maasdorp and John Fremlin, heeded his call and resigned. This did not mean that they renounced their socialist convictions and the ideal of science as an emancipatory force that could be utilised to shape a more equal society. Instead, these ideals had led to the formation of the social relations of science movement and groups such as the CSAWG, and the British Association for the Advancement of Science and the Association of Scientific Workers (A.Sc.W.).

In an article in *Nature*, Steven Rose observed that the group of radically progressive scientists that Maasdorp was a member of was 'a formidable group of intellectuals with a range of talents within their disciplines and political and organisational skills that were to play their part in the total transformation of science in the decade that followed'.[65] Rose included the Fremlins (Maasdorp and John Fremlin) alongside other progressive scientific luminaries such as Bernal, J. B. S. Haldane, Lancelot Hogben, Hyman Levy, Joseph Needham, the Piries and the Woosters. Maasdorp and Fremlin were not the only couple to be active as scientists and political activists at Cambridge at this time. William and Nora Wooster were both crystallographers at Cambridge and active in the CSAWG, as were biochemists Norman and Antoinette Pirie.

Maasdorp left Cambridge to take up the position of national secretary of the A.Sc.W., a post she held from 1937 to 1945. The A.Sc.W. worked for the advancement of scientists and laboratory staff. As secretary, Maasdorp organized campaigns, often writing letters to newspapers and journals, such as promoting 'the proper utilisation of the knowledge and of qualified scientists in time of war'.[66] In a letter published in *Nature* in 1939, Maasdorp advocates for improved salaries for women scientific workers, pointing out that an advertised salary for 'a botanist with an honours degree and at least two years' experience of research' was 'incommensurate with the training required'. In addition, Maasdorp argued that the suggested salary was 'far below the normal salaries obtaining in Government departments for men of science holding equivalent qualifications'.[67]

In the early postwar period, Maasdorp was interviewed for a television programme. Thinking that the discussion would encompass her work with the A.Sc.W., she was dismayed with the angle of the questioning focusing on her gender rather than science: 'Are you married? Do you want children?' A subsequent newspaper article, under the series title 'Odd Occupations', included the following disappointing opening sentence: 'This pretty young woman, who wants children someday, talking abstrusely about science.'[68] Chimba and Kitzinger's study on the gendered reputation of women scientists in British media demonstrates that there has not been much change in the past half-century since Maasdorp encountered sexist questioning. The researchers argue that in the twenty-first century, as was the case previously, 'when women are profiled, the focus is often on their appearance and they may be sexualized ... and that descriptions of them often imply (even as they may seek to address) a contradiction between "airheads" and "eggheads", "bimbos" and "boffins"'.[69] When Maasdorp resigned from the A.Sc.W. on the family's move to Birmingham, she was again the subject of a newspaper article on her work, with a report in the *Daily Mail* under the headline 'Bride won £80,000 pay rise for our scientists.'[70]

In 1953, Maasdorp began teaching secondary-school science, a typical career trajectory for women science graduates. She retired from teaching in 1960. Thane argues that it is crucial to look at the work-life of women such as Maasdorp more broadly than the idea of a conventional career. In addition to paid employment, consideration must be given to 'family life, voluntary work and low-paid work as well as high-flying work'.[71] In this context, reclaiming forgotten women scientists such as Maasdorp writes women back into history in places where they 'existed in the shadows of another's more convenient, accepted, or publicly sanctioned narrative'. Maasdorp's career is typical of her generation. Thane notes that '[b]efore, during and after World War Two, [women graduates] overwhelmingly became schoolteachers, at some point in their lives ... Cambridge and Oxford graduates were more likely than other graduates to obtain posts in higher status independent and grammar schools.'[72] Before World War II, married women faced additional restrictions on employment due to the marriage bar, which necessitated their termination once married. Horrocks argues that only after fundamental legislative changes – the Equal Pay Act (1970), the Sex Discrimination Act (1975) and the Employment Protection Act (1975) – did women start to see more opportunities for scientific careers open up.[73]

Conclusion

A particular challenge in writing histories of women scientists is locating women scientists in the archive. Evidence of women's contribution is often patchy and incomplete, making it difficult to gain a picture of what life was like for women in the laboratory. To achieve a more accurate picture, we must capture the full range of women's participation in science history. This requires some lateral thinking, looking at paid roles such as tutoring and science educators, active involvement in scientific organizations, and unpaid contributions such as performing as a human computer. In this way, we can render women visible contributors to the history of science. Reinet Maasdorp's experience as a postgraduate student at the Cavendish Laboratory and her activism in the social relations of science and the British peace movements during the 1930s may have been lost had it not been for a family history written in 2004.

In 2015, Jones and Hawkins posited that correcting the exclusion of women from the history of science was imperative because the 'continued assumption that women were absent from scientific endeavour, or that they only participated in a secondary capacity … distorts the past and raises obstacles for the future recruitment of women into science'. In addition, the lack 'of female role models helps perpetuate the masculine colouring of science, especially "hard" sciences such as physics, and is a feature of our cultural understandings of science'.[74]

The underrepresentation of women in STEMM, particularly physics, has only slightly improved since Maasdorp's time. In the twenty-first century, science is still a gendered occupation. There are still obstacles for women pursuing a career in science and a lack of gender parity. The gender diversity of applicants for the Royal Society's early-career research fellowship programmes in STEMM continues to show that the number of women applying is not representative of eligible postdoctoral researchers. Women make up only 30 per cent of applicants for physical science fellowships. Women make up only 23 per cent of the academic workforce in physics, with only 16 per cent in senior positions.[75] While statistics have improved since Reinet Maasdorp pursued a career in STEMM, with only 24 per cent of women making up the STEMM workforce, the gender gap is still a significant barrier to women's access, participation and progress in science as a career.[76]

Notes

1 M. Fremlin, 'Reinet Fremlin née Maasdorp 1912 to 1992', p. 5. http://margaret.fremlin.org/ReinetFremlin.pdf (accessed 25 June 2022).
2 Fremlin, 'Reinet Fremlin', p. 9.
3 Fremlin, 'Reinet Fremlin', p. 12.
4 K. Crenshaw, 'Demarginalizing the intersection of race and sex: A Black feminist critique of antidiscrimination doctrine, feminist theory and antiracist

politics', *University of Chicago Legal Forum*, 1: 8 (1989), 139–67, 140, https://chicagounbound.uchicago.edu/uclf/vol1989/iss1/8.

5 Careers Research & Advisory Centre (CRAC), *The Profile of Postdoctoral Researchers in the UK Eligible for Royal Society Early Career Fellowship Programmes*, March 2021, p. 34, https://royalsociety.org/-/media/policy/Publications/2021/trends-ethnic-minorities-stem/Profile-of-postdoctoral-researchers-in-UK-eligible-for-RS-early-career-fellowship-programmes.pdf?la=en-GB&hash=A92E67EA4E2E827907CEA0F195B130B5 (accessed 24 January 2025).

6 L. Guglielmo, 'Introduction: Re-collection as feminist rhetorical practice', in L. L. Gaillet and H. G. Bailey (eds), *Remembering Women Differently: Refiguring Rhetorical Work* (Columbia: University of South Carolina Press, 2019), p. 2.

7 University of Cambridge, 'The History of the Cavendish', www.phy.cam.ac.uk/history (accessed 18 June 2022).

8 P. Gould, 'Women and the culture of university physics in late nineteenth-century Cambridge', *British Journal for the History of Science*, 30 (1997), 128.

9 J. Campbell, 'Rutherford: The road to the nuclear atom', *CERN Courier*, 3 May 2011, https://cerncourier.com/a/rutherford-the-road-to-the-nuclear-atom/ (accessed 24 January 2025).

10 C. Dyhouse, 'The British Federation of University Women and the status of women in universities, 1907–1939', *Women's History Review*, 4: 4 (1995), 469.

11 C. Dyhouse, *No Distinction of Sex: Women in British Universities, 1870–1939* (London and New York: Routledge, 2016), p. 239.

12 Dyhouse, *No Distinction of Sex*, p. 239.

13 W. Ridgeway, 'Women at Cambridge', *The Times* (22 November 1920), p. 8.

14 E. Rutherford and W. J. Pope, 'Women at Cambridge', *The Times* (8 December 1920), p. 8.

15 'First Oxford Women Graduates', *The Times* (15 October 1920), p. 7.

16 'The Newnham Incident', *The Times* (25 October 1921), p. 10.

17 'Our University Women "Aping Man"', *The Times* (22 August 1922), p. 14.

18 See Ogawa's and Edgerton-Tarpley's chapters in this volume.

19 H. Jöns, 'Feminizing the university: The mobilities, careers, and contributions of early female academics in the University of Cambridge, 1926–1955', *The Professional Geographer*, 69: 44 (2017), 673. Maasdorp enrolled in Newnham College.

20 Jöns, 'Feminizing the university', p. 673.

21 'Dr Ann Horton', *Nature*, 215 (1967), p. 1211.

22 Jöns, 'Feminizing the university', p. 674.

23 Dyhouse, *No Distinction of Sex*, pp. 156–57.

24 H. Austin, 'The Cavendish Laboratory, Cambridge', *Nature*, 137 (1936), 766.

25 Dyhouse, *No Distinction of Sex*, p. 249.

26 M. Constable, 'The living past: The Cavendish in 1932', *Cav Mag*, 3 (2010), 3.

27 J. Fremlin and M. Fremlin, 'There isn't a snake in the cupboard: A review of the life of J. H. Fremlin', https://margaret.fremlin.org/book.html (accessed 12 June 2022).
28 Fremlin, 'Reinet Fremlin', p. 13.
29 R. McWilliams Tullberg, *Women at Cambridge* (Cambridge: Cambridge University Press, 1998), p. 178.
30 Constable, 'The living past', p. 3.
31 Austin, 'The Cavendish Laboratory, Cambridge', p. 766.
32 Fremlin, 'Reinet Fremlin', p. 12.
33 Fremlin, 'Reinet Fremlin', p. 12.
34 C. M. Røstvik and A. Fyfe, 'Ladies, gentlemen, and scientific publication at the Royal Society, 1945–1990', *Open Library of Humanities*, 4: 1 (2018), 37.
35 Røstvik and Fyfe, 'Ladies, gentlemen, and scientific publication', p. 45.
36 Røstvik and Fyfe, 'Ladies, gentlemen, and scientific publication', p. 11.
37 Røstvik and Fyfe, 'Ladies, gentlemen, and scientific publication', p. 12.
38 A. E. Kempton, B. C. Browne and R. Maasdorp, 'Transmutation of lithium isotope of mass seven by deuterons', *Proceedings of the Royal Society of London, Series A – Mathematical and Physical Sciences*, 157: 891 (1936), 372–85; A. E. Kempton, B. C. Browne and R. Maasdorp, 'Angular distribution of the protons and neutrons emitted in some transmutations of deuterium', *Proceedings of the Royal Society of London, Series A – Mathematical and Physical Sciences*, 157: 891 (1936), 386–99.
39 Fremlin, 'Reinet Fremlin', p. 13.
40 Røstvik and Fyfe, 'Ladies, gentlemen, and scientific publication', p. 9.
41 Fremlin, 'Reinet Fremlin', p. 4.
42 S. L. Clark, C. Dyar, E. M. Inman, N. Muang and B. London, 'Women's career confidence in a fixed, sexist STEM environment', *International Journal of STEM Education*, 8: 56 (2021), 7.
43 Clark et al., 'Women's career confidence', pp. 7–8.
44 D. M. Young, L. A. Rudman, H. M. Buettner and M. C. McLean, 'The influence of female role models on women's implicit science cognitions', *Psychology of Women Quarterly*, 37: 3 (2013), 283.
45 K. Pickles, 'Colonial counterparts: The first academic women in Anglo-Canada, New Zealand and Australia', *Women's History Review*, 10: 2 (2001), 273.
46 Fremlin, 'Reinet Fremlin', p. 13.
47 Fremlin and Fremlin, 'There isn't a snake in the cupboard'.
48 Tassabehji et al. state that 'until 1945, the term "computer" defined a human (usually female) who carried out calculations – who "computed." Hardware development was men's work; programming was women's.' R. Tassabehji, N. Harding, H. Lee and C. Dominguez-Pery, 'From female computers to male computers: Or why there are so few women writing algorithms and developing software', *Human Relations*, 74: 8 (2021), 1299. See also, P. A. Kidwell, 'Women astronomers in Britain, 1780–1930', *Isis*, 75 (1984), 534–46.

49 R. Howes, *Their Day in the Sun: Women of the Manhattan Project* (Philadelphia, PA: Temple University Press, 1999), pp. 98, 110.
50 J. Light, 'When computers were women', *Technology and Culture*, 40: 3 (1999), 461.
51 Wolbach Library, 'Women at the Harvard College Observatory', https://library.cfa.harvard.edu/glass-plates/women-at-hco (accessed 24 October 2022).
52 T. Buchanan, *Britain and the Spanish Civil War* (Cambridge: Cambridge University Press, 1997), p. 148.
53 A. Brown, *J. D. Bernal: Sage of Science* (Oxford: Oxford University Press, 2006), p. 120.
54 Brown, *J. D. Bernal*, p. 388.
55 Brown, *J. D. Bernal*, p. 122.
56 Fremlin and Fremlin, 'There isn't a snake in the cupboard'.
57 M. Wilkins, *The Third Man of the Double Helix: The Autobiography of Maurice Wilkins* (Oxford: Oxford University Press, 2005), p. 31.
58 R. Davis, 'The British Peace Movement in the interwar years', *French Journal of British Studies/Revue Française de Civilisation Britannique*, 22: 3 (2017), 11.
59 Brown, *J. D. Bernal*, p. 130.
60 Fremlin and Fremlin, 'There isn't a snake in the cupboard'.
61 J. D. Bernal et al., 'Air raid precautions', *Nature*, 139 (1937), 760.
62 C. H. Foulkes, 'The protection of the public from aerial attack', *Nature*, 139 (1937), 608.
63 Bernal et al., 'Air raid precautions', pp. 760–61.
64 G. Werskey, *The Visible College: A Collective Biography of British Scientists and Socialists of the 1930s* (London: Free Association Books, 1988); G. Werskey, 'The Visible College revisited: Second opinions on the Red Scientists of the 1930s', *Minerva*, 45 (2007), 305–19; G. Somsen, 'A history of universalism: Conceptions of the internationality of science from the Enlightenment to the Cold War', *Minerva*, 46 (2008), 369.
65 S. Rose, 'Thirties science movement', *Nature*, 276 (1978), 136.
66 R. Fremlin, 'Scientists in time of war: Their value to the nation', *Manchester Guardian* (8 May 1939), p. 18.
67 R. Fremlin, 'Salaries of scientific workers', *Nature*, 144 (1939), 119.
68 Fremlin, 'Reinet Fremlin', p. 23.
69 M. Chimba and J. Kitzinger, 'Bimbo or boffin? Women in science: An analysis of media representations and how female scientists negotiate cultural contradictions', *Public Understanding of Science*, 19: 5 (2010), 621.
70 Fremlin and Fremlin, 'There isn't a snake in the cupboard'.
71 P. Thane, 'The careers of female graduates of Cambridge University, 1920s–1970s', in D. Mitch, J. Brown and M. H. D. van Leeuwen (eds), *Origins of the Modern Career* (Aldershot: Ashgate, 2004), pp. 207–24, 210.
72 Thane, 'The careers of female graduates', p. 217.
73 S. Horrocks, 'World War II, post-war reconstruction and British women chemists', *Ambix*, 58: 2 (2011), 150–70.

74 C. G. Jones and S. Hawkins, 'Women and science', *Notes and Records of the Royal Society of London*, 69: 1 (2015), 5–9.
75 CRAC, *The Profile of Postdoctoral Researchers in the UK*, p. 23.
76 STEM Women, 'Women in STEM: Percentages of women in STEM statistics', www.stemwomen.com/women-in-stem-percentages-of-women-in-stem-statistics (accessed 12 June 2022).

2

'Your research is crap, do not bother to apply again': Female evolutionary biology theorists as scientific rebels and oppositional scientists

Nuala Proinnseas Caomhánach

Introduction: Can women be rebel scientists?

> *Dick Teresi: Do you ever get tired of being called controversial?*
> *Lynn Margulis: I don't consider my ideas controversial. I consider them right.*
> (*Discover Magazine*, 2011)

In 1967, a twenty-nine-year-old scientist at Boston University, Lynn Sagan, published a paper that exploded the standard evolutionary narrative. Building on earlier, largely speculative, theoretical currents at the margins of evolutionary theory, she put forth the first synthetic hypothesis for endosymbiosis, in which a symbiont dwells within the body of its symbiotic partner. She argued that symbiosis was not only a driving force in the evolution of multicellular life, but the *only* force. Her theory was met with immense resistance; her opponents criticized both the theory and its author Lynn Sagan, soon to be the more-well-known Lynn Margulis.[1] These views have survived in book reviews, obituaries, magazine features, television programmes, articles' comment postings and editorials that commented on issues far beyond the matter of endosymbiosis, the focus of her paper. Placing the archival material we use to recover the role of women in science in conversation with public-facing documents reveals the complexity of being a female theorist. These include the place of the emerging microbial movement in science, the utility of a holistic rather than a reductionist approach to understanding the natural world, the influence of gender in science, questions over who (and what) constituted not only a scientist, but a theoretician in modern biology, and the desire to preserve neo-Darwinism, on which endosymbiosis was an assault. Central to all these controversies was Lynn Margulis herself and how well or poorly she filled the roles of scientist, evolutionary biologist, theorist, thinker, innovator and science popularizer. At the same time her professional qualifications were under fire, gender played a central role in the reception of endosymbiosis theory.

Critics and colleagues often claimed she was 'in your face' as a scientist.[2] As with other iconic figures, such as Richard Feynman or James Watson, it is striking how easily 'Lynn Margulis' and 'endosymbiosis' became a cultural scientific shorthand for a wide range of preoccupations and scientific quandaries. Margulis leaned into the debate head-on to adopt the attitude that she was an outsider and a rebel.[3]

This essay focuses on one particular thread of this multifaceted discourse, Margulis' role as a rebel. Over time, as scientific communities coalesce around particular behavioural patterns and norms, the scientific rebel emerges to counter this performative expectation.[4] A well-known, self-proclaimed rebel was the physicist Freeman Dyson. For Dyson, successful scientists were rebels.[5] They were successful because they did not accept conventional wisdom, whether scientific, political or religious, and rebelled against narrow views of the natural world. Dyson cultivated his image as a maverick genius and Margulis implicitly modelled herself on this bombastic scientist. Margulis as the rebel scientist enables us to ask what we can learn about how male scientists evaluate female scientists who develop scientific theories, and at the same time explore the scientific attitudes towards women in science during a period when science was popularized, televised and brought directly into everyone's homes.[6] Exploring the category of the scientific rebel is not a reification of the concept itself; rather, what this essay aims to do is to build from the concept of the scientific icon, because not all icons are considered rebels, to ponder how female scientists locate themselves within a sometimes hostile intellectual environment.[7] Margulis as an iconic figure and a rebel presents a fruitful opportunity for investigating questions about the creation and perceptions of science and scientists and how scientists create public and professional personas. Rightly or wrongly, roles within science are assigned, both endogenously and exogenously. Perceptions of how fields work are often filtered through (white) iconic individuals. According to the historian David K. Hecht, 'the tendency to individualize a collective endeavor itself is an important feature of public and scientific understanding of science'.[8] Yet, iconic figures reveal much about the expectations, assumptions, values, ideals, excitement and anxieties projected onto them by the scientific community and public audience.

Margulis also offers a window into the scientific debate about whether endosymbiosis theory – and by default the author – should be considered scientific at all. Margulis' theory represented a threat to the dominant neo-Darwinian understanding of evolution – one by a female scientist at that. Critics of Margulis rhetorically dismissed her scientific status and aimed to marginalize the impact of her supporters. They called attention to the lack of scientific evidence in the paper and insisted that she was not a lab-coat-wearing, hands-on researcher, but merely a synthesizer of knowledge produced by others. Gendered rhetoric was a persistent feature of this attempt to distance both the theory and author from the category of 'scientist', 'evolutionary biologist' and 'theorist'. Yet over time, de-gendering and re-gendering enabled Margulis to metamorphose into an acceptable form of a scientist. She was portrayed as a rebellious and obnoxious scientist who threatened the objectivity, order and rationality of scientific advance. She was dismissed as a scientific theorist in part because she was a woman; yet she was criticized as well for violating feminine norms and being a masculine rebel – in your face, obnoxious, insisting on her right to theorize and for being more than a scientist in ways the men were admired. Over time – how much time is not clear – it is precisely these attributes as a masculinized rebel that helped her gain acceptance and enabled her inclusion in the field.

Interestingly, however, it was not only her opponents who conceived of her as something other than a scientist. Margulis' supporters frequently utilized the same attributes – frowned upon by her opponents – as desired traits required to overthrow and subvert the ways in which knowledge was constructed. Her ability to popularize science was praised as being equal to that of astronomer Carl Sagan (even with the complication that they used to be married). Many of these laudatory appraisals specifically commended her for being more than just a scientist or at least for standing outside of the dominant scientific establishment. In other words, many of Margulis' supporters joined her detractors in seeing her in broader terms than scientific ones; they simply differed on the value that they assigned to this interpretation. And they did not feel that this necessarily disqualified her from being a theorist.

These non-scientific (and masculine) attributes helped legitimate Margulis as a theorist and scientist. Her case suggests ways in which such attributes have been fundamental to the assimilation of science and scientists into modern American popular culture more generally. This article first introduces Margulis' 1967 paper and the controversy surrounding it. It then examines the concept of the rebel in science, how her opponents, and Margulis herself, adopted it. The third section discusses the scientific field that Margulis challenged and its central role in helping to fashion her iconic status. The conclusion explores how the discourse on Lynn Margulis reveals how the science-curious public and practitioners understood the boundaries, authority and nature of science – as well as who they would trust to embody it and under what conditions. The article thus contributes to the intellectual project of scholars like Janet Browne, Lorraine Daston, Margaret Rossiter, Declan Fahy, Londa Schiebinger and David K. Hecht, who have treated iconic scientists as an important site of engagement in which Americans have constructed their image(s) of science and the natural world.

Upending competition

In 1967, Lynn Margulis published an article titled 'On the origin of mitosing cells' in the *Journal of Theoretical Biology*. The publication would become as famous for its path to publication as for its scientific content. The article had been rejected by more than a dozen journals before eventually finding a publication outlet. In fifty-six pages, Margulis hypothesized and visualized the evolution of eukaryotic cells and how their internal complex of organelles evolved. Evolution, Margulis argued, did not start with animals, but with microbes. She not only rewound the evolutionary clock to four billion years; she declared war on the view that competition is the force behind evolutionary change. Margulis argued that 'the mitochondria, the (9 + 2) basal bodies of the flagella, and the photosynthetic plastids can all be considered to have derived from free-living cells, and the eukaryotic cell is the result of the evolution of ancient symbioses'.[9] Her proposal that specific, essential-for-life organelles found in the cells of multicellular organisms were of endosymbiotic origin was radical. The prevailing view at the time was that organelles had evolved from within the cell itself. She further argued that the organelles were derived from bacteria that had entered into a series of intimate symbiosis with unrelated hosts. Margulis' theory was not just a thought experiment, but a provocation about the origins of the bifurcation in the Tree of Life between single- and multicellular organisms.

Margulis noted that she was not the first scientist to consider the role of symbiosis in nature. In 1905 and 1907, Russian botanists Konstantin Mereschkowski (1855–1921) and Andrei Sergeyevich Famintsyn (1835–1918) were the first to argue for an endosymbiotic origin of the chloroplast and nucleus.[10] The botanist Boris Kozo-Polyansky (1890–1957) would later argue for the importance of symbiogenesis.[11] Each author emphasized the role of prokaryotic progenitors in evolution before this hypothesis soon fell from the mainstream biological view.[12] Interestingly, one of Margulis' early influences, Edmund Beecher Wilson, America's first cell biologist, considered symbiogenesis as 'entertaining fantasy'.[13] The significance of Margulis' theory was that she tied these concepts together and offered predictions, especially within the burgeoning field of molecular biology, that would eventually support most of her theoretical interventions.

Margulis' tone was earnest and clear as she incorporated a divergent range of fields to present her case, for example, cell biology, ecology, geoscience and ecology. The title, at first glance, seems perfunctory, but it echoed the language of Charles Darwin's *On the Origin of Species* to argue against the prevailing view that evolutionary novelty was derived primarily from natural selection on random mutations. The title itself was bold and as broadsweeping as her core theoretical intervention that evolutionary change often occurs through sudden acts of fusion, synthesis. Serendipitous mergers, more so than competitive and violent struggles, defined the origins of the Tree of Life. From the very beginning of cellular life, then, organisms have been intertwined not only with those of vastly different species, but with entirely different kingdoms of life. Margulis challenged the established binary system of prokaryotes and eukaryotes traditionally considered as taxonomically incongruent. Partnerships, rather than isolated lineages, between fungi and plants enabled life to make its precarious journey from sea to the land, which, 500 million years ago, was an inhospitable place. Cooperation, not competition, was essential to the history of evolution on Earth.

The theory caused quite a stir and soon three scientific camps emerged: those who dismissed it outright and never engaged with it or her again; biologists who critiqued and criticized it and started a long battle with Margulis; and scientists who were captivated by the novel experimental pathways her theory opened. The loudest were her opponents. Some launched their organismal warfare at her choice of symbiont. Heavyweight biochemist Christian de Duve from the Catholic University of Leuven and Rockefeller Institute[14] countered Margulis' bacterial symbiont with a primitive aerobic phagocyte. This imagined phagocyte initially depended on hydrogen-peroxide-mediated respiration during its early evolution. With the adoption of microorganisms that eventually would lose their cell walls, a primitive phagocyte emerged that utilized peroxisomes as the main (aerobic) respiratory organelle. This amitochondriate would later be the host of an aerobic bacterium with oxidative phosphorylation, the ancestor of mitochondria.[15] America's leading microbiologist, Roger Y. Stanier (1916–82) of the University of California, Berkeley, suggested that the first symbiont needed to be an anaerobic, heterotrophic host in the evolution of chloroplasts. He placed the origin of chloroplasts before the origin of mitochondria, arguing that since mitochondria use oxygen, and since eukaryote origin took place in anaerobic times, there must have been first a sufficient and continuous source of oxygen before mitochondria were able to develop.[16] Margulis disagreed.

Most detractors, however, claimed that an evolutionary model grounded in the belief that endosymbiosis was a required step in the evolution of the eukaryotic cell was unnecessarily radical. Zoologist Rudolf A. Raff and chemist Henry R. Mahler of Indiana University wrote:

> In our opinion there is no a priori reason why the eucaryotic [sic] cell, which has proved capable of remarkable evolutionary innovations, should have originated as a collage of procaryotic [sic] cells and part of cells rather than have evolved in a more direct manner from a particularly advanced type of procaryotic [sic] cell. While symbiosis may have been of some evolutionary significance, overdependence on it as an explanation for the origin of the eucaryotic [sic] cell and its organelles may leave interesting questions unasked.[17]

They ended their discussion by stating that, 'while the symbiotic theory may be aesthetically pleasing, it is not compelling'.[18] Raff and Mahler, as with most of her opponents, stood firmly with Stanier.

Stanier, with his colleague C. B. van Niel of Hopkins Marine Station, had earned his reputation based on their famed 1962 paper 'The concept of a bacterium'. They emphasized that the nature and relationships of bacteria, debated since the earliest days of bacteriology, remained unresolved.[19] Furthermore, the prokaryotic–eukaryotic divide was referred to as 'the greatest single evolutionary discontinuity to be found in the present-day world'.[20] The author's frustrations over trying to study bacteria were palpable: '[A]ny good biologist finds it intellectually distressing to devote his life to the study of a group that cannot be readily and satisfactorily defined in biological terms; and the abiding intellectual scandal of bacteriology has been the absence of a clear concept of a bacterium.'[21] Stanier was a force in microbiology and had to 'sign off' on any new developments in the field. He was the intellectual gatekeeper. As legend goes, Stanier, who happened to be in the same department as Margulis during her PhD, met Margulis in an elevator at the University of California, Berkeley, and told her that her strange theories on the origin of mitochondria and chloroplasts would never gain acceptance.[22]

'Strange' was one of many words used to dismiss and deride Margulis' theory. She categorically rejected all arguments that any anaerobic organelle – from the phagocytes to hydrogenosomes – would fit into the underlying concept of endosymbiosis. Her argument was based on the premise that the benefit of mutual symbiosis was founded in shared oxygen utilization. The reaction to Margulis and the endosymbiotic theory seems bloated and dramatic, and hints at a larger system at work in modern biology, based on underlying sexism – a biological boys' club of sorts – and chauvinism towards females who dared to speak up and theorize.[23] Margulis certainly did not shy away from the attention or the scientific battle; if anything, she leaned right into it to construct her own persona as an outsider and a rebel. Margulis was part of a newer generation of American female scientists who knew they had to stand their ground.

The making of a rebel

The appeal of the scientific rebel that would come to characterize Margulis' iconic status has a longer history. The most obvious examples are usually men, such as Richard Feynman, James Watson, Neil deGrasse-Tyson, Kary Mullis and Albert Einstein. The menu of categories of non-scientific attributes these scientists claim is long, including their hobbies, sexual predilections, ethics, drug use, pithy philosophical musings and opinions about issues outside of their own field, and is celebrated as much as their science. Recent scholarship highlights the divergence in this pattern when applied to female scientists; while women participated in research, they have failed to receive both academic and public credit in the same way as their more celebrated male colleagues. Studies of Eleanor Lamson, Rosalind Franklin and Rachel Carson, for example, reveal the ever-shifting parameters and absolute contingency of who and what allowed a woman to be considered a scientist, in the 1920s, 1950s and by Margulis' publication in the late 1960s.

Naomi Oreskes' examination of Lamson highlights the social-class aspect of research science during the 1920s. Fieldwork ranked higher than data analysis. Lamson was an associate astronomer in the US Naval Observatory. She:

> was the person responsible for developing and implementing the procedures necessary to convert the photographic records of the pendulum into measurements of the acceleration of gravity. In the language of Bruno Latour, she was the person responsible for converting instrumental inscriptions into scientific information; Latour and others have emphasized how nontrivial this conversion can be.[24]

According to Oreskes, 'scientists normally make a strong distinction between themselves and their technicians, typically on the grounds of originality'.[25] In this case, Lamson belonged to the category of 'technician' and was, therefore, easy to exclude. Indeed, the actual data were often ranked higher than the person who analysed them, especially if they did not collect said data. By situating the data analysis Lamson conducted, Oreskes established that 'only the men went to sea. Only the men's work could be cast as a heroic voyage to conquer the Earth's secrets. Therefore only the men appeared in the public eye.'[26]

Oreskes' study complicated not only the gendered notion of scientific heroism to show how physical adventurism was more responsible for scientific fame than were scientific achievements. She showed how the stratified social class, reflected in hypermasculine and patriarchal norms, shrouded science research. Fieldwork was more important than the actual scientific data analysis performed; fieldwork seemed dangerous and therefore manly, while data interpretation's tedious and boring nature was more appropriate for females. Oreskes further highlights what happens in the absence of a compelling non-scientific narrative. Fame turned on the presence of non-scientific attributes with a very wide scope of permissible behaviours; without the making of a persona, the scientist remains obscure and invisible.

The story of Rosalind Franklin is significant and has become a relatable gold-standard to understand how invisibility was not only ascribed to female scientists but curated by male colleagues. Brenda Maddox's meticulous account of Franklin would provide compelling evidence of the sexism encountered within scientific culture in 1950s Britain.[27] That Franklin was scooped by James Watson and Francis Crick in the race to elucidate the molecular structure of DNA seems a familiar tale of patriarchal entitlement within academia.[28] But the role and longevity of fame in controlling the narrative of this discovery highlighted the ruthlessness and adroit use of media attention within the systematic sexism at play in this story.[29]

Watson's assault was relentless and boundary-free, as he academically undermined and body-shamed Franklin in a manner he rarely applied to his male counterparts.[30] Franklin's story reveals the chauvinistic culture in scientific institutions towards female scientists in 1950s Britain.[31] Franklin was muted in her contributions, but she was also de-scientized through non-scientific attributes. Maddox's account reveals the force of scientific breakthroughs and modes of self-promotion in this post-World War II scientific culture. Watson's tale is a curious case of fame creating an unfathomable immunity.

Historian David K. Hecht developed the idea of non-scientific appeal more explicitly in his study of Rachel Carson. He demonstrated how the centrality of non-scientific appeal 'is not a quirk of particular, towering figures like Einstein. Rather, it is fundamental to how scientists appear in the public sphere, and particularly to how audiences come to understand them'.[32] In Carson's case, Hecht shows how the convoluted history of non-scientific narratives facilitated the establishment of her credibility as a scientist – at least, among those who chose to grant her such credibility at all. From 'reluctant crusader' to claims that she 'loathe[d] the spotlight', Hecht's rich analysis shows the complicated and contingent path to Carson's credibility as a scientist.

Certainly, the role of the media in constructing Carson's authority as a scientist was paramount. Hecht demonstrates the power of public perception with compelling examples, such as the CBS Reports broadcast on pesticides in April 1963. He argues that this was 'the high point of image dissemination of Carson, as it was viewed by an estimated ten to fifteen million people'.[33] Barrow extended Hecht's argument by examining how newspaper cartoons not only conveyed Carson's central message about the dangers of modern pesticides, but how Carson became an 'inspirational model for how women might play a more active role in the public sphere more generally and in environmental issues specifically'.[34] Both authors demonstrate how the non-scientific appeal was a fundamentally relational concept, and 'scientific' or 'non-scientific' were not static notions, but perceptual categories in the public mind. They provide ample evidence that the persistence of the personal in judging science, inseparable from the research and institution itself, continues today.[35]

By the late 1960s and early 1970s, Lynn Margulis straddled a new world of science as the amount of science reported in the media exploded. In the United States, the 1970s and 1980s saw the creation of science sections in dozens of newspapers, weekly television series, such as *Nova*, and the publication of glossy popular science magazines devoted to science.[36] Margulis constructed the image of the rebel during this period as science began to flow through popular culture, along with environmental consciousness and second-wave feminism. Being an evolutionary theorist during this period was highly problematic, with heavyweights such as Stephen Jay Gould absorbing most of the media oxygen available. But Margulis was ready for battle.

Opponents to her work primarily aimed to question her role as a scientist. They were quick to point out that her paper presented no original research data, as Margulis had drawn on the works of her predecessors to present experimental evidence for her ideas. This tactic wore thin as it was clear to many, including Stanier, that they were the pot(s) calling the kettle black. If the lack of original data did not remove Margulis from the category of science itself, they began to question her scientific abilities. Was she even a microbiologist? Suddenly, stereotypical images of microbiologists emerged highlighting the desperation to locate Margulis. She was criticized for being rarely seen in a laboratory coat or carrying out the manual labour for studying microbes – prepping media cultures, isolating strains for identification. The microbiologist and her colleague Ricard Guerrero, however, defended this aspect of her career as a scientist, stating that, 'her intellectual contributions were essential to many discoveries, reflective of her ability to see "the big picture" … [and she] had an extraordinary ability to interpret micrographs of any kind'.[37]

Unlike her female predecessors discussed above, Margulis' ability to confront any conflict head-on was in part due to the community in which she emerged as a scientist. Margulis and her husband Carl Sagan were part of the new generation of scientists that ushered in the era of science advocacy and popularization.[38] The number of women in science increased during the 1960s for various reasons, including the women's movement and labour shortages.[39] Margulis' conviction about her research enabled her to go on the offensive vocally. Being unabashedly vocal became part of her oeuvre and she became instantly recognizable because of it. For example, Niles Eldredge, a palaeontologist best known for proposing the theory of punctuated equilibrium with Stephen Jay Gould, recalled giving a talk at Amherst College.[40] During the questions-and-answers period, someone in the audience asked if anyone had managed experimentally to produce a new, fully reproductively isolated species in the lab. Eldredge replied that 'Theodosius Dobzhansky had said that he had at first thought that someone had in fact done so with experimental populations of a species of *Drosophila*. [S]o, no, I said no one had managed to produce a convincing, true case of reproductively isolated populations in the lab.'[41] The explosion was instantaneous, Eldredge recalled. When a woman stood up and started 'shouting' he suddenly realized it was Margulis, '*The* Lynn Margulis' (Eldredge's emphasis).[42] Eldredge tried to answer her, and 'though I never shy away from an intellectual argument, there weren't many openings left in Lynn's verbal onslaught'.[43] Years later Elredge stated, 'of course she was right'.[44] The sense of Margulis' always being correct would follow her throughout her career and became a character trait that her colleagues braced themselves for.

The discomfort in dealing with Margulis was not only due to her gender, it was also because she dared to theorize. Theory and experimentation are gender-linked. From the inception of modern science, men theorized to produce scientific interventions and paradigm shifts. Experimentation was broader in scope, more hands-on than intellectual labour, and, therefore, women were allowed.[45] As Margulis was never going to back down, her opponents used non-scientific attributes to de-gender her and create a level of acceptability around her as a theorist. They de-feminized her, and bolstered traits deemed more masculine, such as assertiveness; in so doing, evolutionary biologists were able to enter into intellectual debate with her and at the same time discredit her as well.

A scientific grenade-thrower is born

Literary agent John Brockman, who held numerous scientific salons, stated that '[Margulis] was not shy about expressing her opinions. Her in-your-face, take-no-prisoners stance was pugnacious and tenacious. She was impossible. She was wonderful.'[46] Margulis was often compared to her first husband and science-popularizing astronomer Carl Sagan, for being as charismatic and brilliant as he was. The theme of strength and conviction erupted repeatedly. Biologist Richard Dawkins admitted that he 'greatly admire[d] Lynn Margulis' sheer courage and stamina in sticking by the endosymbiosis theory, and carrying it through from being an unorthodoxy to an orthodoxy ... This is one of the great achievements of twentieth-century evolutionary biology.' The list of terms to describe Margulis was endless, including 'assertive', 'relentless roguishness', 'loud', and yet, unsurprisingly, such images were not wholly accurate. Margulis' own experience as a scientist led to a form of resilience that transformed into a careful curation of her persona as a rebel.

Margulis certainly enjoyed going on the offensive. She often shared the story of how her paper had been rejected by fifteen editors before James

Danielli, co-originator of the model of the lipoprotein bilayer of membranes, decided to publish it. She gloated that one reviewer had written, 'Your research is crap ... Do not bother to apply again.' The term 'crap' implicitly assigned to her not only her research, but her identity as a scientist. If her research was 'crap', therefore she was 'crap'.

Margulis took on her opponents with voracity. In a letter to *Nature* co-authored with Michael F. Dolan, she wrote:

> Sir. Although we agree with William Martin and Eugene V. Koonin's point in Correspondence ... about the validity of the term 'prokaryote,' a term that Norman R. Pace has proposed abolishing ... they have lost sight of the organismic biology forest for the molecular biology trees. The main differences between prokaryotic and eukaryotic cells probably relate to the original symbioses from which eukaryotes evolved.[47]

While Dolan believed that she was 'a bit full of (her)self', Margulis was a fine match to her counterparts.[48] Evolutionary biologists, especially neo-Darwinists, were among her favourite targets. The arguments became so fierce that at one point Richard Dawkins, no stranger himself to fighting, referred to Margulis as 'Attila the hen'.[49] Despite relentless criticism, some themes remained the same for her; she knew her research was solid and she rebelled against all norms of acceptable behaviour within the culture of science to join these men in the academic rabble.

As with all images, the nature of the depiction is more consequential than its precise fidelity to the truth. Such stories gave Margulis an audience to relate to her, a familiar framework through which to understand and interact with her. Non-female imagery provided an interactive structure that helped establish Margulis as a scientist who could be trusted. As botanist Peter Raven put it, 'scientists are mostly instrumentalists, but she's an innovator – often ahead, not always right, but right enough of the time'.[50] These images constituted non-scientific appeal, not in the sense of being unscientific, but rather by facilitating admiration for her person that did not depend exclusively on her science. The rhetoric around Margulis clearly concentrated on de-gendering her or at least focusing on traits more digestible to her male colleagues. They help explain why her personality traits took the form that they did; by diluting or placing her gender to the side, de-gendering of her femininity was more acceptable than any other available approach. But the significance of this discourse goes further still.

Evelyn Fox Keller has noted the degree to which 'our understanding of "feminine" and "scientific" have been historically constructed in opposition to each other'.[51] A focus on gender was also a focus on personal identity, and indeed the conflation of the personal and the intellectual was a central part of Margulis' appeal. Her person was seen as relevant to – and, perhaps, inseparable from – an analysis of her theory. All these comments assigned to Margulis specific values – for example, resilience or courage – that the particular writer held in high regard. Margulis willingly co-constructed these values; she self-described as bossy, rude, hyperactive and self-centred, which is what culminated in her being a rebel. And it was being a rebel and acting with rebellion that defined the legacy of her as a scientist and theorist. But what exactly had Margulis done by casting her theory into the mix of many evolutionary theories? Margulis had challenged their certainty.

Margulis had set the clock back four billion years. Evolution began not with animals but with microbes. Slow change was a part of the evolutionary story, but it was not all of it. Margulis thought that symbiogenesis – a nonviolent, mutually beneficial arrangement – would explain these more immense changes that were necessary. Her theory offered a different vision of evolution, ancient microbes merging for mutual gain. Where Darwin saw 'nature red in tooth and claw', Margulis saw cooperation and networking. Her world was aesthetically calmer than the neo-Darwinists presented.[52] Her world could be interpreted through a feminist or queer lens, but Margulis eschewed this type of gendering of her theory. She was more interested in getting the scientific community to validate her research.

Margulis questioned those neo-Darwinists along two main threads: their theory and their community. She described these male scientists who excessively focused on competition between organisms as 'a minor twentieth-century religious sect within the sprawling religious persuasion of Anglo-Saxon Biology'.[53] She was scathing towards these men, who 'wallow in their zoological, capitalistic, competitive, cost-benefit interpretation of Darwin – having mistaken him ... Neo-Darwinism, which insists on [the slow accrual of mutations by gene-level natural selection], is in a complete funk.'[54] But who exactly were these scientists? Margulis had crossed into the field of evolutionary biology, known as systematics.

In the 1960s, 1970s and 1980s, systematics was fierce. As population geneticist Joseph Felsenstein wrote, 'I used to think that we fought a lot when I worked in population genetics but in that field we used to sit side by side at meetings without growing red-faced, hissing at each other, or spreading scurrilous rumors.'[55] Another biologist, Paul Ehrlich, reported after a conference in the 1960s that it 'went very well – one old-line systematist cried (the meeting would hardly have been a success without that!)'.[56] Systematic biologists became notorious for explosive debates between competing theories of biological classification and phylogenetic inference. Margulis had found her audience – a group of argumentative and arrogant biologists who enjoyed being embroiled in controversy.

Margulis focused on three main theoretical tensions that required endo-symbiotic theory to integrate them: population genetics, the origins of novelty and phylogeny. She accused these scientists of

> codifying ignorance ... I refer in part to the fact that [those from the zoological tradition] miss four out of the five kingdoms of life. Animals are only one of these kingdoms. They miss bacteria, protoctista, fungi, and plants. They take a small and interesting chapter in the book of evolution and extrapolate it into the entire encyclopedia of life. Skewed and limited in their perspective, they are not wrong so much as grossly uninformed.[57]

Their failure was to only see the individual organism and not the collective in their equations and models. As components of her theory began to be empirically supported and understood over time as fundamental to the origins of life on Earth, her persona as a rebel was solidified.

Conclusion: Claiming territory as a theorist

Margulis defined herself by 'oppositional science' and the science community saw her as a modern embodiment of the 'scientific rebel'. She was considered among the elite of 'scientific grenade throwers, and long-entrenched positions shifted when she scored a hit'.[58] Over her fifty-year career, Margulis worked hard to invite a scientifically curious audience into her world. She published popular books, including *Microcosmos: Four Billion Years of Microbial Evolution* (1997), which many scientists saw as an academic jab to her ex-husband's *Cosmos* television series. She was vocal about environmental issues and became revered as a talented and resourceful popularizer of science. Margulis, however, was heavily criticized for being disconnected at times from the scientific achievements of modern biology – and practising her own brand of science, including her collaboration with James Lovelock and the Gaia Theory. She refused simply to doubt her own intuition. Her interview with John Horgan summed up how Margulis interacted with her own persona:

> I asked Margulis if she minded always being referred to as a provocateur or gadfly, or someone who was 'fruitfully wrong,' as one scientist put it. She pressed her lips together, brooding over the question. 'It's kind of dismissive, not serious,' she replied. 'I mean, you wouldn't do this to a serious scientist, would you?' She stared at me, and I finally realized her question was not rhetorical; she really wanted an answer. I agreed that the descriptions seemed somewhat condescending.[59]

But at the end of her life no one questioned her claim as an evolutionary theorist.

How representative was Margulis? Did non-scientific appeal play out in other depictions of female scientists? Is it still worth asking what broader lessons her case might suggest? The gendered nature of the discourse surrounding her – particularly its presence in all kinds of views, not just negative ones – seems likely to be applicable to other iconic female scientists. Margulis, however, stands out as a very atypical scientist too. She was a rebel and a theorist in ways other female scientists were not or could not be. Her generosity to speak publicly about her work, her outsider status, her rebellious nature, controversial stances and chosen field are all factors that would transfer and apply to some, but hardly all, iconic scientists. In other words, not only does Margulis differ in many ways from scientific icons, but the scientific culture that discusses her cannot be considered wholly coterminous with the culture that engaged with Peter Raven, Richards Dawkins, James Watson and others. Margulis' focus purely on the research and not financially viable entrepreneurial side of her discoveries, such as James Watson's directorship at the Cold Spring Harbour Laboratories or Craig Venter and the Human Genome Project, place her as someone who never truly fit anywhere. Known to open lectures by asking 'Any REAL biologists here?',[60] Margulis remains a totally singular iconoclast.

Acknowledgements

For constructive criticism on earlier drafts, I am grateful to Molly Nolan, Myles Jackson and Julie Livingston. I also wish to thank the peer reviewer whose comments and suggestions greatly improved this article.

Notes

1. For continuity and readability, I will use the last name Margulis, as she is best known as Lynn Margulis.
2. J. Brockman, 'Lynn Margulis 1938–2011: "Gaia is a tough bitch"', *Edge Magazine* (23 November 2011) www.edge.org/conversation/lynn_margulis-lynn-margulis-1938-2011-gaia-is-a-tough-bitch (accessed 23 November 2022).
3. L. Margulis, *Lynn Margulis: The Life and Legacy of a Scientific Rebel* (White River Junction, VT: Chelsea Green Publishing, 2012).
4. S. Shapin and S. Schaffer, *Leviathan and the Air-Pump: Hobbes, Boyle, and the Experimental Life* (Princeton, NJ: Princeton University Press, 2011); D. Kaiser, *How the Hippies Saved Physics: Science, Counterculture, and the Quantum Revival* (New York: W. W. Norton & Company, 2011); N. Oreskes, 'Objectivity or heroism? On the invisibility of women in science', *Osiris*, 11 (1996), 91.

5 F. Dyson, *The Scientist as Rebel* (New York: New York Review of Books, 2006); P. F. Schewe, *Maverick Genius: The Pioneering Odyssey of Freeman Dyson* (New York: Thomas Dunne Books, 2013); R. Crease, 'Physics: Rebel without a pause', *Nature*, 494: 311 (2013), https://doi.org/10.1038/494311a.
 6 I. Ockert, *The Scientific Storytellers: How Educators, Scientists, and Actors Televised Science* (Princeton, NJ: Princeton University, ProQuest Dissertations Publishing, 2018).
 7 N. Lutkehaus, *Margaret Mead: The Making of an American Icon* (Princeton, NJ: Princeton University Press, 2008); D. K. Hecht, 'The atomic hero: Robert Oppenheimer and the making of scientific icons in the early Cold War', *Technology and Culture*, 49: 4 (2008), 943–66; C. Bonneuil, 'La Cinquième République des sciences: Transformations des savoirs et des formes d'engagement des scientifiques', in C. Charles and L. Jeanpierre (eds), *La Vie intellectuelle en France*, vol. 2 (Paris: Seuil, 2016), pp. 515–36; L. M. Krauss, 'Scientists as celebrities: Bad for science or good for society?' *Bulletin of the Atomic Scientists*, 71: 1 (2015), 26–32; A. Archer, A. Cawston, B. Matheson and M. Geuskens, 'Celebrity, democracy, and epistemic power', *Perspectives on Politics*, 18: 1 (2020), 27–42; D. K. Hecht, 'Constructing a scientist: Expert authority and public images of Rachel Carson', *Historical Studies in the Natural Sciences*, 41: 3 (2011), 277–302; D. Fahy and B. Lewenstein, 'Scientists in popular culture: The making of celebrities', in M. Bucchi and B. Trench (eds), *Routledge Handbook of Public Communication of Science and Technology*, (Abingdon and New York: Routledge, 2021), pp. 33–35.
 8 Hecht, 'Constructing a scientist', pp. 278–79.
 9 L. Sagan, 'On the origin of mitosing cells', *Journal of Theoretical Biology*, 14: 3 (1967), 226.
10 K. Mereschkowski, 'Über Natur und Ursprung der Chromatophoren im Pflanzenreiche', *Biologisches Centralblatt*, 25 (1905), 593–604; A. S. Famintsyn, 'Die Symbiose als Mittel der Synthese von Organismen', *Biologisches Centralblatt*, 27 (1907), 253–64.
11 B. M. Kozo-Polyansky, *Symbiogenesis: A New Principle of Evolution*, ed. and trans. V. Fet, ed. L. Margulis (Cambridge, MA: Harvard University Press, 2010).
12 See J. Sapp, *The New Foundations of Evolution: On the Tree of Life* (New York: Oxford University Press, 2009). Although I was unable to find evidence, one wonders if the scientists' national origin shaped American views in rejecting the theory of symbiosis.
13 E. B. Wilson, *The Cell in Development and Heredity*, 3rd edition (New York: The Macmillan Company, 1925), p. 278. See also R. Hagemann, 'The reception of the Schimper-Mereschkowsky endosymbiont hypothesis on the origin of plastids – between 1883 and 1960 – many negative, but a few relevant positive reactions', *Annals of the History and Philosophy of Biology*, 12 (2007), 41–60.
14 Now the Rockefeller University.

15 C. de Duve, 'Evolution of the peroxisome', *Annals of the New York Academy of Science*, 168 (1969), 369–81; J. Sapp, 'The prokaryote-eukaryote dichotomy: Meanings and mythology', *Microbiology and Molecular Biology Reviews*, 69: 2 (2005), 292–305.
16 R. Y. Stanier, 'Some aspects of the biology of cells and their possible evolutionary significance', *Symposium of the Society of Genetic Microbiology*, 20 (1970), 1–38.
17 R. A. Raff and H. R. Mahler, 'The nonsymbiotic origin of mitochondria: The question of the origin of the eukaryotic cell and its organelles is reexamined', *Science*, 177: 4049 (1972), 576.
18 R. Y. Stanier and C. B. van Niel, 'The concept of a bacterium', *Archiv für Mikrobiologie*, 42 (1962), 17–35.
19 R. Y. Stanier, M. Douderoff and E. A. Adelberg, *The Microbial World* (Englewood Cliffs, NJ: Prentice-Hall, 1963), p. 18.
20 M. W. Gray and W. F. Doolittle, 'Has the endosymbiont hypothesis been proven?' *Microbiological Reviews*, 46: 1 (1982), 1–42; Stanier, Douderoff and Adelberg, *The Microbial World*; J. M. Archibald, 'Endosymbiosis and eukaryotic cell evolution', *Current Biology*, 25: 19 (2015), R911–21; W. F. Martin, S. Garg and V. Zimorski, 'Endosymbiotic theories for eukaryote origin', *Philosophical Transactions of the Royal Society B: Biological Sciences*, 370: 1658 (2015), 20140330, http://dx.doi.org/10.1098/rstb.2014.0330.
21 Stanier and van Niel, 'The concept of a bacterium', p. 17.
22 R. Guerrero, 'Lynn Margulis (1938–2011): In search of truth', *International Microbiology*, 14 (2011), 183–86.
23 M. W. Rossiter ' "Women's work" in science, 1880–1910', *Isis*, 71: 3 (1980), 381–98; J. Tonn, 'Laboratory of domesticity: Gender, race, and science at the Bermuda Biological Station for Research, 1903–30', *History of Science*, 57: 2 (2019), 231–59; V. Gornick, *Women in Science: Then and Now* (New York: The Feminist Press at CUNY, 2009); S. Rodriguez, 'Watching the watch-glass: Miriam Menkin and one woman's work in reproductive science, 1938–1952', *Women's Studies*, 44: 4 (2015), 451–67; C. Jones, *Femininity, Mathematics and Science, 1880–1914* (Basingstoke: Palgrave Macmillan, 2009); M. Hicks, *Programmed Inequality: How Britain Discarded Women Technologists and Lost Its Edge in Computing* (Cambridge, MA: The MIT Press, 2017).
24 N. Oreskes, 'Objectivity or heroism? On the invisibility of women in science', *Osiris*, 11 (1996), 91.
25 Oreskes, 'Objectivity or heroism?' p. 91.
26 Oreskes, 'Objectivity or heroism?' p. 100.
27 B. Maddox, 'The double helix and the "wronged heroine"', *Nature*, 421: 6921 (2003), 407–8; B. Maddox, *Rosalind Franklin: The Dark Lady of DNA* (New York: HarperCollins, 2002). See also, M. Cobb and N. Comfort, 'What Rosalind Franklin truly contributed to the discovery of DNA's structure', *Nature*, 616: 7958 (2023), 657–60.
28 Franklin's story is also suggestive of the role of antisemitism in post-World War II science.

29 J. Watson, *The Double Helix: A Personal Account of the Discovery of the Structure of DNA* (New York: Atheneum, 1968). Watson admitted that Franklin never gave her data to Crick and himself. He was relentless in ridiculing and disparaging: 'Certainly a bad way to go out into the foulness of a ... November night was to be told by a woman to refrain from venturing an opinion about a subject for which you were not trained' (p. 52). He added that her 'belligerent moods' interfered with Wilkins' ability to 'maintain a dominant position that would allow him to think unhindered about DNA' (p. 17). For that reason, '[c]learly Rosy had to go or be put in her place. ... The thought could not be avoided that the best home for a feminist was in another person's lab' (p. 17).
30 Watson wrote: 'there was never lipstick to contrast with her straight black hair, while at the age of thirty-one her dresses showed all the imagination of English blue-stocking adolescents'. Watson, *The Double Helix*, p. 17.
31 Crick, by no means an innocent bystander, in *Nobel Prize Women in Science* (1993, p. 316), was quoted as saying, 'I'm afraid we always used to adopt – let's say, a patronizing attitude towards her.'
32 Hecht, 'Constructing a scientist', p. 281.
33 Hecht, 'Constructing a scientist', p. 287.
34 M. V. Barrow Jr., 'Carson in cartoon: A new window onto the noisy reception to Silent Spring', *Endeavour*, 36: 4 (2012), 164.
35 S. Shapin, *The Scientific Life: A Moral History of a Late Modern Vocation* (Chicago, IL: University of Chicago Press, 2008).
36 D. Fahy, 'A brief history of scientific celebrity', *Skeptical Inquirer*, 39: 4 (July/August 2015); I. Ockert, 'The scientific storytellers: How educators, scientists, and actors televised science' (PhD dissertation, Princeton University, ProQuest Dissertations Publishing, 2018).
37 C. Chica, *Once Upon a Time Lynn Margulis: A Portrait by Colleagues and Friends* (Editorial Septimus, October 2013) https://issuu.com/estudipuche/docs/once_upon_a_time_lynn_margulis (accessed 30 August 2022).
38 D. Steel, 'Carl Sagan: Practitioner, popularizer and proponent of science', *Contemporary Physics*, 42: 4 (2001), 247–49; T. Caulfield and D. Fahy, 'Science, celebrities, and public engagement', *Issues in Science and Technology*, 32: 4 (2016), 24; A. C. Chambers, '"The handsome astronomer and the yelling lady": Representing scientists and expertise in "Don't Look Up"', *Journal of Science Communication*, 21: 5 (2022), C04; G. Källstrand, 'Warburg's dogs: Nobel laureates and scientific celebrity', *Celebrity Studies*, 13: 1 (2022), 56–72.
39 See L. Schiebinger, 'The history and philosophy of women in science: A review essay', *Signs: Journal of Women in Culture and Society*, 12: 2 (1987), 305–32. See D. Haraway, 'Class, race, sex, scientific objects of knowledge: A socialist-feminist perspective on the social construction of productive nature and some political consequences', in V. Haas and C. Perrucci (eds), *Women in Scientific and Engineering Professions* (Ann Arbor: University of Michigan Press, 1984), esp. pp. 212–13; Gornick, *Women in Science*.
40 The talk was about species, speciation and the fossil record.

41　N. Eldredge, 'The passionate Lynn Margulis', in Margulis, *Lynn Margulis*, p. 47.
42　Eldredge, 'The passionate Lynn Margulis', p. 47.
43　Eldredge, 'The passionate Lynn Margulis', p. 48.
44　Eldredge, 'The passionate Lynn Margulis', p. 48.
45　J. Bangham, X. Chacko and J. Kaplan, *Invisible Labour in Modern Science* (Lanham, MD: Rowman and Littlefield, 2022); S. Shapin, 'The house of experiment in seventeenth-century England', *Isis*, 79: 3 (1988), 373–404; S. Shapin, 'The invisible technician', *American Scientist*, 77: 6 (1989), 554–63; J. S. Light, 'When computers were women', *Technology and Culture*, 40: 3 (1999), 455–83; J. Tonn, 'Extralaboratory life: Gender, politics and experimental biology at Radcliffe College, 1894–1910', *Gender & History*, 29: 2 (2017), 329–58.
46　Brockman, 'Lynn Margulis 1938–2011'.
47　M. F. Dolan and L. Margulis, 'Advances in biology reveal truth about prokaryotes', *Nature*, 445: 7123 (2007), https://doi.org/10.1038/445021b.
48　M. F. Dolan, 'Lynn Margulis and Stephen Jay Gould', in Margulis, *Lynn Margulis*, p. 51.
49　J. Lovelock, 'On Lynn from a close friend and colleague', in Margulis, *Lynn Margulis*, p. 30.
50　Raven is best known for his theory of coevolution with Paul Ehrlich. Quote from J. di Prosperzio, 'Full speed ahead', *University of Chicago Magazine* (February 2004).
51　E. F. Keller, 'The gender/science system: Or, is sex to gender as nature is to science?' *Hypatia*, 2: 3 (1987), 37–49.
52　Throughout her lifetime, Margulis resented and rejected depictions of her as a feminist. While many feminists and female scientists claimed her as their own, she eschewed being pigeon-holed. She particularly rejected the notion that she was trying to replace masculine concepts of nature with feminine ones. She was aware of how her theory seemed to mimic certain societal norms, such as women being cooperative and men competitive by nature. In an interview with John Horgan in 2011, 'she conceded that, in comparison to such concepts as "survival of the fittest" and "nature red in tooth and claw," her symbiosis views might seem feminine. "There is that cultural overtone, but I consider that just a complete distortion."' J. Horgan, 'R.I.P. Lynn Margulis, biological rebel', *Scientific American* (24 November 2011), https://blogs.scientificamerican.com/cross-check/r-i-p-lynn-margulis-biological-rebel/ (accessed 30 August 2022).
53　L. Margulis, 'Kingdom Animalia: The zoological malaise from a microbial perspective', *American Zoologist*, 30: 4 (1990), 867.
54　C. Mann, 'Lynn Margulis: Science's unruly Earth Mother: Lynn Margulis' partisanship of Gaia enrages her colleagues in evolutionary biology, but nobody dismisses her out of hand – because she's been right before', *Science* 252: 5004 (1991), 379, www.science.org/doi/10.1126/science.252.5004.378.
55　J. Felsenstein, 'The troubled growth of statistical phylogenetics', *Systematic Biology*, 50: 4 (2001), 465–67. See also D. L. Hull, *Science as a Process* (Chicago, IL: University of Chicago Press, 2010).

56 K. Vernon, 'A truly taxonomic revolution? Numerical taxonomy 1957–1970', *Studies in the History and Philosophy of Biological and Biomedical Sciences*, 32 (2001), 315–41.
57 Brockman, 'Lynn Margulis 1938–2011'.
58 Brockman, 'Lynn Margulis 1938–2011'.
59 Horgan, 'R.I.P. Lynn Margulis, biological rebel'.
60 Mann, 'Lynn Margulis'.

II

In/visibilities across borders: Scientific collaborations and contestations

3

Inventing a career across borders in the early 1930s: The case of cytogeneticists Eileen W. Erlanson and E. K. Janaki Ammal

Savithri Preetha Nair

Biographical explorations and studies of long-lasting friendships between women scientists offer gender historians a rich resource for unravelling invisible academic networks in operation that helped build careers in science. Historians of biology have begun to shine light on collaborating scientist couples or spouses[1] and on female bonding (covert or overt) at institutions,[2] including research laboratories. Erwin Baur's genetics laboratory in Germany in the early twentieth century employed mostly female staff[3] and included such trail-blazing women geneticists and crop researchers as Elisabeth Schiemann (1881–1972). Her six-decade-long friendship with physicist Lise Meitner (1878–1968), which has received some scholarly attention,[4] is perhaps a comparable case study to ours, but the two shared similar social backgrounds, including a language, lived in the same city and practised two different scientific disciplines (genetics and physics), during singularly volatile historical periods. Although remarkable, their companionship was hardly transnational or cross-border in nature, except perhaps from 1938, when Meitner moved to Sweden, and much later to England.

By contrast, this paper focuses on the as-yet-overlooked case of two women – one British American and the other Asian – bonding over science transnationally and on the fringes of institutions, doing science in each other's company or in close proximity, in a colonial setting.[5] It sheds light on the struggles encountered and strategies adopted by a highly accomplished, but little-known, trail-blazing British American biologist Eileen Jessie Whitehead Erlanson (1899–2001), later Macfarlane, in finding a livelihood commensurate with her education and research experience during the Great Slump. Eileen's life between 1933 and 1936 serves as an object lesson, even a role model, on how forging robust connections across national and disciplinary borders was crucial to 'inventing' a multifaceted scientific career at a time when employment opportunities were especially grim for women, let alone scientists. In her case, the cross-border experiment was facilitated by a close woman contemporary, the Indian plant cytogeneticist

E. K. Janaki Ammal (1897–1984). It was a relationship that would last close to six decades.

The British-born Eileen J. Whitehead obtained her BSc in botany from the University of London in 1919. A year prior to this, she had participated in the Government Potato Disease Survey on the Isle of Wight, and this was perhaps where she first met the soil specialist Earl J. Grimes. She would impetuously marry Grimes (later associate professor of biology at the College of William and Mary at Williamsburg, Virginia, but at the time of the marriage based at King's College, London) to spite her mother, who believed she would end up 'an old maid', being incapable of any 'small talk'.[6] The union would be short-lived, as Grimes died within a year of their marriage. Eileen would however soldier on and publish the *Flora of the Peninsula of Virginia* (1924), interestingly even before she had completed a doctorate, based on the extensive collection they had put together (1921–22). A high-spirited personality, she would soon find love again and marry Carl O. Erlanson, an instructor in botany at the University of Michigan, whom she had met while pursuing doctoral research at the University under H. H. Bartlett – her research was on the American wild rose as Emma Cole Fellowship-holder in botany. On campus, and as a botany instructor, Eileen was very active organizing meetings and botanical seminars, and stood out for her association (as part of the faculty advisory committee) with the Negro-Caucasian Club, the first of its kind anywhere on campuses of American Universities. The Club aimed to 'work towards a better understanding between races and for the abolition of discrimination against Negroes'.[7]

Eileen was perhaps the earliest woman anywhere in the world to be awarded both a PhD and DSc in genetics, and of those born prior to 1900, one of the most adventurous, prolific and public a personality one can find. She ironically remains unknown. With the exception of James Cattell's *American Men of Science*,[8] which mentioned her name, she does not even appear in the classic Rossiter volumes, with their focus on employment issues.[9] This is even more surprising given that she was one of the few biologists to find a place in the Radcliffe 'Women in Science' Exhibit of 1936, part of the Tercentenary celebrations of the Harvard University.[10]

National Research Council Fellow, thrice over

On submission of her doctoral thesis – she was awarded a PhD by the University of Michigan in 1928 for her work on the genetics of the native American rose – Eileen was fortunate enough to be offered employment at the Kent State College in Ohio as a teacher, but when the fruit-fly geneticist (and later Nobel Prize-winner) T. H. Morgan heard of her rose investigations, he invited her to his institute at Pasadena, California. She seized the opportunity and initiated a rose collection. In 1929, Eileen presented a paper at the Botanical Seminar at the University of Michigan on 'Wild roses of the western United States'.[11] A year later, she spoke on 'Sterility in wild roses and in some species hybrids'.[12] *The Michigan Daily* noted that Eileen's work was 'unique' and that she was 'the only botanist whose speciality was roses' and that she maintained 'several acres of roses in the botanical gardens here [the University Botanical Garden, Ann Arbor was the only one 'in which the wild origin of every variety of rose' known could be found], besides extensive rose gardens in Pasadena'.[13] Even after Eileen left Ann Arbor for good, she often visited her alma mater, insisting on her presence in the afternoons in the month of June at the Botanical Garden on Packard Road just outside the city limits, when the wild roses were in bloom, to explain her breeding and cytogenetic experiments to interested visitors.

In early March 1930, twenty-four National Research Council Fellowships for young scientists, 'most prized of all the scientific awards', together with ten extensions of fellowships previously granted were announced. Eileen's name figured in the latter category, alongside the names of the anthropologist Carleton S. Coon and the pioneering woman anthropologist and folklorist Anna Hardwick Gayton (1899–1977) of the University of California.[14] This was Eileen's third National Fellowship – two was customary, but it was unusual for researchers to be awarded a third time – which took her to the John Innes Horticultural Institute (hereafter, the John Innes) in England, founded by William Bateson. At the John Innes in late 1930,[15] Eileen was engaged in 'estimating the number of variations of wild roses and classifying them for the first time' by crossing hybrid seedlings. Cytological research on the rose material was accomplished under the guidance of Cyril D. Darlington. Until her valuable scientific contribution to the genetics of the *Rosa*, much of the classification of wild roses 'was a case of guesswork'.[16]

When Eileen returned to Michigan in the new year (1931), she presented a Botanical Seminar paper titled, 'The Newton and Darlington interpretation of meiotic phenomena'.[17] A few weeks later, she found herself in Pasadena, where she resumed work she had begun three years previously. One of the first things she did was to go on a solitary wild-rose-collecting expedition in her car across the western United States. By this time, Eileen was confident of working out a preliminary classification of the wild roses of North America and had even been invited to publish it in the *American Rose Annual* of 1932. When her paper appeared in the *Annual*, the editor noted that it provided a 'clear look into the make-up of the American native roses from the cytological standpoint, and [set] up a desirable simplification of the species'. Classification of roses was such a vexing taxonomical problem because of the great variation of traits/multiplicity of forms, and the consequent species inflation, which Eileen warned '[got] us nowhere'.[18] It was for synthesizing her *Rosa* research and setting the basis for the taxonomic revision of the rose species, based not only on morphology but also on their cytology and the results of hybrid crossings, and their progeny, besides pollen analyses, that she would be awarded a DSc by the University of London (to which the John Innes was affiliated) in 1934.[19]

By early 1931, her marriage to Erlanson had ended, but their friendship remained unaffected. It was also about this time that she would meet with a horrific motor accident. She had accompanied the geneticist Theodosius Dobzhanksy, his wife and the Austrian microbiologist Karl Bělař (visiting from the John Innes) to the Mojave Desert, when the car they were travelling in crashed perilously, killing Bělař instantly. Eileen was left seriously injured – the Dobzhanskys were very lucky to escape relatively unhurt – and was in hospital at Pasadena for close to three months. Her funding institution, the National Research Council, sent her a cheque of $1,000 to cover expenses, but this would prove insufficient. In early August, while recuperating at the house of the Pasadena geneticist Sterling Emerson, she would write to Bartlett requesting him to persuade the Council to help clear the additional bills incurred during hospitalization; all she had with her at this time, by way of funds, was the Fellowship salary of the last couple of months. She was moreover still 'weak and stiff in the limbs', but because she had no job for the next autumn, she planned to return to Ann Arbor to complete the piece of research that had been interrupted by the accident. For the highly energetic Eileen, the many months of confinement were challenging in the extreme. She was desperate for some normality, which primarily meant engaging in scientific research. 'It will be wonderful to be back to civilian life and to be a useful member of society again', she remarked to Bartlett.[20]

Turning in job applications

These were times when even women with advanced degrees could not find job positions.[21] Among the chief institutions hiring plant geneticists in America were colleges, botanical gardens and horticultural societies. Eileen was alert to every possible opening, if ever there was one, and would tirelessly contact heads of institutions, sometimes well ahead of the position being advertised, and even while laid up and immobile. Within a month's time, she had applied to nearly fifty universities and colleges and had been met with a refusal from most of them. She was particularly disappointed to learn that the College of William and Mary, where her late husband Grimes had been based, did not 'want anyone good [like herself]': 'How do they expect to build up a course in cytology?' she would question Bartlett, clearly peeved. Not one to give up easily, Eileen probed him to ascertain if Ralph H. Cheney, professor of biology at Long Island University, was known to him. She had read in *Science* (10 July 1931) that the Brooklyn Garden was to be affiliated to that university, and that Cheney had been made 'resident investigator (economic plants)'.

> Do you happen to know him? ... The University is new & I would certainly like to get a position in that sort of Institute. It would not matter if one had to begin with a small salary if there was a future. They ought to get good younger people in such departments. The difficulty is that when there is a vacancy it is usually filled by someone known to the head of the department.

Eileen also wondered if the Detroit City College had an opening; this would have enabled her to tend to her rose garden at Ann Arbor on the weekends. In fact, she was even ready to teach at a high school, being 'supremely confident' that she had ability to 'foster enthusiasm in young people', but the fact that she had no degree in education was a major setback.

She was quick to anticipate that a job position would open when she heard of the passing away of Per Axel Rydberg (1931), the first curator of the Herbarium at the New York Botanical Garden, and lost no time influencing Elmer Drew Merrill, the new director of the Botanical Garden, and 'some others', to file the application on her behalf, since she was immobile. She had in fact written to Charles S. Gager, director of the Brooklyn Botanic Garden the previous winter about a position but had been informed in 'a very nice letter' that they were not hiring anyone new. She had also contacted the American Rose Society in the hope of finding a patron. J. Horace McFarland, president of the Society, expressed solidarity with the plan of supporting her research through the Society or some of its wealthy members, such as Captain George C. Thomas of Beverly Hills, California, although he did not suggest how this might be accomplished. Indeed, she had visited Captain Thomas some months previously, and he had happily shared with her details of his rose-breeding experiments. Eileen decided to write to him directly to 'tell him of the situation'. 'He is well known and might be willing to endorse my work, then other wealthy members of the society might realize that there was some use in it – Mrs Ford [Sophie DuPont Ford or Mrs Bruce E. Ford], for instance, who is a trustee of the society', wrote Eileen, revealing her plans to Bartlett. Nothing would come out of all this, but her determination to 'be as active & useful as possible' remained undiminished. On the health front, she was still far from robust.[22]

Invitation from India

Just as she was growing despondent, she received a letter from her Michigan friend and fellow plant cytogeneticist E. K. Janaki Ammal, inviting her to Travancore, a princely state in the south-west of India. The invitation could not have arrived at a better time. Excited at the prospect of going to the 'East' to collect plants, conduct research and reunite with her old friend, she wasted no time informing Bartlett of the new development:

> A letter has just arrived from Janaki [also mentored by Bartlett] saying that she has been appointed head of the Department of Botany at the Maharaja's College of Science, Trivandrum, Travancore. The appointment is for two years, while the regular professor is getting his doctor's degree in England. She says they have fine laboratories & excellent opportunities for collecting in a monsoon area. She has invited me to come out for a year and promises all facilities & no expenses.

Eileen only needed to save about $500 for the journey, which looked quite easily achievable. She wanted to get away from the appalling 'drudgery' of her teaching job at Ohio, which she felt would 'crush her between the professional Educationalists and the ignorance and inertia of the students'. Her years of training, she concluded, were wasted at that place. Eileen was sure there would be others who coveted her present job, which made arranging for a year's leave of absence and finding a temporary replacement relatively easy.

She was keen on accepting Janaki's invitation. 'It [Trivandrum] really is a fine opportunity and I do not see why a year spent in that manner would militate against my getting a position afterwards. If our civilization is going to pieces one would be all well off in the tropics as in a colder region', she rationalized. She wanted Bartlett's opinion on this and asked him to pen a letter to Janaki, who was missing him very much. 'If you will come to India on the way [from his fieldwork in South-East Asia] she will come to meet you with the State Elephants (if you come to Travancore)', Eileen joked all excited, while also stating that her aim in life was to 'do a lot of research but it would help to make enough to defray living expenses'.[23] Janaki would have agreed.

Having earned a doctorate from the University of Michigan in 1931 for her cytogenetic work on the *Nicandra physalodes* under the supervision of B. M. Davis – the first Indian woman to achieve this milestone in the botanical sciences – Janaki embarked on a three-month stint at the John Innes.[24] In late 1931, when Eileen was desperately looking for employment in America, Janaki was already heading home to India to take up a research position in Madras. She had been negotiating with her former teacher, T. Ekambaram at the Madras University, even while in England, about a possible position at the newly established University Botanical Laboratory (UBL) under his charge. Thanks to Ekambaram's recommendation, a research fellowship (1932–33) was awarded to her by the University. This came as a great blessing for Janaki, because she had arrived penniless and had been forced to borrow from an elder brother to make ends meet. Ekambaram had further assigned her the task of developing a university botanical garden. Collecting for it gave her immense joy and even reminded her of the botanical garden at Ann Arbor, but she soon realized the university fellowship was far from sufficient to maintain herself. Moreover, the UBL appointment was only a year-long arrangement. Thus, Janaki began looking out for a position in the agricultural department, the only other place (besides a teaching institution) in India where someone with her qualifications could hope to find decent employment. She had only recently been elected Fellow of the Linnean Society.[25]

Through a timely development, Government Sugarcane Expert T. S. Venkatraman of the Breeding Station at Coimbatore applied to the Imperial Council of Agricultural Research for a cytological assistant, precisely with Janaki in mind, to assist with the cytology of the sugarcane. Even before she could be appointed officially at Coimbatore, Janaki set up a small cytology laboratory at the Presidency College, her alma mater in Madras, to work on some of Venkatraman's sugarcane material. She had also been asked to supervise three male students working on their MSc theses and to encourage cytological studies at Madras University. This was the first time cytology was being taught to university students and none of the male members of the Botany Department of the Presidency College were qualified to do so. The sugarcane material she was studying was producing interesting results, but all the same, Janaki was distressed she was not participating in India's struggle for freedom. She was also becoming increasingly apprehensive about the move to Coimbatore because she dreaded its Brahmanical and patriarchal climate, and preferred cosmopolitan Madras to it. Later that year (1932), Janaki held additional charges as 'Reader in Botany, Chairman of the Board of Examiners in Botany, Member of the Board of Studies and Academic Council and Member of the College Council'. She was still struggling to make a living. Moreover, news about the cytologist's position at Coimbatore was still to be confirmed. Thus, it was mere desperation that had led her to accept the offer of professorship at the Maharaja's College of Science in Trivandrum.

This appointment was not on a permanent footing, but she had created history. It was the first ever time a woman had been appointed to a professorship in what was an all-men's college. News of the appointment was even reported by some Indian newspapers and *The Michigan Daily*. Janaki was however not entirely pleased with her newfound position. Although it made her comfortable financially, she worried her research would be put on the backburner. She was willing to sacrifice any amount of money and comfort for facilities and quiet to do research and become a 'recognized scientist'.[26] Janaki felt 'like a real scientist' only when she sent a research note (on the cytology of the *Cleome viscosa*) to *Current Science*, founded just that year (1932) as India's first journal exclusively devoted to science. She was only the second woman, after Eleanor Dewey Mason of the Women's Christian College, Madras, to have appeared on its pages.[27]

Meanwhile, Janaki had sent Eileen a second letter from Trivandrum, in which she offered to arrange an honorary position for her in Madras, such as 'Head of Herbarium', the present herbarium being 'very poor & must be built up for the Honours Botany classes'. Also on her mind was the directorship of the UBL: Janaki believed that 'biology [was] stagnant in S[outh] India' and that Eileen could 'do a lot for its advance' even in the short span of a year. Janaki was acting on the conviction that the greater the number of applications, the better the chances of appointing 'a good scientist' to such positions. Promptly, Eileen requested Bartlett to write her a recommendation letter for 'the ultimate good of science'. She did not think of herself as a 'brilliant botanist', by which she meant a theoretician, but was certain she had 'a good deal of experience in herbarium, garden & laboratory'. Eileen was certain she 'could do more good for science in Research than in pure teaching'. This was exactly what Janaki hoped for: that Eileen would get across to the authorities in Trivandrum the importance of research over teaching. In her letter to Bartlett, Eileen did not omit to express concern about Ohio (this was the case with almost all teaching institutions in America) going through bad times, with the colleges and universities taking the brunt of it and the salaries of the teaching staff being drastically slashed.[28] Given the situation, the Travancore/Madras job was the best bet, even if it meant leaving America and translocating to an entirely unknown part of the world. Although their friendship would last close to six decades, the focus in what follows is limited to a telling stage in their relationship (1933–34), a period of transition for both women, when employment opportunities were at their lowest.

Preparations for the India sojourn

Eileen was granted a year's leave of absence by the Kent State College authorities, on the recommendation of Bartlett. She was due to sail for India on 22 September 1933 but suffered a major set-back on the health front, a complication that had arisen from the bladder surgery she had undergone in Pasadena after the accident. She had to be admitted to King's College Hospital, London, for an emergency procedure. This disrupted her travel plans and left her with even fewer funds at her disposal. To add to her frustration, recovery was slow, and she had to be on medication (a dose of 'sodium acid phosphate') for the next three months.[29]

Meanwhile, the British marine biologist Geoffrey Tandy of the British Museum (Natural History), an algae specialist, whom she had met on the boat from America, had introduced her to Kenneth de Burgh Codrington, at this time honorary lecturer of archaeology at the University of London. She would have many meetings with Codrington during her stay in London (her mother lived in Putney), which gave her 'several hints' as to how she could use her time in India fruitfully for research. He suggested that she turn to collecting physical anthropology measurements – 'get height-weight ratios', he instructed, for which purpose she got hold of 'a pair of good bathroom scales', besides 'a good pair of second hand calipers [sic]', from the Royal Anthropological Institute, for a guinea. What remained for procurement was a 'head spanner for the auricular radii', which was 'very expensive, and unless a decently priced one could be found' locally, she decided that getting one in India with Janaki's help was more prudent. Indeed, Codrington thought that Janaki 'could do a lot of valuable work [in physical anthropology herself] if she will persevere'. As for J. B. S. Haldane, whom Eileen knew well perhaps from her John Innes days, he suggested that she first learn Malayalam, the language spoken in Travancore. Ironically, a 'First Book' could not be found in London. Before she departed for India, Eileen took care to initiate the necessary proceedings to obtain a DSc, for her work on rose cytotaxonomy[30] – she was one of the world leaders in the field of experimental taxonomy. She also ensured that Bartlett complete the necessary formalities to appoint her an honorary research fellow at the University Botanical Garden, Ann Arbor, which would allow her to continue her connection with Michigan, even while away from the country.

Eileen eventually set sail for India on 3 November and arrived in Trivandrum only a couple of months after the John Innes cytogeneticist, Darlington, had left the city.[31] In an interview with *The Michigan Daily* just prior to her departure, she spoke of the future possibilities Janaki had envisioned for her:

> Dr Janaki expects to obtain some honorary appointment from the Maharani for me, to make my position official. It may be in connection with the herbarium of the College or as the head of a botanical survey of Travancore, which I shall have to organise ... She is eager to advance science and scientific research and appointed my friend, Dr Janaki, to a high position in what had formerly been strictly a men's College.[32]

Sisterhood

Being the only woman professor at the College of Science, Trivandrum, Janaki began a close association with the staff members of the Maharaja's College for Women in the city – women like the Cambridge-educated Louise

Carolina Maria Ouwerkerk, of Dutch origin and British nationality, who was professor of economics, and others like the botanists Mercia Janet and Sosa P. John, who were the first women to research at the UBL. There were also Eunice Gomez and Daisy Muthunayagom, professors of English, alumni of Madras University. Much like Eileen, it was joblessness during the Great Depression that had driven Ouwerkerk, an MA from Newnham College, to accept the offer of Travancore State. Forming an informal sisterhood of sorts, the 'jolly group' enjoyed themselves thoroughly, with frequent visits to the beaches. Sometimes, they had 'an incredibly large tea' on the rocks, after a swim in their saris or swimming costumes. They even 'drew up a Five-Year Plan for Brighter Trivandrum', which involved 'laughter, song and dance – the more mixed the better'.[33] Janaki's presence in Trivandrum was for Ouwerkerk and the others greatly empowering and vice versa. Within only a year of her joining the Women's College, Ouwerkerk had founded the Trivandrum branch of the International Fellowship, which aimed to connect Britishers and Indians. The cosmopolitan Eileen was quick to integrate herself into this group of women academics.

Janaki and Eileen would often reminisce about Ann Arbor, talking through the night sometimes about common friends in Michigan like Frieda Blanchard (in charge of the University Botanical Garden, Ann Arbor, and one of the earliest women anywhere in the world to earn a doctorate in genetics), their mentor Bartlett, the University Botanical Garden and the John Innes. This is not to say that their relationship was always harmonious; like any close relationship, it had its ups and downs. Eileen would in fact remark about Janaki's moodiness to Frieda: her tendency to become 'morbid' when 'moody', leading to much unpleasantness between them. Eileen had even contemplated a return to America after a particularly unpleasant episode, but she dropped the idea when Janaki in a matter of minutes turned volteface and became all caring and warm. Eileen would eventually spend 'a year of research and travel in Travancore and Malabar' in the capacity of honorary professor of botany at the Maharaja's College of Science, Trivandrum.

Making inroads in physical anthropology

Within a short period of her arrival in Travancore, Eileen would convey to Bartlett, sometimes through Frieda, that she was 'very happy' in Trivandrum, and had begun to make forays into physical anthropology, chiefly taking measurements.[34] In line with Codrington's suggestions, her focus would turn to a study of physical growth across castes. She found herself busy measuring fisherfolk besides school students in the city – taking height and weight measurements of 500 boys and girls of different diet groups.³⁵ Eileen's job prospects in Travancore however did not take off as expected, but, being the incredibly dynamic person that she was, she managed to persuade the Travancore government to allow her the use of the state car, surprising even Janaki, 'to make special trips to the High Ranges' to study tribes and also collect plants. She thoroughly enjoyed these excursions despite being physically demanding. The chief secretary was so impressed by her passion that he also allowed her the use of 'a motor launch on a big damned lake in the hills' (the Periyar Lake in what is today the Periyar Tiger Reserve in Thekkady). 'Scarcely anyone gets to use this except the Royal Family, Viceregal Parties & the British Resident', she wrote to Bartlett with obvious pride in having been permitted its use. Accompanied by a forest ranger, she visited the tribal Urali village in the forests; they were, Eileen explained to the very curious Bartlett, 'wandering agriculturalists', and lived in tree houses at night. She took 'still pictures & movies of them', because it would interest him very much.³⁶

On this tour, Eileen was accompanied by the state archaeologist, R. Vasudeva Poduval, who had been directed by the diwan (head of government) to provide her with all assistance. Together, they visited Cape Comorin and several locations in the High Ranges, such as Peermade, Kumili, Vandiperiyar, Devikulam³⁷ and the 'extreme N. E. corner of Travancore', where there were hundreds of dolmens (the megalithic Muniyara dolmens of Marayoor, about 50 km from Munnar, bordering on Tamil Nadu). Another day was devoted to collecting in the forests, this time with the forest ranger and Muduvan trackers, when they came across a dangerous solitary tusker elephant. Eileen had taken some 'nice ciné Kodak pictures of everyday life in Travancore & of some hill tribes', but most of these tribes, she noted, had become 'contaminated or [were] dying out',³⁸ a concern shared by Janaki and the reason she had wanted Bartlett to visit India at the earliest.³⁹ Eileen was able to collect physical measurements of tribal women and this was the first time this had been done. The Madras Museum curator Edgar Thurston, author (with K. Rangachari) of *Castes and Tribes of Southern India* (1909), could not accomplish this because 'the husbands would not allow men to do this'. Even with Eileen, 'the women were very shy', but she had got them to yield to her.⁴⁰

Stumbling upon a collecting career

For both Janaki and Eileen in Trivandrum, holidays were invariably devoted to plant-collecting excursions. Several plant specimens, including marine algae, would be sent in their joint names to the herbaria of Harvard University, the Field Museum of Natural History (Chicago), the University of California, Berkeley and the University of Michigan, Ann Arbor. Among those sent to Harvard, in their joint names, was a *Syzygium caryophyllaeum* from the Pulayanar Kotta hill in the suburbs. In early January 1934, after their return to Trivandrum from Malabar, where they had been to visit Janaki's family and do some collecting, the two were once again engaged in extensive collection of marine algae for despatch to some of the world's leading herbaria.

An outgoing person, sometimes to the point of annoyance to Janaki, Eileen began exploring the city and its surroundings on her own, and at other times, when Janaki was busy, in the company of her women friends attached to the Women's College. She had also hired a young boy to carry the plant press for her, when she went about collecting. Eileen would despatch 'one lot of seeds' to the United States Department of Agriculture (USDA), and spend several days identifying the plants and fixing labels; one set was to reach the University of Michigan. She assured Bartlett that the transportation charges would be taken care of, as she had arranged for a loan from England to tide over the expenses. Janaki had also single-handedly built up a large collection. They worked on it together, sometimes through the night. If Janaki was too busy correcting examination papers or preparing for classes, Eileen worked on it herself. The Forest Office Herbarium in Trivandrum was 'good', with many of the sheets determined not very long ago by experts at Kew, which helped with identification.[41] Eileen would occasionally make short visits to Madras, to the Presidency College, to have specimens identified.[42] She had also been introduced to Janaki's friends at Queen Mary's College and was thus very much at home even without her friend in the city.

Eileen would also build up a large collection of 'anthropological articles', including 'bows and arrows, traps, mats etc made by the Hill tribes', which went on to become part of the Museum of Anthropology of the University of Michigan. These objects, all labelled (along with the plant specimens – over 600 accession numbers in newspapers with her field numbers written on the margin) were sent to the despatch agent of the USDA in New York; they went by a barge from Trivandrum to 'Quilon c/o Harrison & Crossfield Ltd'.[43] In so doing, Eileen had turned into a veritable collector, a woman counterpart of her mentor, Bartlett, or her exact contemporary, Walter Norman Koelz (1895–1989), the Michigan zoologist, collector of natural history and artefacts and museum curator,[44] even if she had not collected on that scale. But this was only the beginning. Incidentally, Koelz was in India on a two-year collecting mission for the University of Michigan around the same time (1932–34). Eileen seemed to have stumbled upon a new career, or sub-career, as a collector of things 'Indian', a prospect that might not have happened had she not seized the cross-border opportunity and bonded with a close woman contemporary from the 'East', however different by way of personality from herself.

Marriage as panacea for all ills

By about May 1934, Eileen had travelled to Madras, well ahead of her departure to America via Japan. She inspected the university herbarium at Janaki's suggestion, to figure out how it could be improved, and rendered herself useful for the honours botany classes. It was also during her stay in Madras that she met a Scottish engineer, James B. Macfarlane, employed with Burmah-Shell Oil Co. They soon got engaged, much to the surprise of all her friends. They planned to marry in London after six months. While her contemporaries in science like Janaki, C. K. Kausalya, M. M. Mehta, McClintock, Rosalind Franklin and several others chose to remain single to focus on their careers, she thought otherwise. Eileen believed that she was a very 'fortunate person to get another chance at a normal, happy family life, after [her] various troubles of the past twelve years'. She was convinced that science was 'a good field for women', if they did not 'mix up love with emotions and become vegetables'. The only problem, she stated, was that 'women can't concentrate on their work once they fall in love. Then by the time they've got their heads out of the clouds, they're 10 years behind time in their field!' Eileen was clearly in the minority among successful women scientists of the times for her view that marriage did not necessarily get in the way of a woman's career in science.

In the early years of their friendship, the compulsion to make close female allies to bolster one's scientific practice was far more crucial to the unmarried Janaki than for Eileen, who was twice married by the time they had met each other at the University of Michigan. Moreover, Janaki hailed from a family dominated by women, even matriarchs, for that matter; she strongly bonded with her mother, sisters (she had five, four elder and one younger) and several nieces, was schooled at a convent-run girls' school, graduated from a women's college and taught immediately after at an exceptional all-female higher-education institution in Madras, the Women's Christian College. She was most happy mingling with women, and sisterhood mattered much to her, when it came to practising science; moreover, her decision to remain single had been taken very early in life. However, being single by itself did not predicate strong female bonding, as far as women practitioners of science went, the case in point being the maize-geneticist Barbara McClintock (1902–92), a close geneticist contemporary of Janaki and Eileen.[45] In the case of Eileen, it was after she had divorced her second husband, Carl Erlanson, that she felt a greater need to bond with Janaki and, later, undertake the cross-border adventure. Quite expectedly, when she found her third husband in Madras, Eileen moved away once again; it was not as if she lost contact with Janaki entirely, but that strong need to bond was no more conspicuously present.

In an interview given to a local American newspaper in her later years, Eileen would claim that Macfarlane, whom she had met while doing physical anthropology in India, was always interested in her work, 'not jealous' about her achievements.[46] In fact, she thought marriage was a solution to life's several problems, which comes as something of a surprise. Speaking of Janaki, on an earlier occasion, Eileen commented: '[She] thinks that her work (research) is the only joy she has in life and in it she feels very much alone since there is nobody to discuss it with. I feel very sorry for her ... If she could only find some Indian biologist and could marry and help him I think it would solve all her difficulties.'[47] Eileen's finances continued to be poor and the Travancore stint had not turned out as expected, making a return to the Kent College, at least for a semester, inevitable. She would however return to India in a matter of months, this time as Mrs Macfarlane, and settle down in Cochin, a port city on the Malabar Coast of India, where her husband, an engineer with Burmah-Shell, was to be based.

Redrawing disciplinary boundaries

As soon as she landed in Cochin in March 1935, despite the trying weather, Eileen began an anthropometric study of the 'small, ancient colony of white Jews' of Cochin, whom she found exceedingly fascinating because they were all inter-related and were only seventy in number. She was also interested in an 'outcast group in South Malabar'. All physical anthropology until this point in India was a prerogative of male practitioners, the likes of Thurston and Rangachari in South India, or a B. S. Guha in Bengal, and limited to male subjects and physical measurements.[48] On the contrary, Eileen's focus, as a biologist/geneticist, was invariably in the outliers – outcasts, and mixed-race populations, which she found aplenty in India thanks to her cross-border adventure – and relied on blood groups (besides physical measurements) as a guiltless and objective basis for ordering the human races.[49] Through her pioneering serological researches among these people, she had extended the frontiers of biology and contributed to the making of seroanthropology, a new field of knowledge. Her work would however come to an abrupt halt when she was accidentally poisoned with mercuric chloride (exactly how is unknown) and had to be hospitalized for three weeks. She would soon have to leave Cochin for Madras, because her husband had been transferred to that city.

Disenchantment with India

Going to Madras was 'another upheaval' for her and she was far from comfortable this time. There was a plan to offer her an honorary fellowship at Madras University, something there was no precedent for, but her husband was unimpressed and insisted that a salaried position was preferred. He had begun to get the feeling that she was spending 'too much time and energy, not to mention money, for those and not those resident in India'. As for Madras University, it was hiring directly from Britain, not from European residents in India, and this deeply disappointed Eileen. To keep her American citizenship from lapsing, she ensured Frieda Blanchard renewed her honorary appointment as fellow at the University Garden at Michigan.[50]

If previously Eileen had felt her talents were being wasted in America, she now complained to Bartlett that she was feeling 'cramped' in Madras and 'that the world [was] not getting the benefits it could receive from [her] training and natural gifts'. She was convinced that she had 'more enthusiasm, ability, personality and worth-while experience than most people' and that her talents were being wasted in India, 'an exasperating place', which had not improved 'under the Indianization policy of the British'. It was safe to look for a suitable position in America at the earliest, she decided, while she was still physically able. She commented,

> Probably you still have several poor deserving botanists on your hands, but I hope you will agree with me that I have something unusual to give any college or research institution or botanic garden. I think that the idea that a job is a charity affair and should be given to the person who is poorest, or has the most dependants is pernicious as bad as the communal idea in India.

Eileen felt being 'happily married', and the fact that she was not physically fit to have her own children (owing to the serious injuries sustained in the motor accident), would make her 'a better teacher'. Her desperation was even more evident when she wrote to Bartlett: 'there must be a place for me somewhere, after all the pains I have been to get the training and degrees. Will you please keep me in mind until 1938 if you hear anything?' She was willing to return to America and take up a teaching position at Howard University or, even better, at the Brooklyn Botanic Gardens. Eileen was ready to send out her curriculum vitae to a few institutions but waited to hear from Bartlett.[51]

Engaging in diverse research

The study of the blood groups of the 'white' Jews would be resumed upon return to Cochin in the new year (1936).[52] Invariably, as a trained academic, she did a thorough job of whatever the subject of her research – reading all the pertinent literature, accessing a network of experts and sometimes writing book-reviews, besides conducting original research in the field. In this case, the expert was the South Indian anthropologist L. K. Ananthakrishna Iyer (1861–1937), who described the Malabar Jews in his *Castes and Tribes of Cochin* (vol. 2, 1912) from an ethnographical point of view.[53] She made friends wherever she went. Some among them were highly influential members of society, who would go on to play important parts in forwarding her research, thanks to her capacity for negotiation and an eye and respect for local specificities, including wearing the Indian sari (see Figure 3.1). In Cochin, it was S. S. Koder, the wealthy 'white' Jew, a younger contemporary; it was through Koder that she was able to gain access to this otherwise-closed community of Jews, also called Paradesi (meaning foreign) or Malabar Jews, and collect blood samples, the first time anyone had done this. Her presence in Cochin and the association with someone as important as Koder would promptly be reported by the local English newspaper: 'Dr (Mrs) Macfarlane, who is making a research study of the various races in Malabar by an examination of their blood, is now in Cochin testing the blood of White and Black Jews. She is a guest of Mr. S. S. Koder, Cochin.'[54]

As noted above, Eileen was rarely at a loss for biological subjects to research, even in a place like Cochin, which could not boast of anything like a botanic garden, laboratory or biological field station. She would in fact undertake a diverse set of research problems, and conterminously at that, during her residence in this port city. For instance, in April 1935, she embarked on a survey of marine boring organisms and an analysis of the flora of the largest man-made island by dredging (called the Willingdon Island), even while researching the blood groups of the 'white' Jews of Cochin and continuing her study of Malabar flora which she had begun in Travancore. That the harbour could be approached as a biological field station, offering singular opportunities to the biologist, was what Eileen ingeniously demonstrated in Cochin. In the artificial island formed through dredging, she discovered an ideal subject for understanding the process of plant colonization in the tropics. As for the survey of marine boring organisms that destroyed wooden crafts, which she undertook in parallel, it was a pioneering project, from the point of view of both taxonomy and economics, so much so that it was reported in the newspapers.[55] She would make a collection of these destructive crustaceans and send them to experts to confirm their identification and names. The British and European contacts

Figure 3.1 Eileen Erlanson (later Macfarlane), dressed in Indian sari, undated (courtesy of the McCamic family)

chosen for this purpose were her exact contemporaries, the brilliant woman zoologist/carcinologist Isabella Gordon of the British Museum (Natural History), and the French naturalist, Africa explorer and future anti-nuclear activist Théodore Monod, who was also an expert on isopods. Locally, she relied on the expertise of biologists like M. O. Parthasarathy Iyengar, who had been introduced to her by Janaki, besides R. Gopala Aiyar of Madras University, to resolve the many scientific issues involved in the project.

Imperative to publish and communicate science to the public

It was vitally important for Eileen to convert any and every piece of research accomplished, however preliminary or modest in scope, either as academic papers, and/or as communications to the public in the form of illustrated lectures at scientific and other institutions, locally and abroad, and/or as articles in popular magazines and newspapers. These were indispensable components of her scientific practice and something she had already demonstrated in America and Britain, in the context of the American wild rose. The cross-border quest would expand the breadth of her research interests and make her more prolific – there was much that was new and exciting to investigate and communicate to the world from the biological point of view. Whether it be the jottings of her impressions after six weeks of life in Trivandrum, which she wished Frieda Blanchard in Michigan would send to a university newspaper or journal for publication, or her preliminary investigations on heredity and racial difference, based on blood-group analysis, which would appear as 'some special articles ... in the foremost South Indian newspapers',[56] her eagerness to reach out to the public was clearly evident. The driving force behind her research was usefulness. Even while confined to a hospital bed in Cochin, she wrote a short note for the local English newspaper, on the state of care provided by the government hospital and how it might be improved.[57]

In late 1935, Eileen wrote to Bartlett asking him for advice about her anthropological data: whether the Michigan Academy of Sciences was a good place for publishing it, or whether a British journal would suit it better. Eventually, her first note on the 'white' Jews – an ethnographic study of a Malabar Jewish wedding – would appear in a college magazine from Cochin[58] and her preliminary findings based on the serological data collected from the community in *Current Science*. She was only the third woman, after Mason and Janaki, to publish in it.[59] Eileen had gone one step further however: she found herself on the journal's board of editors and wrote the editorial for the issue devoted to human genetics (March 1936), in which number her paper on the 'white' Jews had appeared, besides summaries of lectures delivered on cytogenetics in the princely state of Mysore. She would go on to publish several more articles in the journal over the next few months and years. Her preliminary reflections on the marine boring organisms of the Cochin harbour would also find a place in *Current Science*.[60] As for the paper on plant colonization, it would appear in the *Journal of the Indian Botanical Society*, thanks to Janaki, who was secretary of the Society at this time.[61]

On her return to America after her first stint in India, Eileen confidently delivered at the University of Michigan an illustrated lecture on the botany and anthropology of southern India,[62] marking the beginning of a parallel career in America as an 'India Expert'. In fact, on one occasion, she wondered if she should register with some agency as a public lecturer, which she would indeed eventually do. In London too, just after her marriage to James Macfarlane was solemnized, Eileen made a similar presentation at the Royal Anthropological Institute, complete with lantern slides and a film shot with the Kodak camera, on the hill tribes and fisherfolk of Travancore.[63] When she delivered a public lecture on the subject at the Madras Rotary Club, it aroused much public interest. Madras Surgeon-General Cuthbert Allan Sprawson wanted her to do a similar project of physical anthropology in Madras, and Eleanor Mason, a fellow woman of science, teaching at one of only two women's colleges in the city after having obtained a doctorate in physiology from Harvard Medical School, invited her to collaborate on 'a study of growth in relation to nutrition among the natives'.[64]

Eileen had been trying to land a job with Mysore State, but nothing came out of it. Filled with anxiety at being jobless, she concluded that India would never be 'a White Man's country and [was] even harder on white women', and that her talents and education were 'too good to waste'. She also feared her husband's life span would be drastically cut short if they stayed on and was determined not to remain in India after his (expected) retirement in 1938. In any case, she decided 'not to get rusty' until then.[65] Ironically, this was the very time – while unemployed and despondent in Madras – that she would receive news of being showcased as one of the world's outstanding women in science, in an exhibition held as part of Harvard University's tercentenary celebrations!

Concluding remarks

Both cytogeneticists, Eileen Erlanson and Janaki Ammal were acutely aware of their vulnerability in the face of shackling social and cultural values, the lack of employment opportunities for high-achieving women like themselves, and the need to build robust local and even transnational networks and be surrounded by women who affirm, if they were to negotiate social hurdles and shape their varied scientific careers. Access to Janaki's academic position, her generosity and her social networks in India were crucial to Eileen in making a success of her cross-border quest, and the bonding was equally fruitful and empowering for Janaki, while she began a new phase of life in sedate Trivandrum, both socially and from the point of view of so new a science as genetics.

Eileen's Travancore and Cochin sojourns were a demonstration of how a creative, alert and persevering, even persistent, biologist could never be at a loss to find subjects for research. She was in fact setting a role model for others to follow: she valued mobility, was unafraid of crossing borders – national and disciplinary – quickly assimilated local specificities, tapped into and expanded her social networks, seized every opportunity that came her way to explore and investigate and insisted on being self-supportive. Marriage, for her, was not a hurdle to doing science, but it was crucial to be financially independent. Also central to her practice was the refashioning of every piece of research accomplished, however modest in scope, as publications and/or lectures, both academic and popular. It was important for her to leave her mark everywhere she lived and, equally, the fact of having lived everywhere. Among the strategies she adopted to subsist as a scientist (if not thrive) was to 'invent' a varied career, as against a unidimensional one; in her case, in addition to being a rose geneticist, by virtue of having crossed borders, she had also become a seroanthropologist, human geneticist, collector, popular science writer, traveller, public lecturer and an 'expert on India' in America, albeit in the making.

This biographical case study reveals how the cartography of a discipline (in this case, genetics) sometimes changed, when practised in complex social and political settings, resulting in new fields of knowledge such as human genetics and seroanthropology. Eileen's multifaceted practice was fired by a sense of usefulness and involved surveys and the use of biological techniques to address social problems. If genetics, her field of specialization, was crucial to comprehending human heredity, physiology and evolution, it also opened a window into public health, racial and gender differences, blood ties and the very fabric of local life. Her anthropological research, although in a preliminary stage, pointed to the gender gaps extant in South Indian society (with respect to growth and nutrition, for instance) and also in the very practice of the discipline of physical anthropology until then, which had been an exclusively male domain. Not least, Eileen's anthropological interventions of the early 1930s in Travancore and Cochin provide us with that early and rare glimpse of a (Western) woman anthropologist at work in a colonial setting.

Acknowledgements

I would like to thank the family of Eileen Erlanson Macfarlane for giving me access to archival material in their collection.

Notes

1 M. L. Richmond, 'South American fieldwork/cytogenetic knowledge: The cytogenetic research program of Sally Hughes-Schrader and Franz Schrader', *Perspectives on Science*, 28: 2 (2020), 127–69; M. L. Richmond, 'A model collaborative couple in genetics: Anna Rachel Whiting and Phineas Westcott Whiting's study of sex determination in Harbrobracon', in A. Lykknes, D. L. Opitz and B. Van Tiggelen (eds), *For Better or Worse: Collaborative Couples in the Sciences* (Basel: Birkhäuser, 2012), pp. 149–89. P. G. Abir-am and D. Outram (eds), *Uneasy Career and Intimate Lives: Women in Science, 1789–1979* (New Brunswick, NJ: Rutgers University Press, 1987).
2 For a paper that examines female subcultures and the making of physiology, see T. A. Appel, 'Physiology in American women's colleges: The rise and decline of a female subculture', *Isis*, 85: 1 (1994), 26–56, https://doi.org/10.1086/35672.
3 See I. H. Stamhuis and A. B. Vogt, 'Discipline building in Germany: Women and genetics at the Berlin Institute for Heredity Research', *British Journal for the History of Science*, 50: 2 (2017), 267–95.
4 E. Scheich, 'Science, politics and morality: The relationship of Lise Meitner and Elisabeth Schiemann', *Osiris*, 12 (1997), 143–68.
5 For a sample of publications adopting a transnational approach to writing the history of twentieth-century biology, see A. Barahona, 'Transnational science and collaborative networks: The case of genetics and radiobiology in Mexico, 1950–1970', *Dynamis*, 35: 2 (2015), 333–58; A. Barahona, 'Local, global, and transnational perspectives on the history of biology', in M. R. Dietrich, M. E. Borrello and O. S. Harman (eds), *Handbook of the Historiography of Biology*, vol. 1 of the Historiographies of Science Series (Cham: Springer, 2021), https:/doi.org/10.1007/978-3-319-74456-8_17-1. For a collection that explores how transnational connections shaped women's lives outside of science, see C. Midgley, A. Twells and J. Carlier (eds), *Women in Transnational History: Connecting the Global and the Local* (London: Routledge, 2016); S. Mukherjee, *Indian Suffragettes: Female Identities and Transnational Networks* (New Delhi: Oxford University Press, 2018); C. von Oertzen, *Science, Gender and Internationalism: Women's Academic Networks, 1917–1955*, trans. K. Sturge (New York: Palgrave Macmillan, 2014) focuses on the International Federation of University Women (IFUW), which played a major role in connecting academic or intellectual women in the first half of the twentieth century. For a biographical exploration of an Asian woman scientist adopting a transnational approach, see S. P. Nair, *Chromosome Woman, Nomad Scientist:*

E. K. Janaki Ammal, A Life 1897–1984 (New York and London: Routledge, 2022).
6 Interview with Eileen Macfarlane, *The Cincinnati Post* (3 February 1958), p. 9.
7 *Michiganensian*, 32 (1928), p. 298.
8 *American Men of Science, A Biographical Dictionary*, suppl. The Physical and Biological Sciences (L-O) (New York and London: R. R. Bowker Company, 1966), p. 3336.
9 M. W. Rossiter, *Women Scientists in America, Struggles and Strategies to 1940* (Baltimore, MD: Johns Hopkins University Press, 1982); M. W. Rossiter, *Women Scientists in America: Before Affirmative Action, 19401–972* (Baltimore, MD: Johns Hopkins University Press, 1995).
10 Unpublished finding aid, Radcliffe College Archives, Schlesinger Library, Harvard Radcliffe Institute, Cambridge, MA, B-6. Also, C. A. Elliott, 'The tercentenary of Harvard University in 1936: The scientific dimension', *Osiris*, special issue, *Commemorative Practices in Science: Historical Perspectives on the Politics of Collective Memory*, 14 (1999), 153–75, even if this does not discuss the 'Women in Science' Exhibition, and only makes a brief mention of it (p. 160).
11 *The Michigan Daily* (13 January 1929), p. 8.
12 *The Michigan Daily* (29 January 1930), p. 8.
13 *The Michigan Daily* (8 March 1930), p. 5.
14 *The Ithaca Journal* (7 March 1930).
15 She would also attend the Fifth International Botanical Congress at Cambridge during this visit. The John Innes Horticultural Institution (John Innes), *Annual Report*, 1930, p. 4.
16 *The Ithaca Journal* (7 March 1930).
17 *The Michigan Daily* (18 January 1931), p. 8.
18 E. W. Erlanson, 'American wild roses', *American Rose Annual* (1932), 85.
19 W. H. Lewis, A. A. Reznicek and R. K. Rabeler, 'Identifications and typifications of Rosa (Rosaceae) taxa in North America described or used by E. W. Erlanson, 1927–1934', *Novon*, 22: 1 (2012), 41–42 (a biographical note on Eileen Erlanson).
20 Bentley Historical Library, University of Michigan (hereafter, BHL, UoM): University Herbarium Records (University of Michigan) H. H. Bartlett series, Box 12, Eileen Erlanson to H. H. Bartlett, letter dated 5 August 1931.
21 See S. G. Kohlstedt, 'Sustaining gains: Reflections on women in science and technology in 20th-century United States', *NWSA Journal*, 16: 1 (2004), 9.
22 BHL, UoM, Eileen Erlanson to H. H. Bartlett, letter dated 5 August 1931.
23 BHL, UoM, Eileen Erlanson to H. H. Bartlett, letter dated 19 November 1932.
24 For biographical studies of E. K. Janaki Ammal, see S. Kedharnath, 'Edavaleth Kakkat Janaki Ammal (1897–1984)', *Biographical Memoirs of Fellows of the Indian National Science Academy*, 13 (1988), 90–101, https://insaindia.res.in/BM/BM13_8808.pdf; C. V. Subramanian, 'Edavaleth Kakkat Janaki Ammal', *Resonance* 12: 6 (2007), 4–9; V. Damodaran, 'Gender, race and science in twentieth-century India: E. K. Janaki Ammal and the history of science',

History of Science, 51: 3 (2013), 283–307; V. Damodaran, 'Janaki Ammal, C. D. Darlington and J. B. S. Haldane: Scientific encounters at the end of empire', *Journal of Genetics*, 96: 5 (2017), 827–36.
25 Nair, *Chromosome Woman*, p. 104.
26 Nair, *Chromosome Woman*, p. 115.
27 Nair, *Chromosome Woman*, p. 123.
28 BHL, UoM, Eileen Erlanson to H. H. Bartlett, letter dated 27 January 1933.
29 BHL, UoM, Eileen Erlanson to H. H. Bartlett, letter dated 24 October 1933.
30 BHL, UoM, Eileen Erlanson to H. H. Bartlett, letter dated 22 September 1933.
31 For the circumstances of his visit, see Nair, *Chromosome Woman*, pp. 124–27.
32 Cited in Nair, *Chromosome Woman*, p. 141.
33 Nair, *Chromosome Woman*, p. 127.
34 BHL, UoM, Eileen Erlanson to Frieda Blanchard, letter dated 23 January 1934.
35 BHL, UoM, Eileen Erlanson to H. H. Bartlett, letter dated 18 April 1934.
36 BHL, UoM, Eileen Erlanson to H. H. Bartlett, letter dated 18 April 1934.
37 Proceedings of the Government of HH the Maharaja of Travancore, Archaeological Department, Administration Report 1109 ME (1933–34 AD), p. 1. See also pp. 4–5, which describe the dolmens they found in Marayur.
38 BHL, UoM, Eileen Erlanson to H. H. Bartlett, letter dated 18 April 1934.
39 Nair, *Chromosome Woman*, p. 52.
40 BHL, UoM, Eileen Erlanson to H. H. Bartlett, letter dated 18 April 1934.
41 BHL UoM, Eileen Erlanson to Frieda Blanchard, letter dated 23 January 1934.
42 BHL, UoM, Eileen Erlanson to H. H. Bartlett, letter dated 18 April 1934.
43 BHL, UoM, Eileen Erlanson to H. H. Bartlett, letter dated 18 April 1934.
44 For a biography of W. N. Koelz, see C. M. Sinopoli, *The Himalayan Journey of Walter Norman Koelz: The University of Michigan Himalayan Expedition, 1932–1934*, Anthropological Papers Series, no. 98 (Ann Arbor, MI: Museum of Anthropology, 2013).
45 For book-length biographies of McClintock, see E. F. Keller, *A Feeling for the Organism: The Life and Work of Barbara McClintock* (New York and San Francisco, CA: University of California Press, 1985): N. Comfort, *The Tangled Field: Barbara McClintock's Search for the Patterns of Genetic Control* (Cambridge, MA: Harvard University Press, 2003).
46 *The Cincinnati Post* (3 February 1958).
47 BHL, UoM, Eileen Erlanson to H. H. Bartlett, letter dated 27 November 1926.
48 An exception to this was the Indian anthropologist Irawati Karve (1905–70), a younger contemporary of Eileen Erlanson. For a biographical study of Karve, see N. Sundar, 'In the cause of anthropology: The life and work of Irawati Karve', in P. Uberoi, N. Sundar and S. Deshpande (eds), *Anthropology in the East: The Founders of Indian Sociology and Anthropology* (New Delhi: Permanent Black, 2007), pp. 360–416; U. Deshpande and T. P. Barbosa, *Iru: The Remarkable Life of Irawati Karve* (New Delhi: Speaking Tiger, 2024).
49 See Nair, *Chromosome Woman*, pp. 179–80.
50 BHL, UoM, Eileen Erlanson to H. H. Bartlett, letter dated 29 October 1935.
51 BHL, UoM, Eileen Erlanson to H. H. Bartlett, letter dated 29 October 1935.

52 For Eileen's contributions to seroanthropology, based on her published papers, see P. B. Mukharji, 'From serosocial to sanguinary identities: Caste, transnational race science and the shifting metonymies of blood group B, India *c.* 1918–1960', *The Indian Economic and Social History Review*, 51: 2 (2014), 143–76.
53 For a biographical essay, see K. Ram, 'Anthropology as "Ananthropology": L. K. Ananthakrishna Iyer (1861–1937), colonial anthropology, and the "native anthropologist" as pioneer', in Uberoi, Sundar and Deshpande (eds), *Anthropology in the East*, pp. 64–105.
54 *The Malabar Herald* (15 April 1936), p. 5.
55 *The Malabar Herald* (12 October 1935), p. 4.
56 *The Kent Stater* (5 December 1935).
57 *The Malabar Herald* (17 August 1935), p. 3
58 'A white Jew's wedding', *Maharaja's College Magazine*, Cochin State, 19: 1 (1937), 1–7.
59 E. W. Erlanson Macfarlane, 'Preliminary note on the blood groups of some Cochin castes', *Current Science*, 4: 9 (1936), 653–54.
60 E. W. Erlanson, 'A preliminary survey of marine boring organisms in Cochin Harbour', *Current Science*, 4: 10 (April 1936), 726–32.
61 E. W. Erlanson, 'Plant colonisation on two new tropical Islands', *The Journal of the Indian Botanical Society*, 15 (1936), 193–214.
62 *The Michigan Daily* (9 August 1934), p. 3.
63 E. J. Macfarlane, 'Hill tribes and fisherfolk of Travancore', *Man*, 38 (March 1938), 43–44.
64 *The Kent Stater* (5 December 1935).
65 BHL, UoM, Eileen Erlanson to H. H. Bartlett, letter dated 9 April 1936.

4

Vlasta Kálalová Di-Lotti in Iraq: Medical practice and scientific research

Adéla Jůnová Macková

Vlasta Kálalová Di-Lotti, a physician who attempted to combine her medical education with scientific research in the field of medicine and entomology, is a very important, though widely forgotten, figure in the history of Czechoslovakian science.[1] Women's education and opportunities for further employment in science were very scarce in the Czech lands at that time. Although they could join graduation examinations at male grammar schools from 1878 onwards, the first grammar school for women in Central Europe was founded only in 1890, six years before Kálalová was born. By the time she was four years old, the first eight women had graduated from the Faculty of Arts. After the establishment of Czechoslovakia in 1918, women's access to university positions increased. Between 1909 and 1939, 163 women worked as assistants at the Medical Faculty in Prague, thirty at the Czech Technical University and four in the Faculty of Science. It was only in 1925 that historian Milada Paulová became the first female assistant professor and later, in 1935, the first female extraordinary professor in history at the Faculty of Arts of Charles University in Prague.

Vlasta Kálalová thus belonged to the second generation of university-educated women. Education for women was available only in big cities, but Vlasta's elder sister was a dentist who worked in Prague, which enabled her to attend the Prague Girls' Grammar School and live with her sister even while studying at the Faculty of Medicine. She graduated with the highest grades but, being a woman, was expected to work as a physician in her hometown. She defied that fate and decided to act on her decision to establish an institute for tropical disease research in the Middle East and immerse herself in the scientific study of tropical diseases. Her life and scientific activity were remarkable for the first half of the twentieth century, as well as her chosen location of research – the Middle East – and her background as a Central European woman physician and scientific researcher. For all these reasons, her case is an important addition to the history of women in science, which also demonstrates how she attempted to reconcile her scientific work with her responsibilities as a wife and mother.

In this chapter, I will focus mainly on Kálalová's work in Iraq, where she founded her own small hospital. Her intention was to secure funding for the establishment of an institute of tropical diseases research, to be linked to the hospital, which was to serve as a research base for parasitologists from the National Institute of Health. In addition to work in the hospital, she was also involved in research on the transmission of leishmaniasis and the prevention of the disease through skin graft. This aspect of her work is very difficult to trace in the archival sources of official institutions and needs to be examined in conjunction with other material. Here, I will concentrate both on her manuscript *Across the Bosphorus to the Tigris*, which covers her time in Turkey and Iraq, and her correspondence with the first Czechoslovak president, Tomáš Garrigue Masaryk,[2] whose support enabled her to establish a hospital in Baghdad.

Vlasta Kálalová Di-Lotti: A biography

Vlasta Kálalová Di-Lotti was born on 26 October 1896 in Bernartice, a small town near Tábor.[3] Her father, Karel Kálal, a teacher at a primary school, took care of his children's education and Vlasta began her secondary-school studies in Prague at the Girls' Grammar School in Královské Vinohrady. In 1916, she passed her maturity exam and enrolled in the Faculty of Medicine of the Czech University in Prague. She attended common medical lectures – histology, anatomy, pathology – but also more specialized courses in pharmacology, dental medicine, ophthalmology, gynaecology, basics of microscopy and electrotherapy. Students also took lectures in chemistry and physics, especially in modern disciplines like roentgenology and radiology.

At the same time, she also studied Arabic and Turkish, including the history of literature, Islam and the modern history of the Middle East, at the Faculty of Arts of the Czech University under Professors Rudolf Dvořák[4] and Rudolf Růžička.[5] Later she attended the Arabic lectures taught by Professor Alois Musil.[6] She also made lifelong friends there, among them Marie Tauerová,[7] her Orientalist brother Felix Tauer[8] and zoologist Jiří Baum.[9] These lectures were very important to her, because knowledge of local languages enabled her to communicate directly with people in Baghdad.

In 1919, she attended Professor Jaroslav Hlava's lecture on tropical parasitology, which discussed the topic of little-known and difficult-to-treat diseases among patients who had returned from travel abroad. To tackle such ailments, it was proposed to establish an Institute of Tropical Diseases somewhere in the Middle East, where samples could be obtained and sent to laboratories in Prague for further study. This proposal met with almost no response, so Vlasta resolved to go to Baghdad or Jerusalem in future and put it into practice.

Her graduation[10] three years later was followed by two further years of study, during which time she specialized in surgery, a field she considered very important to achieve her professional goals. From spring 1922 onwards she worked in a hospital in Louny and from mid-1922 she spent a short period of time at the Veterinary Clinic in Brno working under Professor František Král,[11] eventually completing a nearly two-year internship in Brno at the surgical clinic of Professor Petřivalský, St Annes Hospital.[12] During her medical work, however, she continued her language education: 'Prof Petřivalský's surgical clinic is sufficiently furnished and there is a lot of work from the beginning, so I hope I will be able to benefit a lot here. ... Of the oriental languages, I am now doing regularly only Persian and reading the journal *Stambul Seriyati*. I have absolutely no time, only for the clinical work.'[13]

The first mention of her attempts to travel to the Middle East dates to 1921, when she asked Alois Musil, her Arabic teacher, to help her be admitted to an American college in Istanbul,[14] where foreign students were studying medicine. She hoped to secure a job in a Turkish hospital and advance her knowledge of tropical diseases. This turned out to be impossible, but Musil promised his student to support her future endeavours in the Middle East. During that time, Kálalová continued to work in hospitals in various towns, while Musil helped her with her Arabic grammar and encouraged her to continue her studies. In 1924, she decided to pursue her goal of creating a Czechoslovak institute for tropical disease research in the Middle East. The first step towards this was a study trip:

> Now I would like to spend some time in the clinic of Haidar Pasha,[15] to improve my Turkish and to master the Arabic medical terminology, which the Turks actually use. After that I would like to continue my study trip to some Arab cities, especially to Baghdad. I am planning to leave for Constantinople around 29–30 September.[16]

In the end, Musil was able to help Kálalová only indirectly. From 1920 onwards, when he moved from Vienna to the Czech University in Prague, he promoted the idea of a Czechoslovak cultural and economic penetration into the so-called Orient. The 'Oriental' market was seen, together with the Balkan Peninsula, as a perfect opportunity for exporting Czechoslovak industrial products after the dissolution of the Habsburg monarchy and the disappearance of the tariff-protected market. Czechoslovak companies produced about 60 per cent of the industrial products under the monarchy, especially heavy industrial products such as armament and plant units destined for export. The Oriental Institute, established under Musil's leadership in the 1920s, benefitted from the support of the first Czechoslovak president, Tomáš Garrigue Masaryk. Its aim was to advance Czechoslovak economic and cultural interests in the Middle East. Alois Musil intended to use the financial budget from the Cultural and Economic Department of the Institute as well as from companies that were seeking to export to this region to finance periods of study for Orientalists abroad, so that they could conduct research and provide information about potential trade opportunities. Upon return, the Orientalists were expected to train people to represent these companies abroad. Since Alois Musil had spent many years in the Middle East studying and travelling – his most important work took place in today's Jordan, where he explored Quseir Amra – he wanted to create a network of friends, Czechoslovak compatriots, commercial agents and scholars working abroad to promote relations between countries in the Middle East and Czechoslovakia.

It was this idea, which also appealed to President Masaryk, that Vlasta took up. She decided first to establish a hospital in Baghdad, which would then financially support the Institute for Tropical Diseases Research. Prof Petřivalský from the Brno clinic advised her to apply for a travel grant from the Ministry of Education and the Ministry of Foreign Affairs. Neither of those institutions supported her plan: the Ministry of Foreign Affairs argued that it was a scientific, not promotional, trip, while the Ministry of Education was influenced by the negative opinion provided by the management of the Brno clinic. The Ministry of Health also refused to help, pointing out that it lacked the necessary funds for this kind of undertaking.

Vlasta's cooperation with the Czechoslovak Red Cross led her to Alice Masaryk,[17] daughter of the Czechoslovak President Tomáš Garrigue Masaryk, who arranged for her to meet directly with her father.[18] On 15 September 1924, Vlasta arrived at Topoľčianky, the president's summer residence, where she stayed for four days. On the first evening, she was invited to an audience with Masaryk. We know of this audience from two sources, the manuscript *Across the Bosphorus to the Tigris*,[19] and reports held by the Archive of the Office of the President of the Republic. Masaryk was very supportive of the whole venture and promised to finance a study trip to the Middle East, during which Vlasta was to decide on the best location to practise medicine and gain further experience.

The total amount was probably agreed very loosely, as the minutes of the audience only give the approximate costs for the first year of her activities, estimated at CZK60,000.[20] Kálalová herself had originally estimated her expenses at CZK25,000 for the travel costs[21] and the stay in Istanbul, and CZK60,000 for basic expenses; Masaryk's financial assistance thus contributed significantly and enabled her to stay for several months in Istanbul, where she gained practical and linguistic knowledge at the gynaecological clinic of Prof Besim Ömer.[22]

From her stay in Istanbul in 1924–25 dates the first preserved letter from her correspondence with Masaryk and his daughter Alice. The information in it refers mainly to opportunities for medical practice in Iraq. Thanks to a letter of recommendation[23] from the Office of the President of the Republic, she received a warm welcome from the Czechoslovak Embassy in Istanbul. Through the Czechoslovak representative Rudolf Světlík,[24] she met the director of the Baghdad branch of the Ottoman Bank (Banque impériale ottomane), Mr Reid.[25] Reid used his connections in Baghdad to enlist the support of a Dr Dunlop,[26] director of the Royal Hospital Baghdad.[27] Dunlop tried to obtain a post for her in the hospital, but this proved impossible due to financial considerations. He advised her to set up a private medical practice, which he believed would be very difficult, and estimated she would not have many patients initially. The advice encouraged Vlasta. She decided to stay in the Middle East and set up a small private practice to save the expenses of travelling to and from her home country. As she put it, she wanted to 'establish an institute that would provide us with a culturally colonizing commercially important point in the near Orient'.[28] Correspondence with Masaryk indicates that, from the very beginning, she was not only interested in medical practice and the establishment of an institute for tropical disease research, but also in securing a foothold in Iraq for Czechoslovakia's science (e.g., study stays for Orientalist scholars, natural scientists) as well as its commercial interests, exactly as Alois Musil had intended when he founded the Oriental Institute in Prague as an institution that would promote the country's cultural and economic interests in the Middle East.[29]

Vlasta was able to form a more accurate idea of the situation in Iraq, both financially and medically, only in March 1925, when she arrived in Baghdad. According to a letter to Masaryk in March 1925, ninety-six male doctors[30] were in private practice.[31] With regard to women medical practitioners, she only mentions an English doctor who left Baghdad in 1924 and a French doctor, Pin,[32] who ran a medical practice there. However, her position was very difficult because she did not speak Arabic and had no surgical training. She was forced to give up her practice only a year after Kálaová's arrival in Baghdad. The report of the British Colonial Office includes a list of female doctors working in Baghdad. Dr Pin and Vlasta Kálalová Di-Lotti are mentioned, but Vlasta is mistakenly identified as an Austrian doctor:

> There is an opening for a few lady doctors in the large towns but this is in the nature of a gamble. There are no appointments in Government service but permission to practice might be granted to an applicant of suitable attainments, who should not however come to Iraq without first obtaining this permission.

> Two lady doctors (French and Austrian) are already in practise in Baghdad. One is doing well and the other, though competent, is doing badly.[33]

Because of her connection to Dr Dunlop, Vlasta was well received by the British doctors and it was even thought initially that she might head the newly established gynaecology and obstetrics ward of their civilian hospital, to gain experience before becoming independent. In the event, she failed to secure the job. This was probably due to Oriental Secretary Gertrude Bell,[34] whose meeting with Vlasta turned out rather awkwardly:

> A native servant in the commissariat is a courtesy in itself. Does he think I am English too? The clerk came out to find out why I was coming. He is respectful and helpful, but remarks that the lady is busy. He left and came back. She has agreed to see me after all. The thin old lady with a sharp nose. Only the brocade of her unstitched shoes reveals her predilections, which in her travelogues are reflected in her connoisseur's love of flowers. 'I address you as a European and an experienced expert on the condition of Iraqi women. Do you think there is a need for female doctors in Baghdad?' 'I don't know. The health inspector or the hospital director can say.' The political discretion of the Oriental Secretary is greater than the humane relations with Arab women. So I won't detain her. 'Would you be so kind as to give me an introductory letter to the health inspector?' 'I cannot introduce you. I don't know you.' I prefer blunt frankness to false excuses. But perhaps even Miss Bell is ashamed of the level of her humanity. 'You have a letter of recommendation to the Director of the Royal Hospital. Contact him first. And let me know the outcome at the end.'[35]

Vlasta arrived in Baghdad on 15 March 1925 and, in view of the rejection from the British medical services and the forthcoming law that would make it difficult for foreign doctors to establish themselves in Iraq through fees and complicated paperwork, she immediately applied for permission to practise medicine. She received it ten days later. Pondering whether to open a small practice or a clinic, she eventually settled for the latter:

> I could only open an office and operate on surgical cases in the Armenian sanatorium, newly established here on the outskirts of the city. But I think that would be a mistake for two reasons. First, it would miss the most opportune time to set up an institute, which will be set up today in a form that really meets a local need. The Institute will gain the gratitude of the population and will be able to easily continue in future in any other branch of medicine as a scientific institute; on the contrary, if I wait, there will be no natural grounds for the establishment of an institution later on, which will then certainly meet with many obstacles that don't exist now.[36]

Vlasta finally decided to open a small practice, for six months, to adjust to the new environment and have enough time to order equipment from the Czechoslovak Republic. In autumn, she was joined by nurse Ruth

Tobolářová.³⁷ In September 1927, she was replaced by another Czech nurse, Marie Marianinová,³⁸ who remained in Baghdad for nearly three years.

The financial cost of such an undertaking was, of course, extensive. Without Masaryk's help, Vlasta Kálalová Di-Lotti would have never had the opportunity to rent a house and equip the small clinic in Baghdad with all the necessary equipment. Apart from her own family and the money she had saved throughout her university studies and during her internships in Czech hospitals, Masaryk was the only one who helped Vlasta Kálalová Di-Lotti finance her dream.³⁹ We have detailed information about her finances from her correspondence with President Masaryk. She had CZK50,000 for her stay in Baghdad, which the Presidential Office sent to her via the Ottoman Bank in Istanbul, and CZK50,000 of her own savings, with which she intended to buy equipment.⁴⁰ Altogether, CZK300,000 was spent on all the expenses related to the trip, the stay in Turkey, the move to Baghdad and the establishment of the clinic. President Masaryk paid her CZK10,000 while still in Bohemia,⁴¹ and the Presidential Office sent CZK50,000 to Istanbul and CZK100,000 to Baghdad.⁴² She then received less than CZK94,000 to pay for the necessary medical equipment which she had been sent from Bohemia. Masaryk provided these funds as a scholarship, but Vlasta always considered them only as a loan (except for the CZK10,000 that was intended for the study trip to Istanbul) and she managed to repay the whole amount within two years.⁴³

Medical practice in a hospital

We know very little about the beginnings of Vlasta Kálalová's medical practice. All the work was her responsibility, she had no qualified help and she got by with the help of her servant Mahdy. As a woman, she needed him as a necessary companion in public. In addition to the servant, she also employed a cook. She bought the most essential equipment for her medical practice relatively cheaply in the Baghdad market, from a Jewish pharmacist who had bought British medical supplies from army surplus after the war, including an examination table. Thanks to him, she was also able to rent a suitable house that met both sanitary and space requirements:

> The green balcony looked out onto Palace Avenue. The house formed a corner a few steps from the family property of my Muslim friends. I couldn't have asked for better. The very next day, April 7, Mahdy and I moved into number 2/17 on Džádet es-Seraj. The tailor working in the arcade of Mahmood's house had sewn white linen curtains for the glass walls of the surgery. Mustafa Kamil, with the help of his friends, helped me furnish the waiting room with new purchases. The young Russian Resler, who oversaw the chemical

department of Baghdad's largest department store, granted my request and ordered medicines from several Czechoslovak pharmaceutical factories. The surgery was not even finished yet, and already patients were coming in. Or their family members with requests for medical consultation.[44]

It is obvious that after the opening of the clinic in the spring of 1925, Vlasta Kálalová was quite busy with various arrangements for her apartment and office, so that she did not have much time to write letters, but the letters that have been preserved are an important source to understand her life there: 'I just want to keep busy now. There's not that much work yet. The average daily income just covers the expenses (which are quite considerable here, at least CZK150 a day), only the operations are a welcome plus.'[45]

In July 1925, the necessary surgical equipment arrived from Prague and in September, Ruth Tobolářová arrived in Baghdad. In October 1925, when Masaryk received a further report on the development of the Czech 'enclave', Vlasta already had her regular patients and an important assistant in the person of nurse Ruth Tobolářová and could therefore afford to rent a larger house in a better area of the town. She found a suitable house that was larger than the previous one and could also accommodate rooms for post-operative patients. The house was situated in the centre of the city and was easily accessible for her patients:

> I rented a spacious house (10 rooms on the first floor with a kitchen, cool rooms for summer and servants' quarters on the ground floor) for Rs. 3000 per year. My house is still being repaired and we will be able to move in about a week. It will have, besides the consulting room, rooms for diathermy and a waiting room, quite a nice operating theatre and two large rooms for patients, which can accommodate 8–10 beds, and a third room can be added later![46]

The new house, known as the Burazanli House, was repaired during the autumn (she moved in on 7 November 1925) and adapted to the needs of a small clinic. On 15 December, by a decree of the Baghdad director of health, the activity of the 'Czekoslovak Mustausaf'[47] was authorized. 'Mustausaf' indicated, according to Kálalová, a dispensary or small private clinic, and was a general term used for a medical facility with one doctor. 'Mustashfa',[48] the hospital, had to have at least two doctors, twenty beds and its own pharmacy.

In January 1927, when the small clinic had already been receiving Iraqi patients for a year, Vlasta Kálalová sent the first comprehensive report to Masaryk, detailing her time in Baghdad and containing information about the country in which she lived, including climate, natural resources, industry, trade, agriculture, population, religious groups, education and politics. The report was carefully divided into headings and the text was accompanied by illustrative photographs.[49]

In explaining her practice, Vlasta wrote in great detail about the Iraqi health system, which had been the responsibility of the Health Secretariat since 1919, followed by the Ministry of Health and, after its abolition, the Ministry of the Interior. For example, during this time a water supply system had been established in Baghdad, river water was disinfected, a decree was issued on the reporting of infectious diseases and vaccination became compulsory during epidemic outbreaks. In 1927, the whole of Iraq had only twenty-six hospitals and 200 doctors, half of whom were based in Baghdad. The leading posts were, of course, held by British doctors.[50]

Although Vlasta Kálalová always stressed the importance of her small clinic as the basis for the establishment of a large scientific institute for tropical disease research, her medical work enabled many Muslim women to undergo examinations and operations that a male doctor was not allowed to perform. The importance of the small Czech clinic in Baghdad can also be gauged by the statistics that Vlasta Kálalová kept:

> Of the surgeries performed last year, 70 percent were women and 30 percent were men. From the beginning of my stay in Baghdad, from April 1925 to the end of 1926, I treated 1522 patients, of whom 690 in 1925 and 823 in 1926. Since the opening of the dispensary (1 January 1926) 36 patients have been admitted, with 345 days of treatment and 98 operations (15 laparotomies) have been performed.[51]

Other surviving correspondence between Kálalová and Masaryk, which testifies to the operation of the Czech hospital, dates from the year 1929 and suggests that some letters might have been lost. The emerging economic crisis caused Kálalová to fear that all her efforts would be destroyed by the lack of funds to build an institute for tropical disease research:

> Yet, in spite of my own dissatisfaction, in many ways, I have the impression in my conscience that my stay here was not in vain, even though the purpose for which I came – the establishment of an Oriental exponent for the Czechoslovak Chair of Tropical Diseases – is uncertain as a result of the present political situation.[52]

Kálalová tried to further develop her small clinic; her practice expanded and patients, especially women, found their way to her. The information contained in the report indicates both an increase in the number of patients and surgeries: 'In the thirty-two working months from the beginning of 1927 to the end of 1929, 1700 new outpatients were treated and 172 operations were performed. Of these, sixty-five inpatients were admitted (690 treatment days).'[53]

In the autumn of 1929, the Czechoslovak clinic moved once more, this time to the so-called Ant Street, and again to a larger house located in a more affluent neighbourhood. The reason for this was the ever-increasing rent and deteriorating condition of the building:

> In the autumn of 1929, the Czechoslovak dispensary moved to another location in the central district of Baghdad. For Mr. Burazanli was raising the rent by ten percent annually from the original three thousand rupees, and at the same time denying any building repairs. So, it came to pass that we exchanged the ancient house for a neat new building in nearby Darbúnet-en-Nemle, Ant Street. A quiet link between the noisy arteries of the capital.[54]

Vlasta's patients included Muslims, Jews and Christians from all social circles. She also treated the family of King Faisal I of Iraq. In addition to the British Royal Hospital, where she assisted in operations, she had one of the best-equipped small hospitals in Baghdad and one of the best-equipped operating theatres. She also used electrotherapy, high-frequency electromagnetic currents in surgical procedures, to cauterize blood vessels and to destroy neoplasms (tumours). Other examples of her work, documented in her notes, manuscript and letters, included operating a cleft palate on a three-day-old baby girl, the spindle-cell sarcoma on the head of a forty-year-old woman, reconstructing the nose of a boy injured by a horse, performing hymenoplasties (a procedure undertaken to reconstruct the hymen) and complicated deliveries in the hospital, as well as in private homes.

Scientific work

In addition to her medical practice, Vlasta Kálalová Di-Lotti also devoted herself to scientific research on tropical diseases, which was to be the first small contribution before a large scientific institute could be established. She focused on cutaneous leishmaniasis,[55] also known at the time as 'Oriental' or 'Baghdad lump', which was caused by protozoa of the genus *Leishmania* and left permanent scars in the form of large ulcers all over the body. She set herself the task of finding a natural vector, while trying to prevent the disease from developing by transplanting a small graft of skin from a person who had already had the disease, which provided immunity. In her report to Masaryk, she stated:

> In the field of medical research, cutaneous leishmaniasis (the so-called Baghdad lump) has been investigated. To prove its natural vectors, the cooperation of Dr Drbohlav of the State Institute of Health was requested for microbiological investigations on a large scale. Personally, I am trying my method of immunisation against the Baghdad lump (by skin transplantation, with very good results so far).[56]

Vlasta Kálalová Di-Lotti was interested in the latest research in the field of tropical medicine, which since the beginning of the twentieth century had strongly emphasized research on leishmaniasis. We can only assume that she was led to it by the doctors at the British Royal Hospital, who dealt with the disease on a daily basis and also had the latest scientific articles at their disposal.[57] She shared her research with scientists from the State Institute of Health in Prague, Jaroslav Drbohlav (head of the Laboratory for Entomology and Parasitology) and Heřman Šikl, founder of the field of histology in Czechoslovakia. She also sought advice from colleagues from the University of Jerusalem, Oskar Theodor and Saul Adler, who worked in the Department of Parasitology and visited Iraq in 1928. During the 1940s, Saul Adler developed a vaccine against leishmaniasis using living parasites. Vlasta used electrocauterization of the ulcer shortly after detection but tried as well to find a prevention method. Her immunization method was based on transplantation of a skin graft collected from an area near the ulcer. At first, she tried to apply her own skin graft that she transplanted to her first nurse Ruth Tobolářová. Later, she used the same method on her two children. In these three cases she had a positive outcome, and all of them became immune to leishmaniasis. She practised this method while using the parent's skin graft for their children. In cases where these parents had been infected in their childhood (i.e., a long time ago) the immunization failed, but when they had been infected in adulthood, she reported positive outcomes. Given the short time Vlasta spent in Baghdad and the small sample she was able to test her method on, it is difficult to evaluate her work. However, a paper on her scientific work was published in the prestigious Czechoslovak periodical *The Journal of Czech Physicians*.[58]

Across the Bosphorus to the Tigris records scepticism about her ability to understand the aetiology of leishmaniasis, as well as cooperation with scientists in Czechoslovakia, which was to be a major contribution of the scientific institute in Baghdad:

> I had a bad experience with our parasitologists. I was interested in the spread of the Baghdad blight. A little girl with a disintegrated leishmaniasis ulcer moved into Rustemia, where there was no endemic cutaneous leishmaniasis before. An opportunity arose to demonstrate a natural mode of transmission. A rare opportunity for our investigators, I thought, and we began to carefully collect and send, with meticulous records, the flies from near and far around the child who might have been the source of infection. 'In Paris they envy the Iraqi material,' wrote Assoc Prof Drbohlav, but he buried it for ever. I asked in vain for confirmation of the shipments and for the results of the investigation into the phlebotomus.[59]

Kálalová Di-Lotti's promising scientific cooperation with Czechoslovak scientists, which was to be the main focus of her future work in Iraq, failed. Dr

Drbohlav probably did not communicate with her any further, as we learn nothing about further shipments or potential cooperation from subsequent sources. Despite these disappointments, the transmission of leishmaniasis by sandflies was eventually proven.

Attempts to establish a radiotherapeutic institute

The failed collaboration with the National Institute of Health and the end of hopes for the establishment of the institute for tropical disease research in Baghdad were perhaps the main reason she decided to venture into a new project, the establishment of a radiotherapy institute in Jerusalem. The plan to leave Baghdad might have also been related to her personal life. Vlasta married an Italian, George Silvio Di-Lotti, in June 1927, she was already raising a family – her son Radbor and daughter Lydia were born in Baghdad (see Figure 4.1) – and might have believed, as other colonial accounts also suggest, that the Iraqi climate was unsuitable for European children. We do not know exactly why she settled upon the idea of a radiotherapy institute but given her physics lectures at the University of Prague and her work in hospitals in Czechoslovakia, it appears that she had a fairly broad understanding of radiology and the use of radiation in medicine. She also planned to establish a hospice as part of the institute.

In summer 1927, during her stay in Bohemia, Kálalová Di-Lotti was granted an audience by Masaryk to report on her work and introduce her husband. At the same time, she approached the Foreign Ministry about the post of honorary consul in Iraq for her husband, who had applied for Czechoslovak citizenship. Other sources suggest that she was dealing with the establishment of the radiotherapy institute in Jerusalem, especially its funding. Masaryk made sure that the idea was positively received at the Ministry of Foreign Affairs, which handled the whole matter through the Czechoslovak consulate in Palestine. Kálalová Di-Lotti set off for Jerusalem. The Office of the President of the Republic paid for the family's travel, during which time Vlasta managed to find a suitable plot of land for the building.[60]

When the Czechoslovak consul Vladimír Fric returned to Jerusalem in November 1927, another plot was chosen based on the applicable sanitary regulations, including the need to have a plot of at least 900 m^2 for a hospital with twelve beds. The estimated costs increased due to this (the land with basic improvements would have cost CZK2.5 million) and in 1928 documents from the Office of the President of the Republic already show that they considered buying or renting the house. The radiotherapy institute was regarded as a good investment that would earn its own money, but the initial investment for the house, rent, doctors, nurses and all the equipment

Figure 4.1 Vlasta Kálalová Di-Lotti with nurse Marie Marianinová and son Radbor, Baghdad, 1928. SOA Třeboň, SOkA Písek, fond Rodinný archiv Di-Lotti (1887–1982), box 2, inv. no. 193 (courtesy of the State Regional Archives in Trebon, Czech Republic)

(partly counting on donations from Czechoslovak companies) was so high that by the end of 1928 any information regarding the enterprise had disappeared from the archival documents.[61] It is therefore likely that the idea failed to materialize.

Conclusion

Vlasta Kálalová Di-Lotti ended up staying in Baghdad for almost seven years, working tirelessly throughout her stay on all the tasks she set out to accomplish. Her large medical practice speaks of the success she achieved – the only setback was the failure to find a successor to take over and maintain such a practice (she offered it for free) after her illness prevented her from continuing her work.

In autumn 1931, Kálalová Di-Lotti decided to return to Czechoslovakia. She feared not only for the health of her children, but also for her own health. She suffered from high fevers, most likely the outcome of dengue fever. In the spring of 1932, after futile attempts to find a successor among Czechoslovak or foreign doctors to continue her work, she left for her homeland and her clinic closed its doors.[62]

Vlasta Kálalová Di-Lotti also worked to improve health conditions for the Iraqi population, especially women, who could not be treated by a male doctor. She sought to extend compulsory education for midwives, who were to be allowed to practise their profession with only basic reading and writing skills. We cannot document today how much influence she had on the founding of the first medical school in Iraq, but at its inauguration she took a place of honour next to King Faisal I.[63]

The Office of the President of the Republic has preserved a copy of a certificate issued to Vlasta Kálalová Di-Lotti in January 1933 as an evaluation of her medical practice in Baghdad:

> We confirm in addition to your request that you established a surgical institute (Mustausaf Čechoslovak) in Baghdad in 1925 with the moral and material support of the President of the Republic and that you ran the same as an independent surgeon until 31 March 1932. Your clinic was equipped according to the model of our hospital surgical departments. We have been following your activities all the time and we like to express our satisfaction, especially because you have been successful in a foreign environment and alongside competing English institutes.[64]

Vlasta Kálalová Di-Lotti's intention to establish and run a successful medical institute was therefore successful from the perspective of the Office of President Masaryk, who financially supported the project. Though it is nearly impossible to imagine her hospital competing with the Royal British Hospital in Baghdad, as a single woman she managed to build a financially self-sufficient hospital that attracted clients from different social strata and was also popular in the highest circles. However, this was not enough to further her ambitious plans. Since financial problems were affecting the then-established British Health Service, which had the apparatus of the mandate system behind it, Kálalová Di-Lotti had little hope of being able to establish a Czechoslovak research institute in Baghdad in a situation where there was not even a fully equipped Czechoslovak embassy in Iraq. Her sudden departure for Czechoslovakia, which made it impossible to choose a successor, marked the end of her ambitious plans.

After returning to Czechoslovakia, Vlasta Kálalová Di-Lotti was treated in her hometown of Bernartice for several years. Dengue fever together with the exhausting work in Baghdad took its toll. It was not until 1936 that she moved with her family to Prague and, despite persistent health problems, began working for the Czechoslovak Red Cross. The family remained in Prague until the German occupation but lived through World War II in Bernartice. On 8 May 1945, SS troops were withdrawing through Bernartice into the American war zone and during a firefight, Vlasta's entire family was killed. She herself was shot in the shoulder and bore the consequences for the rest of her life. After a year, she returned to her post in Prague and in 1946 accepted an invitation from her old friend from the Czechoslovak Red Cross, Mary Harrison, to join her in the United States.[65] She stayed in the USA for several months and actively participated in the International Women's Congress. There she met Norwegian writer Ingeborg Refling Hagen,[66] with whom she visited Norway on her way back to Czechoslovakia, where she held lectures for physicians in the Norwegian language. After returning from the US, she worked for some time at Professor Diviš's surgical clinic[67] at the Prague General Hospital, but she never returned to the scientific research she had begun in Baghdad. Her only published scientific study, documenting her research in Baghdad, is an article based on a lecture given at the Society of Czech Physicians on 16 January 1933 and published in the 1930s in the prestigious *Journal of Czech Physicians*.[68] After retirement, she moved to her native town of Bernartice and died almost forgotten in 1971 in a hospital in Písek.

Vlasta Kálalová Di-Lotti's attempts to establish an institute for tropical disease research in the Middle East and study 'tropical diseases' in Iraq have been largely forgotten in the history of science. Her case demonstrates how she attempted to reconcile her scientific work with her role as a physician and surgeon and responsibilities as a wife and mother. Notwithstanding the fact that she benefitted from the support of President Masaryk, as a woman researcher and practitioner of medicine, Vlasta's efforts were often thwarted by the male-dominated establishment in her home country, as her interaction with Dr Drbohlav suggests. Although relegated to the margins of history, her research and, in particular, her medical practice that provided Iraqi women with access to medical treatment, including complex surgical interventions, was very important. In that respect, her years in Iraq also presented opportunities for her to try to assert her value as a medical practitioner and scientist, by acquiring knowledge that was not easily accessible to her male peers, either because they did not have the experience of having lived in Iraq or because they could not treat women patients. This situation is not dissimilar to that of cytogeneticist Eileen W. Erlanson, discussed by Nair in this volume, who undertook a similar journey to India, eventually fashioning herself as an 'India expert' upon her return to the US.

At the same time, it is important not to lose sight of the political contexts in which the lives of these women unfolded. A female physician from Central Europe, Vlasta Kálalová Di-Lotti worked and conducted research in the Middle East, where Czechoslovak political, economic and cultural ambitions had to contend with the more established imperialist ambitions of Britain and France. From the beginning, her project was planned as a cultural project, with the hospital and institute envisaged as a centre that would enable scientists to conduct their research. Financial considerations played an important part. The project initiated and financially supported by the president of the Republic was successful in the case of the hospital, but the attempt to extend it to scientific research, whether as an institute for tropical disease research or radiotherapeutic institute, eventually failed due to the lack of finances and geopolitical barriers.

Notes

1 With the institutional support of the Czech Academy of Sciences, Masaryk Institute and Archives, RVO: 67985921.
2 Tomáš Garrigue Masaryk (1850–1937), first Czechoslovak president, professor of philosophy at the Faculty of Arts, Czech University, Prague.
3 See Státní okresní archiv (State Regional Archives; hereafter, SOkA), Písek, fond Family Archive Di Lotti 1887–1982, box 1, inv. no. 3, biography of Vlasta Kálalová. Another important source for tracing Vlasta Kálalová's life are documents from the National Archives in Prague. See Národní Archiv (National Archives; hereafter, NA), fond Police Headquarters Prague II – general office 1941–1951, box 6755, sign. L/2186/11. A novel that documents Kálalová was authored by journalist and novelist I. Borská, *Doktorka z domu Trubačů* [*The Doctor from the Trumpeters' House*] (Prague: Mladá Fronta, 1978). These materials include, for example, passport and gun licence applications and permanent residence reports.

4 See Marie Tauerová, *V duchu s MUDr: Vlastou Di-Lottiovou Kálalovou* [*In the Spirit with MUDr: Vlasta Di-Lotti Kálalová*]. Private archive of Hana Tauerová.
5 Tauerová, *In the Spirit with MUDr: Vlasta Di-Lotti Kálalová*.
6 Alois Musil (1868–1944), a prominent Austro-Hungarian and Czechoslovak Orientalist, Bible scholar and traveller, spent several years in the Middle East at the end of the nineteenth and beginning of the twentieth century, where he lived with several Bedouin tribes. From 1920 onwards, he was a full professor at the Faculty of Arts of Charles University in Prague.
7 Marie Tauerová (1896–1981), librarian, sculptor, translator from Persian, worked at the Moravian Library in Brno.
8 Tauerová, *In the Spirit with MUDr: Vlasta Di-Lotti Kálalová*.
9 Jiří Baum (1900–44), Czech zoologist, traveller and writer, received his doctorate at the Faculty of Science of the Charles University in 1928 and worked for the National Museum, whose zoological collection he expanded through travels around the world.
10 SOkA Písek, fond Family Archive Di Lotti 1887–1982, box 2, inv. no. 210, medical diploma V. Kálalová.
11 MVDr. František Král (1892–1980), professor of special pathology and therapy of internal diseases of domestic animals, whose scientific works dealt with radiology. For Vlasta's work at the veterinary clinic, see Literární archiv Památníku národního písemnictví (Museum of Czech Literature; hereafter, LAPNP), fond Alois Musil, box 169, inv. no. 57, V. Kálalová to A. Musil, 28 December 1922, Brno. 'Before it opened [here, V. K. means Prof Petřivalský's clinic], I accepted the assistantship offered to me at Prof Král's internal clinic, at the veterinary school, and I am glad I did so, for I had the opportunity to learn in this way things that may be useful to me one day and which I would otherwise have found difficult to learn. I have also benefited with respect to human medicine, paradoxical as it may seem. Especially in parasitology.'
12 Július Petřivalský (1873–1945), founder of the Moravian Surgical School, studied in Prague and Innsbruck. In 1919, he was appointed full professor of pathology and therapy of surgical diseases and became the first head of the surgical clinic of the medical faculty of Masaryk University in Brno (part of St Ann's hospital). LAPNP, fond Alois Musil, box 169, inv. no. 57. V. Kálalová mentions her medical practice in her letters from 21 March 1922, 2 May 1922, 10 May 1922 and 28 December 1922, Prague. See also NA, fond Ministry of Social Welfare, box 3963, file Palestina, scrapbook. V. Jedlička, 'Česká chirurgická klinika v Bagdadu' [Czech surgical clinic in Baghdad], *Písecký kraj* (9 November 1929).
13 LAPNP, fond Alois Musil, box 169, inv. no. 57, V. Kálalová to A. Musil, 28 December 1922, Brno.
14 Üsküdar American Academy was founded in 1876 in Bahçecik. The school, whose mission was to provide quality education for girls, moved to Istanbul in the early 1920s.
15 Turkish Haydarpaşa, the Asian part of Istanbul.

16 LAPNP, fond Alois Musil, box 169, inv. no. 57, V. Kálalová to A. Musil, 29 August 1924, Bernartice.
17 Alice Masaryková (1879–1966), daughter of the first Czechoslovak president Tomáš Garrigue Masaryk, founder and first chairwoman of the Czechoslovak Red Cross. After the death of her mother Charlotte, she took over the role of First Lady.
18 President Tomáš Garrigue Masaryk awarded scholarships from the Masaryk National Fund to distinguished scientists. Professor Lexa, an Egyptologist, went on a scientific trip to Egypt in 1930–31 thanks to his help.
19 See manuscript *Přes Bospor k Tigridu* [*Across the Bosphorus to the Tigris*], pp. 10–13. LA PNP, fond Family Archive Di Lotti 1887–1982, box 1.
20 CZK60,000 amounted to approximately £376 (£1 = CZK160 in the 1920s–1930s).
21 The journey from Prague to Baghdad cost approximately CZK4,000 at that time.
22 Tauerová, *In the Spirit with MUDr: Vlasta Di-Lotti Kálalová*.
23 *Archiv kanceláře prezidenta republiky* (Archive of the Office of the President of the Republic, hereafter AKPR), fond *Kancelář prezidenta republiky* (Office of the President of the Republic, hereafter, KPR), inv. no. 807, file MUDr. Vlasta Kálalová Di-Lotti, activities abroad, Vladimír Kučera, Secretary to President Masaryk, to Czechoslovak Embassy in Istanbul, 18 September 1924.
24 The first Czechoslovak embassy in the Ottoman Empire after the establishment of Czechoslovakia in 1919 was the Office of the Czechoslovak Delegate in Istanbul, led from 1920 to 1924 by lawyer and diplomat Rudolf Světlík (1869–1934).
25 I have been unable to trace this person.
26 William Dunlop (1886–1967), physician, worked at the Royal Baghdad Hospital from 1919 to 1935. He was professor of clinical treatment at the Royal Baghdad Medical School, where he lectured on childhood diseases. He worked in Iraq until 1935. For more details on his person, see 'Obituary Notices: W. Dunlop', *British Medical Journal*, 3: 5561 (5 August 1967), 375–76.
27 The Royal Baghdad Hospital was established in 1897 as an Ottoman military hospital under the name Madžīdīja chastachāne-si. After the occupation of Baghdad by the British Army in 1917, it was converted into a permanent military hospital, No. 23.
28 AKPR, fond KPR, inv. no. 807, file MUDr. Vlasta Kálalová Di-Lotti, activities abroad, V. Kálalová to T. G. Masaryk and A. Masaryková, 24 January 1925.
29 This is a complex issue which this chapter only touches upon marginally, in connection with the practice of Vlasta Kálalová Di-Lotti, who mediated the export of Czechoslovak pharmaceutical companies to Iraq and ordered most of the medicines and medical supplies from Czechoslovakia. However, mutual trade between Czechoslovakia and Iraq was very low in the 1920s and 1930s.
30 Kálalova Di-Lotti compares this figure with the number of four doctors before World War I. This figure is undoubtedly a gross underestimate. There was a state hospital in Iraq at the time, later the Royal Hospital, and private doctors

also practised. Longrigg reports that the state health service alone employed twenty-five doctors in 1921. See S. Longrigg, *Four Centuries of Modern Iraq* (Beirut: Librairie du Liban, 1968), p. 170.
31 AKPR, fond KPR, inv. no. 807, file MUDr. Vlasta Kálalová Di-Lotti, activities abroad, V. Kálalová to T. G. Masaryk, 30 March 1925, Baghdad.
32 I have not been able to find out the name of the British doctor; the French doctor is only named by her surname Pin. In her 1929 report to President Masaryk, Kálalová mentions a French doctor, Pin, who had settled in Baghdad a few months before her arrival, probably at the end of 1924, and remained there for less than three years (probably until the spring of 1927). At the time of Kálalová's arrival in Baghdad, Dr Pin was the only female doctor in the city. She had no surgical training and did not speak Arabic, which Kálalová believes was the reason her practice did not take off. At the same time, however, Kálalová noted in her report that Pinová remained in the Middle East and opened a practice in the Iranian city of Kermanshah. See Masarykův ústav a Archiv AV ČR, Archiv Ústavu Tomáše Garrigua Masaryka (Masaryk Institute and Archives of the CAS, Archive of the Institute of Tomáš Garrigue Masaryk; hereafter, MÚA, AÚTGM), fond T. G. Masaryk, box 505, inv. no. 8, file Iraq, Report, Baghdad, 23 December 1929.
33 See National Archives (Kew), Colonial Office (CO) 730/119/9. Report of the British High Commissioner to the Colonial Office, 15 February 1927, No. CO/230, p. 25.
34 Gertrude Bell (1868–1926), a prominent British traveller, writer and archaeologist who became involved in British wartime activities in the Middle East from 1915. From 1917, she held the post of oriental secretary to the British occupation (later mandate) administration of what was later Iraq, a position roughly equivalent to that of a political adviser, a post to which she was entitled by her superior knowledge of the region gained from previous travels and archaeological excavations. Bell was one of the figures within the British administration sympathetic to the moderate Arab national movement; the following negative, even caustic, assessment of Kálalová therefore seems rather biased.
35 *Across the Bosphorus to the Tigris*, pp. 162–63. LA PNP, fond Family Archive Di-Lotti 1887–1982, box 1. At the time of her meeting with Kálalová, Bell was younger than fifty-seven years old (if it occurred in the spring of 1925, shortly after her arrival, as the following text shows). Bell was suffering from several physical and mental health problems towards the end of her life and may therefore have given an older impression.
36 AKPR, fond KPR, inv. no. 807, file MUDr. Vlasta Kálalová Di-Lotti, activities abroad, V. Kálalová to T. G. Masaryk, 30 March 1925, Baghdad. Probably the sanatorium of the Armenian physician Topolian mentioned in V. Kálalová's letter to T. G. Masaryk. MÚA, AÚTGM, fond T. G. Masaryk, sign. Kor-II-63, box 710, V. Kálalová to T. G. Masaryk, 12 January 1927, Baghdad.
37 Emilie Ruth Tobolářová (1895–1976) was the first Czechoslovak nurse who worked with Vlasta Kálalová in Baghdad in 1925–27. She worked as a laboratory nurse for Professor Ladislav Syllaba at the Faculty of Medicine of Charles

University in Prague (he was also Masaryk's personal physician), where Vlasta met her. For her nursing work in Iraq, she was granted a leave of absence from the State Nursing School, where she worked as deputy director.
38 Marie Marianiovou probably worked as a nurse at Kálalová's clinic between 1927 and 1930.
39 AKPR, fond KPR, inv. no. 807, file MUDr. Vlasta Kálalová Di-Lotti, activities abroad, V. Kálalová to T. G. Masaryk, 30 March 1925.
40 The receipt of this amount was confirmed in December 1924. See AKPR, fond KPR, inv. no. 807, file MUDr. Vlasta Kálalová Di-Lotti, activities abroad, V. Kálalová to the Office of the President of the Republic, 23 December 1924.
41 AKPR, fond KPR, inv. no. 807, file MUDr. Vlasta Kálalová Di-Lotti, activities abroad, Minutes of the audience of Vlasta Kálalová with T. G. Masaryk in Topolčianky, 15 September 1924.
42 AKPR, fond KPR, inv. no. 807, file MUDr. Vlasta Kálalová Di-Lotti, activities abroad, the loan and repayment schedule of V. Kálalová.
43 'I hope that it will be possible for me to repay, within the next year or two at most, the other part of my debt, which I have always considered your kind scholarship to be.' See MÚA, AÚTGM, fond T. G. Masaryk, box 505, inv. no. 8, file Iraq, V. Kálalová to T. G. Masaryk, 28 October 1925, Baghdad. The total amount of CZK244,000 was at that time approximately £1,525.
44 *From Bosphorus to the Tigris*, p. 184. LA PNP, fond Family Archive Di-Lotti 1887–1982, box 1. Palace Avenue is one of the alleys in old Baghdad near ar Rashīd Street; its Arabic name at the time was Džāddatu's sarāj. Mustafā Kāmil was Kálalová's friend, advocate from Baghdad. Resler could not be traced.
45 Archiv Ústavu dějin lékařství a cizích jazyků, 1. Lékařská fakulta Univerzity Karlovy (Archive of Institute for History of Medicine and Foreign Languages, 1. Faculty of Medicine, Charles University), fond Vlasta Kálalová Di-Lotti, No. 1711/7/72, V. Kálalová to M. Tauerová, 20–22 May 1925, Baghdad.
46 MÚA, AÚTGM, fond T. G. Masaryk, box 505, inv. no. 8, file Iraq, V. Kálalová to T. G. Masaryk, 28 October 1925, Baghdad. Diathermy is a type of therapy that uses high-frequency electric current to deep-heat areas of the body.
47 MÚA, AÚTGM, fond T. G. Masaryk, sign. Kor-II-63, box 710, V. Kálalová to T. G. Masaryk, 7 April 1926.
48 From Arabic, *mustawSaf* (clinic, dispensary) and *mustashfaa* (hospital).
49 See MÚA, AÚTGM, fond T. G. Masaryk, box 505, inv. no. 8, file Iraq, V. Kálalová to T. G. Masaryk, 26 March 1927. Kálalová also described her medical practice and health conditions in Iraq in the medical journal *Praktický lékař*. V. Kálalová Di-Lottiová, 'Cařihradské a bagdádské kapitoly' [The Constantinople and Baghdad chapters], *Praktický lékař*, 13: 15 (1933), 436–40; 13: 16 (1933), 467–70; 13: 17 (1933), 492–95; 13: 18 (1933), 526–28; 13: 19 (1933), 550–51; 13: 20 (1933), 573–74.
50 MÚA, AÚTGM, fond T. G. Masaryk, box 505, inv. no. 8, file Iraq, V. Kálalová to T. G. Masaryk, 26 March 1927.

51 MÚA, AÚTGM, fond T. G. Masaryk, sign. Kor-II-63, box 710, V. Kálalová to T. G. Masaryk, 12 January 1927, Baghdad.
52 AKPR, fond KPR, inv. no. 807, file MUDr. Vlasta Kálalová Di-Lotti, activities abroad, V. Kálalová to T. G. Masaryk, 23 December 1929.
53 MÚA, AÚTGM, fond T. G. Masaryk, box 505, inv. no. 8, file Iraq, Report, Baghad 23 December 1929, p. 6.
54 *From Bosphorus to the Tigris*, p. 324. LA PNP, fond Family Archive Di-Lotti 1887–1982, box 1. The so-called Ant Street was located in the central part of old Baghdad.
55 Leishmaniasis is the collective name for a parasitic disease caused by protozoa of the genus *Leishmania*, transmitted to humans by the bite of insects of the genus *Phlebotomus* or *Lutzomyia*. This infectious skin disease causes purulent lumps that leave visible scars. Healed leishmaniasis results in lifelong immunity.
56 MÚA, AÚTGM, fond T. G. Masaryk, box 505, inv. no. 8, file Iraq, Report, Baghdad 23 December 1929, pp. 5–6. See also V. Kálalová Di-Lottiova, 'O chorobách a zdravotních poměrech v Íráku' [On diseases and health conditions in Iraq], *Časopis lékařů českých*, 1933, 72: 14 (1933), 430–33; 72: 15 (1933), 460–63; 72: 16 (1933), 495–98. Jaroslav Drbohlav (1893–1946) graduated from the Faculty of Medicine of Charles University in 1915–17. He worked as an assistant in the bacteriological institute of Professor Honl. In 1921, he was a Rockefeller Foundation scholarship-holder and went to the US, where he worked on the cultivation of amoebae causing tropical dysentery. After the establishment of the State Institute of Health in 1925, he worked in the Department for Microbial Diagnostics, which he headed from 1928 onwards. In 1931, he became associate professor of pathological microbiology.
57 On the history of the leishmaniasis research, see D. G. Jogas Jr., 'The tropics, science, and leishmaniasis: An analysis of the circulation of knowledge and asymmetries', *História, Ciências, Saúde – Manguinhos*, 24: 4 (2017), www.scielo.br/hcsm.
58 Kálalová Di-Lottiova, 'On diseases and health conditions in Iraq'.
59 *From Bosphorus to the Tigris*, p. 309. LA PNP, fond Family Archive Di-Lotti 1887–1982, box 1. Ar Rustumīja, then a village, now a suburb of Baghdad. The site is located about 10 km south-east of central Baghdad, on the Dijālā River, about 5 km from its confluence with the Tigris. Phlebotomus is a mosquito of the genus *Phlebotomus* that transmits leishmaniasis. Kálalová sometimes refers to this insect by its English name 'sand fly'.
60 AKPR, fond KPR, inv. no. 807, file MUDr. Vlasta Kálalová Di-Lotti, activities abroad. Reports from July and August 1927.
61 AKPR, fond KPR, inv. no. 807, file MUDr. Vlasta Kálalová Di-Lotti, activities abroad, Vladimír Fric to the Office of the President of the Republic, 26 January 1928, 1 February 1928 and 13 June 1928.
62 Vlasta Kálalová offered the health authority the facilities of her sanatorium free of charge if they were used to set up a gynaecological clinic run by a doctor of any nationality. Kálalová, 'Cařihradské a bagdadské kapitoly' [The Constantinople and Baghdad chapters], p. 28; SOkA Písek, fond Family

Archive Di-Lotti 1887–1982, box 1, inv. no. 4, biography of Vlasta Kálalová. Here she states that she stayed in Baghdad from 15 March 1925 to 1 March 1932.

63 Manuscript of a lecture held by Jaroslav Slípka (Society for the History of Science), 'MUDr. Vlasta Kálalová Di-Lottiová (1896–1971)', private archive of Professor Jaroslav Slípka.

64 AKPR, fond KPR, inv. no. 807, file MUDr. Vlasta Kálalová Di-Lotti, activities abroad, Office of the President of the Republic to V. Kálalová, 28 January 1933.

65 About her stay in the USA, see Borská, *Doktorka z domu Trubačů* [*The Doctor from the Trumpeters' House*], pp. 237–50.

66 Ingeborg Refling Hagenová (1895–1989), Norwegian writer and teacher, whom Vlasta Kálalová met at the International Women's Congress in New York.

67 Jiří Diviš (1886–1959), Czech physician, follower of the surgical school of Rudolf Jedlička, specialist in thoracic surgery. He worked at various Czechoslovak clinics. In 1926, he obtained his *Habilitation* in pathology and the therapy of surgical diseases. In 1946, he was appointed full professor.

68 Kálalová Di-Lottiová, 'On diseases and health conditions in Iraq'.

5

Early years of the International Conference of Women Engineers and Scientists: Shaping transnational collaboration in the Cold War era, 1964–1975

Emily Rees Koerner and Graeme Gooday

> Gone were the many diverse people from all corners of the world. Now, we were a Conference.
>
> Ruth Schafer, The Conference Story: Recollections of the First International Conference of Women Engineers and Scientists, June 1964.[1]

> It is also important that our conference is held in Poland, a socialist country. It justifies the need for international cooperation among countries of different social and economic systems.
>
> Maria Milczarek, welcoming address at the fourth ICWES, 1975.[2]

Introduction

In studying women's role in the history of twentieth-century science, we cannot limit our scope to either the abstract academic or fundamental natural sciences, or to just the familiar countries of our own origins. This paper examines instead the historical value of seeking out women's roles in the important intersection between the more practical sciences and the world-building profession of engineering within a transnational context. As we have argued elsewhere, this domain of useful techno-sciences was both less exclusionary of women and manifested the kinds of transnational collaboration previously seen in women's peace movements of the early twentieth century.[3] In this chapter, we go further to explore – as the above quotations indicate – how such collaborations could cross all the extreme political polarities of the Cold War: from the military–industrial complex of the capitalist USA, to the socialist domains of the Soviet bloc that emerged from the ashes of World War II.[4]

So how did women in engineering and science collaborate across borders in the second half of the twentieth century? As previously discussed by Gooday and Rees Koerner, one of the most distinct modes of this kind of collaboration was – and continues to be – the International Conference of Women Engineers and Scientists (ICWES), an ongoing international gathering to discuss contemporary issues in the fields of engineering and applied science, and the promotion of women's place within them.[5] Instigated by the USA's Society of Women Engineers (SWE), the conference first took place in 1964, where nearly 500 women with backgrounds in science and engineering, from around thirty-five countries across the world, came together in New York.[6] Although not originally envisaged by the USA hosts as a transnational series, since 1964, a total of eighteen meetings have been held in different continents every three to four years, with attendees from multiple countries, constituting a crucial meeting point for women in engineering and applied science.[7] Yet only a small amount of academic attention has been paid to either the initial formation or the long-term continuation of these conferences, despite their status as exemplars of transnational collaboration and science and engineering diplomacy led by women during the Cold War.[8]

This chapter examines the first four conferences held in New York, USA, in 1964; Cambridge, UK, in 1967; Turin, Italy, in 1971; and Krakow, Poland, in 1975 – the largest of the four gatherings, with 600 (mostly Polish) attendees.[9] These spanned what has too often been over-simplistically represented as the 'Iron Curtain' to demonstrate women's peaceful and productive collaboration during the most intense Cold War competitions over nuclear arms and Space Race technology. Specifically, we suggest that it was by the third and fourth meetings that the three characteristic themes of ICWES meetings were established: i) the enhancement of women's position in science and technology; ii) the deployment of engineering and applied science to promote equality and well-being in all nations, especially for food supply and infrastructure; and iii) an initial welcome that highlighted the host nation's accomplishments on these first two points, and sponsorship by leading political figures and national industries.

On average, women from thirty different countries attended each of the first four conferences, although which thirty countries participated varied with each meeting: South American nations were most visible in ICWES 1 (USA); African nations were first apparent at ICWES 2 (UK); and only at ICWES 4 (Poland) was there broad participation from Eastern European nations, with the USA constituency (44 attendees) for the first time overshadowed by a large majority of native (Polish) participants (423 attendees).[10] Given such variation of constituencies between meetings, we cannot glibly describe their geographical scope as global. Yet these meetings certainly brought into dialogue a multiplicity of national perspectives on women's prerogatives in engineering and applied science, guided by the overarching goals mentioned above. They thus bring more explicitly into relief the cultural variations of women's standing in STEMM than previous nationally focused stories.[11] Our aim is to analyse the first four ICWES conferences as an example for how we can situate such national histories in both local and global contexts.[12]

This chapter will show how the organization and running of those first four ICWES meetings relied not just on the host nation's resources but both on transnational cooperation and, especially from the second meeting onwards, the enabling impetus of individual women (e.g., Ira Rischowski). Specifically, we show that the planning and programming of successive meetings relied on the national-level organizations for women in science and engineering most prominently in eight countries: Brazil, France, Italy, Japan, the Philippines, Poland, the UK and the USA. This transnational networking became an integrative mechanism for the continuation of the conferences in the Cold War context, until it was succeeded in 2002 by a new oversight organization: the International Network of Women Engineers and Scientists, which was set up (in Canada) to manage future ICWES gatherings. While our chapter does not take the ICWES story into the twenty-first century, nor does it cover the entire Cold War period, it certainly follows recent scholarly trends in challenging the dichotomous 'Iron Curtain' discourse, to show that ICWES enabled more free exchange of STEMM knowledge between 'East' and 'West' than in the fraught espionage-prone domains of nuclear weaponry and space rocketry.[13]

Finally, we argue below that ICWES's aims of peace and progress, as well as its focus on women as professionals, situated ICWES within a broader historical trend for women's organizations to pursue irenic goals, such as the Woman's International League for Peace and Freedom, which was founded in 1915.[14] And in more concrete ways, we can note that the early ICWES conferences made ties with other comparable global women's learned organizations, such as the International Federation of University Women (IFUW), founded in 1919; indeed representatives from the International Federation of Business and Professional Women and the International Council of Women attended the first four ICWES conferences. Alliances between international women's organizations were thus a crucial part of ICWES's history; it cannot be told as a story of ICWES emerging as if in isolation.

It is important to emphasize that many of the national organizations active in peacefully sustaining ICWES – such as SWE and the UK's Women's Engineering Society (WES) – were created for women in engineering rather than scientists. Yet since histories of women in engineering are often marginalized in favour of women scientists (or elided in histories of science without proper demarcation), another key mode of collaboration that we consider is that between scientists and engineers, extending the work carried out by the national networks of women STEMM practitioners that emerged in the early to mid-twentieth century.[15] As we have explored elsewhere, it is in the arenas of engineering and applied science that evidence of the labour of women – and, more specifically, women of colour – can be more readily found, thus making it all the more important to remedy this marginalization within the scholarship.[16]

In that regard the ICWES meetings are particularly revealing; while ICWES 1 (USA) was established to look at the deficit of women engineers and scientists in Western nations in particular, it was clear at ICWES 4 (Poland) that the Eastern European socialist/communist nations sought to represent themselves as having 'solved' the challenge of recruiting and sustaining women in STEMM through their rigorous socialist principles of workplace gender equity and collective childcare. Thus, while the agenda of the first meeting was initially set by the USA's Society of Women Engineers, the character and prerogatives of ICWES gatherings developed a pattern by the Polish meeting of 1975 showing how nation-specific agendas of technoscience could nevertheless be harmonized into a broader statement of women's understated role in technoscience, without any single Cold War nation or ideology being dominant – exemplified by the fact that the USA conceded its early steering of ICWES to a politically heterogeneous alliance of nations.

Responding to the Soviet Union: The first ICWES in New York, USA, 1964

The first ICWES in 1964 was hosted by the USA's Society of Women Engineers at a historical juncture when more women than ever before, in a range of different countries, were entering into the fields of engineering and applied science. As Laura Puaca's work has shown, one key context for this recognized by SWE was women's extensive participation in communist Russia: by 1963, 30 per cent of the Soviet Union's engineers were women, an order of magnitude greater than most Western nations.[17] And in that same year, Russia's first female astronaut, the engineer Valentina Tereshkova, demonstrated that communist technocracy could extend women's prerogatives beyond the Earth's surface too (continuing the USSR's startling lead in the 1957 launch of *Sputnik*, the world's first artificial satellite).[18] Although the proportion of women was still relatively low in Western nations, both that Cold War context and the processes of decolonization and modern industrialization brought opportunities for women's training and education, as well as initiatives to change the perception of which careers were open to women.[19] From a variety of causes, the second half of the twentieth century saw a growing need for trained engineers and scientists to assist in the (re-) building of many different nations. This ranged from the provision of basic infrastructure pertaining to water and electricity to the building of rockets to be sent into space.[20]

Where there were difficulties in recruiting men to engineering positions, women were called upon to fill the skills gaps. In many countries there were recruitment drives to get more girls and women into training programmes in these fields to create workforces fit for national needs. The Cold War context certainly galvanized more activity: Puaca has argued that SWE actively capitalized upon associated 'emergency mentalities' to engage more government organizations and educational authorities in their quest for the US government to encourage more girls and women into engineering.[21] And thus it is no surprise that the US president actively welcomed the first ICWES gathering in 1964 (see below). Despite such drives, however, the reality of discrimination faced by women who aspired to be astronauts and engineers in NASA has been documented by Marie Lathers, while the invisibility of the African American women employed by NASA in crucial calculations of rocket trajectories has been brought to prominence in the popular book *Hidden Figures: The American Dream and the Untold Story of the Black Women Who Helped Win the Space Race*.[22] While it was the Space Race with the USSR which prompted this less-visible form of technocratic inclusivity in the USA, neither the Space Race nor the internal US politics of race were mentioned at ICWES 1. As will become apparent, it was left instead to the communist countries participating in ICWES meetings to highlight their more developed capacity to educate, equip and enrol women for roles in engineering and applied science, with British commonwealth initiatives to support women's training in former African and Indian colonies relatively modest by comparison.[23]

Having only previously held national conferences for women engineers in the USA since its foundation in 1950, it is within this multifaceted global context of growing women's participation that SWE's leaders decided to hold an international conference for women engineers and scientists in 1964.[24] At a time when new hard Cold War and post-imperial borders were being enacted and new forms of national sovereignty re-established for many countries, it is noteworthy that the ICWES 1 organizers aligned this event with the growing irenic culture of internationalism; subsequent organizers continued this focus on cross-cultural exchange by holding the fourth conference in Poland in 1975, a then-communist-run country behind the so-called Iron Curtain.[25]

The date of the first ICWES was chosen by SWE to coincide with an event far from the hostilities of the Cold War: the 1964 World Fair in New York. Ruth Schafer, the chair of the first ICWES, later wrote that the aims of the conference spoke to a USA-centric modus operandi: the organizers 'wanted to further the aims of the SWE' by alerting 'guidance counsellors, educators, and the public' of the imperative to encourage women into engineering. And to do so SWE revealingly now had 'to learn from the International Delegates' what conditions in their own countries had provided 'an encouraging atmosphere for women to obtain and use technical educations'.[26] As then-President of the USA Lyndon B. Johnson echoed in his telegram to the conference, printed at the start of the proceedings, the USA had 'lagged behind many European countries in its utilization of the abilities of women in the professional fields, particularly scientific research, engineering, law and medicine'.[27] This patriotic 'utilization' rhetoric to recruiting skilled labour helped to ensure that the SWE conference organizers received funding from both government organizations and businesses. Funding for ICWES 1 was provided by the National Science Foundation, the USA's Asia Foundation and the Engineers Joint Council, among others.[28]

To challenge the conservative gender norms of 1960s America, the conference had to be recognized by 'high-level representations from the United States Government, the state of New York and the city of New York'.[29] Thereafter, it was common for high-ranking figures from government to provide an opening speech at ICWES meetings or to attend parts of the conference, as in the case of the third conference in Italy, where the undersecretary to the Ministry of Public Works attended the final part, while government representatives were present in Poland too.[30] Such appeals to the bastions of traditional institutions of governance show the ICWES 1 organizers' wish for the conference to sit within the mainstream. This is a contrast perhaps to the newly emerging Women's Liberation Movement in the USA, which came to be associated with second-wave feminism, which often took a more grassroots approach.

Such overtly feminist campaigning is rarely visible in the publicly accessible records of the ICWES meetings. Indeed, the only exception we have thus far located is Rita Levi-Montalcini's paper to ICWES 3 (Turin, 1971) on 'Women scientists and the Women's Liberation Movement'. On that occasion this Jewish Italian erstwhile refugee from fascist persecution in World War II outlined what she saw as the 'depressing' persistence of sexist bias against women's participation in STEMM. The most successful female practitioners had declared to her their 'total indifference and almost annoyance' towards the Women's Liberation Movement, claiming that they had never suffered 'discrimination', and thus shunned any radical moves to eliminate sex discrimination in science.[31] Nevertheless, it was thus very revealing that a subsequent male speaker at the same conference session, the Italian undersecretary of the Ministry of Public Works, Vincenzo Rosso, felt compelled to conclude his account of the importance of female emancipation for human progress with a stark warning to 'women against being dragged away into extreme movements which can often lead to a fierce feminism'.[32]

Still, it is important to note that in the 1960s and 1970s even the act of holding a conference dedicated to women in science and engineering could be read by those attending as a radical act, since at the time such spaces primarily dedicated to discussion of women's agency and achievements in STEMM were rare. This is a point highlighted in a post-conference article in *The Woman Engineer* reporting on the third ICWES in Turin, 1971: 'This conference was attended by delegates from more than 30 nations from all over the world: nothing unusual for an international congress except that nearly all these delegates were women, highly qualified in some branch of science and technology.'[33]

Another relevant point to remark on is that the context of the women's movement/third-wave feminism is only broadly applicable in certain countries, with other countries undergoing their own transformations in terms of gender relations and gender in the workplace. ICWES's international nature perhaps made it a forum for discussion and comparison, rather than a platform for one version of feminism. This is particularly relevant for the fourth ICWES, held in socialist Poland in 1975, where organizers were keen to emphasize that equal opportunities for women were embedded within the country's governing principles, as will be explored in more detail in a later section.

Continuing the conference: The second ICWES in Cambridge, UK, 1967

Despite the importance of the first ICWES for initiating and laying the pattern for future conferences, SWE did not, in fact, initially conceive of it as the beginning of an international series of conferences, rather as a one-off gathering. Reports at the time suggest that the idea for a second conference was only born during the first conference, with delegates from the UK offering to host the next gathering. As Ruth Schafer later reported, several days into the evidently successful first conference there was 'eager whispering' about possible future developments. Several 'enthusiastic people' were already asking, 'When and where will there be a Second Conference, Ruth?' Her invariable answer was, 'Patience, let's have this one first.' But towards the end of the conference, as the Friday lunch neared, such tension could be contained no longer: 'Isabel Hardwich, bless her, grabbed the mike and shouted, "You're all invited to England in' 67." Pandemonium!'[34]

In contrast to the initial US-centric aims of the conference's beginnings, the planning notes for the second conference, which was organized primarily by the UK's Women's Engineering Society (WES), showed a more overtly internationalist approach: 'By holding another conference within such a comparatively short time, it is anticipated that a genuine international movement among women engineers may be set in motion' and 'as an isolated occurrence this [ICWES] could pass away and no more need have been heard, but, as the beginning of a movement, its effect could be continuing'.[35] The commitment of the WES delegates to hold a second conference meant that an International Continuing Committee with ten area representatives was formed to decide where later conferences would take place.[36] The proceedings from the first four conferences document the countries that put themselves forward to host the subsequent meeting, which were then voted on by country representatives, deciding the outcome of the location.

WES's ability to lead a second conference was, arguably, not surprising. As one of the older organizations committed to women and engineering – it was founded in 1919 – it was also, always, an outward-looking organization, with members from around the world and frequent reports from overseas about women in engineering.[37] Indeed, one of the key members of the second ICWES organization committee was German Jewish refugee Ira Rischowski, who found her home in the UK and in WES after she fled Nazi Germany in the 1930s. Furthermore, the second conference was better attended by women engineers and scientists from African nations: at ICWES 1, only Morocco, until 1956 a French colony, had been represented among African countries. Largely due to the persistence of Commonwealth-based links, following the independence of several nations from British colonial rule, many more were in attendance at ICWES 2: Ghanaian zoologist Leticia Obeng, Ugandan mechanical engineer Miriam Muwanga and Nigerian physicists Ebun Adegbohungbe and Deborah Ajakaiye were all listed as delegates and speakers, offering their perspectives on their work in their respective countries. The ICWES meeting in Cambridge could well be viewed as a mode of post-Empire diplomatic relations between Britain and the Commonwealth, and it is notable that all these women had received part of their educations at British universities, marking a further point of interaction between the former colonies and Britain in the form of educational exchange.

Broadening the management of ICWES

While it may seem that SWE and WES – two well-established organizations for women in engineering from then-leading global powers – dominated the early history of ICWES, this did not last. In fact, the subsequent ICWES proceedings and other documents show that the continuity and success of the conference depended upon the participation of several other organizations from around the world in ways that gradually shifted control of ICWES away from SWE and WES. The organization of the first two ICWES meetings did indeed require close cooperation with national organizations across the globe, including le Cercle d'études des femmes ingénieurs de l'Association des Françaises diplômées des universités, the Society of Japanese Women Scientists and the Women Chemical Engineers of the Philippines.

The transnational ICWES Continuing Committee scrutinized a range of bids after each conference on who might run the next international gathering in three or five years' time. Each ICWES meeting was then run by the successful host country's national organization for women in engineering and/or science. For example, the Associazione Italiana Donne Ingegneri e Architetti (AIDIA) took over the running of the third conference in Turin, Italy, while the Central Technical Organization (NOT) and the National Council of Polish Women managed the fourth conference in Krakow, Poland. The proceedings from the third conference in Italy indicate that there were fourteen national organizations represented, from Brazil, France, Germany, Greece, India, Italy, Japan, Mexico, Poland, Portugal, Switzerland, the UK, the USA and the USSR. The transnational nature of ICWES appears to have depended upon the collaboration of various national organizations, each of which brought its own representation and national perspective into dialogue with its counterparts from around the world.[38]

Such collaboration was not only transnational but also interdisciplinary. While women's organizations for engineers were more prevalent, the ICWES meetings were also a place where women from across engineering and scientific fields – and their various organizations – came together. As noted in the introduction, engineering and applied science were the focus of ICWES gatherings, not the abstract and typically theoretical 'pure science' long privileged within histories of science. Indeed, the practical forms of 'applied science' are often where the work of women and other marginalized groups was typically called upon by the industrial or industrializing states, as evidenced at early ICWES meetings.[39] Unlike international initiatives in, for example, high-energy physics, that had no direct humanitarian applications,[40] the transcultural, discursive world of ICWES centred on creating and sharing knowledge that would be directly useful for human well-being.

Formulating long-term ICWES goals for peace: ICWES 3 and 4

Having been established as a conference over the eleven years covering the first four meetings, how did the leaders of the ICWES gatherings arrive at some continuing long-term goals for it? To some extent, the first conference in the USA set the internationalist tone; whereas Ruth Schafer privately focused on SWE's inward-looking motives, the SWE president in 1964, Aileen Cavanagh, offered a more global approach to focusing the work of male and female engineers and scientists on the 'future needs of humanity'. Alluding to the Space Race and projects of global exploitation so obviously shunned at ICWES 1, Cavanagh noted that while some 'are striving to conquer outer space and attempting to unlock the mysteries and vast resources of the oceans and the bowels of the earth', others were taking a very different approach: ICWES's core constituency was instead 'working daily to nurture the hopeful spirit of life among the world's people'. It was in that vein that Cavanagh argued that 'technical progress properly applied' could still be a 'strong force for good'; the goal was to 'increase human dignity and to ease the burdens of hunger, ignorance, and despair' that afflicted so many people throughout the world in the 1960s.[41]

This tone is then echoed at the subsequent meetings in a variety of ways: in the conference proceedings for the first four ICWES meetings certain concepts come to the fore in the opening addresses, illustrating an investment in progress, lack of selfishness, public service and an orientation towards a better future. In her opening address at the third ICWES in Turin, Italy, in 1971, for example, the lead organizer Anna Amour contrasted these guiding concepts to the (implicit harms of) so-called technological progress wrought in the Space Race and nuclear warfare:

> We are here planning for progress, and not just the scientific and technological progress which has launched men on the moon or in the infinitesimal secrets of the atoms, but a real full progress that will help men and women to proceed along the way of civilisation and to eliminate dangers of destruction that we are facing because of some imbalanced scientific and technological achievements.[42]

It is interesting to note here Amour's contrast between perceived successes of human ingenuity, such as landing a man on the moon, and the more genuine idea of 'real full progress'; the aim of the conference was not gratuitous scientific and engineering progress but rather a practicable project that could eliminate the pressing global problems of that era. It suggests that they see women as able to play a mitigating role against 'imbalanced scientific and technological achievements' (perhaps largely conducted by men), to achieve something different, pertaining to a different model of progress. Similar sentiments were repeated at the fourth ICWES in Poland, which was centred on technological achievement and the welfare of all nations. The keystone message was expressed as 'the conviction that science and technology should serve mankind in satisfying its needs' and that 'the contribution of women to the successful solution of this problem seems to be of paramount importance'.[43]

The proceedings of ICWES 3 cite the final conference resolutions made in Turin, stating the now-developed themes for the conference going forward, marking out the importance of this meeting for the cementing of a particular format. This included the two strands 'planning for progress' and 'women's professional and family duties', framing the two main prerogatives of what the conferences had thus far and would go on to explore. It ends with a pledge from the 'women engineers and scientists from 32 countries' gathered in Turin to 'dedicate their efforts' to improving 'the quality of human life' through 'the responsible use of scientific and technological achievements', once again reiterating the above message.[44]

As alluded to earlier, by the third and fourth ICWES meetings, clear links also became apparent between ICWES and contemporary conferences working towards similar global aims. For example, the third ICWES set its resolutions to align with the 1972 Stockholm Conference on the Human Environment, while the proceedings of the fourth ICWES make mention of the UN's International Women's Year in 1975, the World Women Congress in Berlin, the World Conference of Women in Mexico and the Conference of Security and Peace in Europe (held in Helsinki). With these multiple modes of collaboration, the organizers – and the organizations involved – might be seen to be enacting women-led Cold War science and engineering diplomacy, with a clear set of diplomatic principles to advocate for, which, unlike more conventional modes of science diplomacy, were not directed by one national agenda.[45] Ito and Rentetzi, for example, recognize a recent focus on the 'important roles played by non-governmental actors' in Cold War technoscience.[46] As the next section will demonstrate, ICWES's transnational mode of diplomacy manifested itself in practical ways across Cold War divides.

Crossing the 'Nylon Curtain' into Poland: ICWES 4

As highlighted in the introduction, ICWES came into being at a time of shifting geopolitical terrain, particularly in light of the Cold War, thus we might see these connections to other international organizations and events as one of the ways in which the ICWES conference organizers saw the conference as situated within a wider global picture. It begs the question, therefore, to what extent ICWES was an active response to some of these challenges and how they were defined and expressed in the conference discourse. From analysing the proceedings and other available conference documents, a defining feature of the ICWES conference discourse appears to be a non-direct approach to politics, what we might instead interpret as diplomacy. There is very little mention of contemporary political situations, except in an oblique way, for example, through the mention of poverty, hunger, lack of infrastructure or non-specific references to space travel and the potential vagaries of technological development when carried out with the wrong intentions.[47] Instead, as already alluded to, there is an emphasis on progress, as well as a commitment to peace.

The discourse around peace and progress is noticeably prevalent at the fourth ICWES in Poland. As the opening quotation indicates, there was a conscious recognition of Poland's status as a communist nation, and the need for cooperation between countries with different economic and social systems. In opposition to the hostile rhetoric apparent on both sides of the Cold War divide, the president of the Polish Women's Council, Maria Miczarek, called for collaboration and unity. The final words of her opening speech quoted the motto of the 1975 International Women's Year: 'Development–Progress–Peace'.[48] As suggested in the introduction, this speaks to recent research on the Cold War which has argued that there was much more exchange between East and West than the hostile rhetoric would suggest, leading some scholars to reframe the overly rigid framing of the 'Iron Curtain' in more porous, flexible terms as the 'Nylon Curtain'.[49] Concurrently, there was a concerted effort to draw attention to women's established place within the social and working fabric of the country: 'Polish women participated actively in the rebuilding of our country, developing the socialist economy, science and culture.'[50]

We can see her speech as a response to the need expressed by Ruth Schafer and Lyndon B. Johnson in 1964 for the USA to learn from other nations on how to include more women in engineering and science: Miczarek said directly that 'the participants of the conference will have the opportunity to find out for themselves how far we have succeeded in Poland'.[51] The contrast between Western nations and those under socialist governments in Eastern Europe was sharply noted in a report from the fourth ICWES on the numbers of women engineers in different countries: it stated that 16 per cent of engineers in Poland were women, while in the USSR it was 39 per cent, compared to 1–4 per cent in USA, Canada and the UK.[52] Revealingly, at ICWES 4, the high importance of women's participation was brought up front to be the first major topic: section 1 of the conference was dedicated to 'contemporary sociological problems of women engineers and scientists', ahead of sections on the environment and transport, nutrition, supply and equipment problems, computer science and materials and engineering.

It could be argued that, from ICWES's outset, the Soviet bloc was keen to use the conference as a platform for promoting its socialist egalitarian version of progress. At all four of the first conferences there were papers from speakers representing the USSR, including from the Soviet Women's Committee, commenting on the USSR's success in integrating women into engineering and applied science. At the third ICWES, for example, one paper focused on how the USSR provides women with the ability to work and to be mothers. In contrast to papers from other nations, largely highlighting what was preventing women taking part in the professional workforce, these papers aimed to show other nations what communism could do for women.

While many of the statistics and parts of the presentations might have held true, it is important to note that the Soviet Women's Committee was, according to the *Encyclopedia of Russian Women's Movements*, a 'government mouthpiece' and 'propaganda tool' that was actively involved in international and regional women's organizations and conferences to highlight women's success and happiness in the USSR.[53] Whether or not women in the USSR fully experienced the transformations propounded by such governmental mouthpieces has been called into question, with research indicating that the lived reality of women was at times very far removed from the rhetoric.[54] Despite the intentions of ICWES to be a non-political space, there is evidence that it, at times, served the political agendas of certain nations, something that could bear greater scrutiny in further research.

Conclusion

The wider purpose of ICWES as it reached maturation in the mid-1970s can be found encapsulated in the resolutions formulated at the end of the fourth ICWES in Poland:

> Despite the fundamental differences in life-style of the peoples, and despite the basic dissimilarity of the actual tasks which are performed under different social and political conditions, there are several common, important problems to be solved by mankind.
>
> Among these problems are: the prevention of war and, in particular, of worldwide nuclear war, the protection of the natural environment, reasonable utilization of natural resources, the struggle against starvation, destitution and maladies suffered in many parts of the world, the elimination of disproportional economic and cultural development.
>
> The methods of solving these problems must vary from country to country. However, under the conditions of progressive détente and peaceful technical knowledge is always in keeping with her human respect for society and we ask you to propagate this idea wherever may be. While working for the development of her own country, she must also consider the good of all mankind.[55]

Within the allusions to the Cold War, the message is clear: that nations have to work together and that difference can, and should be, for the benefit of all human life. This chapter set out to examine how women in engineering and applied science collaborated across borders in the second half of the twentieth century, locating the study in the previously overlooked International Conference of Women Engineers and Scientists, positioning the conferences as an example of women-led science and engineering diplomacy.

The conference, though situated within a wider context of women's professional and peace-oriented organizations, formed something rare in its focus on women in the engineering and scientific professions, drawing together various national organizations set up during the twentieth century to support women in their endeavours in these fields. As we have seen against the backdrop of the Cold War, the conference sought – albeit within a relatively mainstream manner and with the support of traditional bastions of power – to champion the role that applied science and technology could play in solving the various global issues, including hunger, lack of infrastructure and gender inequality. As the chapter has also drawn attention to, this was a double-way utilization, with the 'traditional bastions of power' opportunistically using the conferences as spaces to promote their own agendas or learn from other countries how best to tap into women as an underused labour resource. However, despite this utilization approach, the early conferences' dominant focus on peace and progress meant that they offered an alternative discursive space within the science and engineering domain, going against an approach to science and engineering that furthered hostile agendas in the form of the Space Race and the production of military equipment.

Traces of the geopolitical context are evident in the creation and maintenance of the first four ICWES meetings. The first conference was managed and held in the USA and was defined by a somewhat USA-centric agenda to improve the US's own practices by learning from others, perhaps to ensure that it did not lose competitive edge. The more outward-looking WES from the UK ensured the conference continued for a second time and onwards to the current day, but only with the support and cooperation of several other national organizations. It was at the third and fourth ICWES meetings, in Italy and Poland respectively, that the values and ideals of ICWES were formulated, and more alignment with the broader international situation is evident. This chapter has highlighted the significance of the fact that the fourth ICWES took place in a socialist country, and it has suggested that, as a group of women, the ICWES organizers and attendees sought to offer an alternative narrative to conflict and division, superseding imposed nationalistic barriers, aligning with other concurrent internationalist efforts happening in other disciplines/spheres.

It also became apparent that the alignment of ICWES with governing bodies and institutions set it apart from other women-led movements. Later in the 1960s, and throughout the 1970s, many explicitly feminist organizations took an active form of protest against the kinds of traditional institutions that were welcomed at the ICWES gatherings, as these were often seen to be the oppressors of women. In this way, ICWES differs from other (Western-led) feminist movements; while it was governed by principles that can be understood as feminist – furthering the cause of women and promoting their equality – it was not an early part of the feminist movements, nor did it express its cause in the language of feminism that was developing at the time.[56] This aligns with Pamela Mack's contention that 'the fight for equality [in the USA] succeeded in bringing more women into engineering, but those women did not bring many feminist ideas with them'.[57] Instead, the conference and its organizers appear to have chosen a more 'mainstream' approach, aiming to work with governments and institutions, rather than seeking to disrupt, as would become the norm with rallies, sit-ins, protests and marches organized by women as part of the burgeoning civil rights movement.

The extent to which the discourse of ICWES matched up to its effects has been beyond the scope of this chapter to explore, but questions can be raised in this direction: how effective was ICWES as a mode of transnational science and engineering diplomacy? There are further questions about the limitations of ICWES, especially in terms of who was and was not included and why some countries were more actively involved than others. In its arguably traditional approach, could it be said to have served a hegemonic, Western agenda, rather than challenging it? Clearly there are other modes of transnational collaboration between women in engineering and applied science that can be explored alongside ICWES. Further research could, for example, look for points of transnational interaction that were not centred in Western locus points, as ICWES was, such as those that have been found between the USSR and Africa in the field of architecture in the twentieth century.[58]

Notes

1 R. I. Schafer, *The Conference Story: Recollections of the First International Conference of Women Engineers and Scientists*, June 1964, 20 (Society of Women Engineers archives, Wayne State University, LR001689/2/3).

2 M. Milczarek, 'Welcoming address', *Proceedings of the Fourth International Conference of Women Engineers and Scientists*, 1975 (Institution of Engineering and Technology archives, NAEST 132/4.1.5).

3 G. Gooday and E. Rees Koerner, 'Formulating a transnational history of women in engineering and applied science', *Women's History Today*, 3/4 (Summer 2022), 4–13.

4 For the historical literature on (Western) women as peacemakers, see note 14 below. For the anti-nuclear 'Pugwash' movement launched in 1957 and led by (mostly) male scientists from both sides in the Cold War, see J. Rotblat, *Scientists in the Quest for Peace: A History of the Pugwash Conferences* (Cambridge, MA: The MIT Press, 1972).
5 Gooday and Rees Koerner, 'Formulating a transnational history'. See also I. Vardi and L. Smith-Doer, 'Women in the knowledge economy: Understanding gender inequality through the lens of collaboration', in D. L. Kleinman and K. Moore (eds), *Routledge Handbook of Science and Technology Studies* (Abingdon: Routledge, 2014), pp. 388–405.
6 *Proceedings of the First International Conference of Women Engineers and Scientists*, 15–21 June 1964, I-4, https://uihistories.library.illinois.edu/REPO SITORYCACHE/156/d5ne15BSVY0hV4ogbAmN6p592koYodSl607Z2in6k m8aHI1y4S15vb3R58qwP34ra4tJ4nx1TJgHk55V7f17K3l8FSw6XNK5m jz39ERv2o9_3409.pdf (accessed 21 April 2022).
7 While the conference name suggests ICWES meetings were for scientists and engineers, the majority of papers and discussion points concerned the practical applications of science, rather than relating to 'pure science', thus applied science is perhaps the more accurate terminology.
8 Most references to ICWES relate to the first conference in the USA in 1964, in relation to histories of the USA's Society of Women Engineers, for example, M. W. Rossiter, *Women Scientists in America: Before Affirmative Action 1940–1972* (Baltimore, MD and London: The Johns Hopkins University Press, 1995). A more comprehensive, descriptive overview of the management and content of the conferences, and the formation of INWES, can be found in M. Frize, C. Deschênes and R. Heap, *Women's Contribution to Science and Technology through ICWES Conferences* (Cham: Springer Nature, 2023).
9 Proceedings of ICWES 6 (unpaginated) give estimates for the overall attendance at successive meetings: ICWES 1 (USA): 500; ICWES 2 (UK): 300; ICWES 3 (Italy): 200; ICWES 4 (Poland): 600; ICWES 5 (France): 150; ICWES 6 (India): 300.
10 Attendee information is available in the conference proceedings for ICWES 1–6, which were accessed at the IET archives, London. For ICWES 4, see data in *Proceedings*, pp. 69–99.
11 See, for example: L. M. Puaca, *Searching for Scientific Womanpower: Technocratic Feminism and the Politics of National Security, 1940–1980* (Chapel Hill: University of North Carolina Press, 2014); A. S. Bix, *Girls Coming to Tech! A History of American Engineering Education for Women* (Cambridge, MA: The MIT Press, 2014); A. Canel, R. Oldenziel and K. Zachmann (eds), *Crossing Boundaries, Building Bridges: Comparing the History of Women Engineers, 1870s-1990s* (Amsterdam: Harwood Academic, 2005); B. Zengin, *Women Engineers in Turkey: Gender, Technology, Education and Professional Life* (Saarbrücken: Lambert Academic Publishing, 2010); C. Franchini, 'Women pioneers in civil engineering and architecture in Italy: Emma Strada and Ada Bursi', in M. Groot, H. Seražin, E. M. Garda and C. Franchini

(eds), *MoMoWo: Women Designers, Craftswomen, Architects and Engineers* (Ljubljana: Založba ZRC, 2017), pp. 82–101; M. Hicks, *Programmed Inequality: How Britain Discarded Women Technologists and Lost Its Edge in Computing* (Cambridge, MA: The MIT Press, 2017).

12 G. Y. Shen, 'Women and the transnational dynamics of science education in early twentieth century China: A quiet revolution', *Chinese Annals of History of Science and Technology*, 3: 2 (2019), 62–93; C. Midgley, A. Twells and J. Carlier, *Women in Transnational History: Connecting the Local and the Global* (London and New York: Routledge, 2016); C. von Oertzen, *Science, Gender and Internationalism: Women's Academic Networks, 1917–1955* (Basingstoke: Palgrave Macmillan, 2021). This is also part of a wider trend in science and technology studies; see, for example, J. Krige (ed.), *How Knowledge Moves: Writing the Transnational History of Science and Technology* (Chicago, IL: University of Chicago Press, 2019).

13 See, for example: P. Babiracki and K. Zimmer (eds), *Cold War Crossings: International Travel and Exchange across the Soviet Bloc 1940s–1960s* (Arlington: Texas A&M University Press College Station, 2014); P. Babiracki and A. Jersild, *Socialist Internationalism in the Cold War: Exploring the Second World* (Basingstoke: Palgrave Macmillan, 2016); and J. Mark and P. Betts, *Socialism Goes Global: The Soviet Union and Eastern Europe in the Age of Decolonization* (Oxford: Oxford University Press, 2022).

14 For more information, see: L. K. Schott, *Reconstructing Women's Thoughts: The Woman's International League for Peace and Freedom Before World War II* (Stanford, CA: Stanford University Press, 1997); J. Blackwell, *No Peace Without Freedom: Race and the Women's International League for Peace and Freedom, 1915–1975* (Carbondale: Southern Illinois University Press, 2004); C. C. Confortini, 'Doing feminist peace', *International Feminist Journal of Politics*, 13: 3 (2011), 349–70.

15 Though there have been notable studies of women in engineering in a national context (see note 10 above), conferences – including the one this volume was born out of – and oft-cited literature tend to focus more on science. See, for example: R. Watts, *Women in Science: A Social and Cultural History* (London and New York: Routledge, 2007); C. G. Jones, A. E. Martin and A. Wolf (eds), *The Palgrave Handbook of Women and Science since 1660* (Basingstoke: Palgrave Macmillan, 2022), N.B., this volume does contain one chapter on women in engineering; P. Fara, *A Lab of One's Own: Science and Suffrage in the First World War* (Oxford: Oxford University Press, 2018). There are some exceptions, for example, M. W. Rossiter, *Women Scientists in America* (which contains chapters on women in engineering and SWE); Bix, *Girls Coming to Tech!*; and N. Kodate and K. Kodate, *Japanese Women in Science and Engineering: History and Policy Change* (London and New York: Routledge, 2016).

16 Gooday and Rees Koerner, 'Formulating a transnational history'. This builds on earlier research by Steven Shapin into the 'invisible technicians' working in scientific laboratories in the eighteenth century. See S. Shapin, 'The invisible

technician', *American Scientist*, 77 (1989), 554–63. Women, however, can be seen to be doubly invisible, since the assumption has tended to be that historically women are largely absent as scientific and/or technological workers.

17 Puaca, *Searching for Scientific Womanpower*, p. 121.
18 B. Evans, *Escaping the Bonds of Earth: The Fifties and the Sixties* (Cham: Springer Science & Business Media, 2009), pp. 52–58.
19 A report printed in the proceedings of the first ICWES indicated the following numbers of women engineers in European countries: England, 149; France, 1,500; Ireland, 3; Italy, 313; Norway, 900; Switzerland, 3 (based on self-reporting from these countries). The report states that the 1960 census showed that 7,000 women were working as engineers (in 1962, there were 615,400 engineers employed in industry in the USA according to an Engineers Joint Council study). For the contrast with Soviet republics, see later in the chapter.
20 Gooday and Rees Koerner, 'Formulating a transnational history'.
21 L. M. Puaca, 'Cold War women: Professional guidance, national defense, and the Society of Women of Engineers, 1950–60', in A. M. Knupfer and C. Wayshner (eds), *The Educational Work of Women's Organizations, 1890–1960* (Cham: Springer 2008), pp. 57–77; Puaca, *Searching for Scientific Womanpower*.
22 M. L. Shetterly, *Hidden Figures: The American Dream and the Untold Story of the Black Women Who Helped Win the Space Race* (London: Harper Collins, 2017); M. Lathers, ' "No official requirement": Women, history, time, and the U.S. Space Program', *Feminist Studies*, 35: 1 (2009), 14–40.
23 Gooday and Rees Koerner, 'Formulating a transnational history'.
24 Bix, *Girls Coming to Tech!*; Puaca, 'Cold War women'.
25 This was part of a growing trend: the 16th International Congress for the History of Science took place in Bucharest in 1981, for example, and this was where the first meeting of the Commission of Women in the History of STM also took place.
26 Schafer, *The Conference Story*, pp. 1–2.
27 Telegram from Lyndon B. Johnson, 17 June 1964, *Proceedings of the First International Conference of Women Engineers and Scientists*, 15–21 June 1964, p. vii.
28 T. Eller English, 'Finding aid for the International Conference of Women Engineers and Scientists Records', SWE archives, Wayne State University, https://reuther.wayne.edu/files/LR001689_guide.pdf (accessed 11 May 2022).
29 Schafer, *The Conference Story*, p. 21.
30 *Proceedings of the Fourth International Conference of Women Engineers and Scientists*, p. 5.
31 R. Levi-Montalcini, 'Women scientists and the Women's Liberation Movement', *Proceedings of the Third International Conference of Women Engineers and Scientists*, 1971, p. 13.
32 V. Rosso, 'Women engineers and scientists for human progress', *Proceedings of the Third International Conference of Women Engineers and Scientists*, 1971, p. 20.

33 R. West, 'An opinion of the purpose and achievements of TICWES', *The Woman Engineer*, 11: 3 (Winter 1971), 22.
34 Schafer, *The Conference Story*, p. 26.
35 Executive Committee Minutes for the Second International Conference of Women Engineers and Scientists (Institution of Engineering and Technology archives, London, NAEST 92/15/2/1).
36 Eller English, 'Finding aid'.
37 G. Gooday, 'Internationalism and the UK's Women's Engineering Society (WES)', Electrifying Women blog, 21 April 2020, https://electrifyingwomen.org/internationalism-and-the-uks-womens-engineering-society-wes/ (accessed 21 April 2022).
38 It is worth noting that Isabel Hardwich from WES was a leading figure in the early ICWES meetings and became the president of the International Organizing Committee of ICWES.
39 For more detailed discussion of the speakers and topics of papers at the early ICWES conferences, see Gooday and Rees Koerner, 'Formulating a transnational history'.
40 Shiv Visvanathan's description of the founding of the high-energy physics laboratory CERN – a far more widely known international postwar trans-European project – may well have epitomized what he calls 'the internationalism of pure science'. S. Visvanathan, *Carnival for Science: Essays on Science, Technology and Development* (Oxford and New York: Oxford University Press, 1997), p. 176.
41 A. Cavanagh, 'Official opening', *Proceedings of the First International Conference of Women Engineers and Scientists*, 15–21 June 1964, I-4.
42 A. Amour, 'Opening address', *Proceedings of the Third International Conference of Women Engineers*, 1971.
43 *Proceedings of the Fourth International Conference of Women Engineers.*
44 *Proceedings of the Third International Conference of Women Engineers.*
45 For recent literature on science and engineering diplomacy, see, for example: M. Adamson, 'Science diplomacy at the International Atomic Energy Agency: Isotope hydrology, development, and the establishment of a technique', *Journal of Contemporary History* 56: 3 (July 2021), 522–42; B. Amadei, 'Engineering for peace: Challenges and opportunities', in *2018 World Engineering Education Forum: Global Engineering Deans Council (WEEF-GEDC)* (Albuquerque, NM: IEEE, 2018), pp. 1–6; S. Arapostathis and L. Laborie, 'Governing technosciences in the age of grand challenges: A European historical perspective on the entanglement of science, technology, diplomacy, and democracy', *Technology and Culture*, 61: 1 (2020), 318–32; S. Kunkel, 'Science diplomacy in the twentieth century: Introduction', *Journal of Contemporary History*, 56: 3 (July 2021), 473–84; Z. Li, F. Cui and Z. Wang, 'A discussion on practices and characteristics of science and technology diplomacy in twentieth-century China', *Cultures of Science*, 6: 2 (June 2023), 186–98; S. Robinson, M. Adamson, G. Barrett, L. Lund Jacobsen, S. Turchetti, A. Homei, P. Marton, et al., 'The

globalization of science diplomacy in the early 1970s: A historical exploration', *Science and Public Policy*, 50: 4 (14 September 2023), 749–58.

46 K. Ito and M. Rentetzi, 'The co-production of nuclear science and diplomacy: Towards a transnational understanding of nuclear things', *History and Technology*, 37: 1 (2 January 2021), 10.

47 At a meeting of the national representatives on 9 September 1975 where thirty-two countries were represented, chaired by Isabel Hardwich, a special statement from the representatives of the United Kingdom said: 'The Representative of the U.K. expressed very strongly her belief that the conference papers should be non-sectarian and non-political'; *Proceedings of the Fourth International Conference of Women Engineers*, 1975, p. 47.

48 Miczarek, *Proceedings of the Fourth International Conference of Women Engineers*, p. 33.

49 György Péteri refers to the divide as a 'nylon curtain' rather than an iron one, while Michael David-Fox writes of it as a semi-permeable membrane. See G. Péteri, 'Nylon Curtain: Transnational and transsystemic tendencies in the cultural life of state-socialist Russia and East-Central Europe', *Slavonica*, 10: 2 (2004), 113–23; M. David-Fox, 'The Iron Curtain as semipermeable membrane: Origins and demise of the Stalinist superiority complex', in Babiracki and Zimmer (eds), *Cold War Crossings*, pp. 14–39.

50 Miczarek, *Proceedings of the Fourth International Conference of Women Engineers*, p. 32.

51 Miczarek, *Proceedings of the Fourth International Conference of Women Engineers*, p. 33.

52 *Proceedings of the Fourth International Conference of Women Engineers.*

53 N. N. Corigliano and C. Nechemias (eds), *Encyclopedia of Russian Women's Movements* (Westport, CT and London: Greenwood Press, 2001), p. 174.

54 See, for example: J. Laycock and J. Johnson, 'Creating "New Soviet Women" in Armenia? Gender and tradition in the early Soviet South Caucasus', in C. Baker (ed.), *Gender in Twentieth-Century Eastern Europe and the USSR* (Basingstoke: Palgrave Macmillan, 2017), pp. 64–78.

55 *Proceedings of the Fourth International Conference of Women Engineers*, pp. 50–51.

56 For example, Betty Friedan's seminal feminist text *The Feminine Mystique* was published in 1963, one year before the first ICWES was held.

57 P. E. Mack, 'What difference has feminism made to engineering in the twentieth century', in A. N. H. Creager, E. Lunbeck, C. R. Stimpson and L. Schiebinger (eds), *Feminism in Twentieth Century Science, Technology and Medicine* (Chicago, IL: University of Chicago Press, 2001).

58 Ł. Stanek, *Architecture in Global Socialism: Eastern Europe, West Africa and the Middle East in the Cold War* (Princeton, NJ: Princeton University Press, 2020). See also C. Katsakioris' chapter 'The Soviet-South encounter: Tensions in the friendship with Afro-Asian partners, 1945–1965', in Babiracki and Zimmer (eds), *Cold War Crossings*, pp. 134–65.

III

In/visibilities in medicine and care:
Treating, teaching, reforming

6

'A model of devotion to the school': Female doctors in secondary schools in interwar Romania

Camelia Zavarache

Women have been linked historically to medicine and medical practice, not necessarily as active producers of knowledge, but more as objects of scrutiny and medical intervention.[1] The complex relationship between women and medical science has long been acknowledged by historians, sociologists and feminist scholars, who have approached the topic from a highly critical, gender-dominated perspective.[2] Investigations have focused on women's contributions to the history of medicine,[3] including the professional paths of female practitioners, their interaction with the male-dominated establishment and their role in defining and fighting 'women's illnesses'.[4] In a Central European context, the intersectionality of race and gender has also received attention as the basis of discrimination against Jewish women attending medical schools.[5]

In the Romanian Principalities of Wallachia and Moldavia, women were part of the group of healthcare professionals paid by the state.[6] By 1852, two schools for training midwives had been created in Bucharest and Iași.[7] Their establishment was important at a time when women gave birth at home, with no medical assistance. By the turn of the century, following women's admission into medical and pharmaceutical schools, graduates started working in other institutions. For instance, in 1910, female professionals were working in the laboratories of the Pharmaceutical School in Bucharest and the Obstetrics and Paediatric Clinics attached to the Medical School in Iași.[8] However, it was World War I that made women more visible in hospitals, mostly as nurses and, to a lesser extent, as doctors, as was the case with the first female surgeon, Marta Trancu Rainer.[9] Records from the archives of the Ministry of Public Health, Labour and Welfare show that in the early 1920s, due the loss of experienced doctors during the war, women also started being appointed as doctors in rural and smaller urban hospitals.[10]

This chapter focuses on female doctors in secondary schools in interwar Romania, thus adding another site of knowledge-making and practice to the ones already discussed by scholars with an interest in women and medical science. In particular, it will examine the multiple roles school authorities assigned to women medical graduates in the 1920s, when new legislation was introduced to provide medical assistance to secondary-school students. Under that legislation, female doctors were given priority to work and teach hygiene in secondary girls' schools all over the country. However, their involvement in teaching and treating female students was a fraught process that generated strong resistance from the male doctors who tended to monopolize such positions and from school principals and local revisers who often regarded women professionals with profound mistrust. From 1928 onwards, the law generated a massive process of substitution, with more and more female doctors demanding the right to access such teaching positions. This was made possible by women's broadening access to medical schools post-World War I. Although women's involvement in science was not research-related, by screening students' health and promoting hygiene practices, female doctors acted as agents of modernization who made medical knowledge accessible to younger generations.

The recruitment of female doctors into Eastern European healthcare administrations and schools became a general trend during the 1920s–1930s. Increased access to medical training and difficult socio-economic conditions seem to have been important factors in convincing authorities to enlist female doctors, thus making them more visible as social actors. In this regard, the chapter speaks to Edgerton-Tarpley's and Corbi's discussions of Chinese women in this volume, which highlight their agency as both healthcare agents and teachers.

The chapter is divided into eight sections: the first introduces the situation prior to the war, discussing hygiene in secondary schools' curriculum, in connection with the access of female students to medical schools at the Universities of Bucharest and Iași, while also analysing the new political configuration after the war. The next two sections discuss the 1928 secondary law on medical assistance for secondary-school students and the tendency for principals to work primarily with male doctors. The fourth section focuses on the professional requirements which doctors working in schools had to observe, while the following two dive into the actual process of substitution, analysing the diversity of contexts in which female doctors pursuing teaching positions in schools found themselves. The last two sections document the procedure for securing tenure of two female school doctors during the 1930s as the final stage in a series of developments which allowed them to establish themselves as respected professionals and modernizing agents. The chapter ends with several conclusions.

In writing this chapter, I have relied heavily on the archives of the Ministry of Cults and Public Instruction,[11] since that institution was responsible for overviewing the entire process of assigning school positions to female doctors. The correspondence between the Ministry's representatives and the doctors in local schools reveals the many struggles younger doctors, men but especially women, faced as they tried to carve a path for themselves.

Medical science: A beacon of light in difficult times and an instrument of modernization

In nineteenth-century Romania, hygiene was incorporated into the two education laws passed in 1896 and 1898 and consequently in the 1898 curriculum for secondary schools, targeted at both boys and girls.[12] In the case of girls' schools, it was taught alongside medicine and domestic pharmacy, in five-year schools (*școli de fete de gradul I*), as well as four-year schools (*școli de fete de gradul II*),[13] the latter corresponding to the superior stage of high school, which allowed graduates to enter university. This was the outcome of a top-to-bottom dissemination of health culture, from the medical elite towards the lower social strata, the emerging process of medicalization and the popularization of social hygiene.[14]

However, to teach such classes, secondary-school principals employed male doctors, the only ones available at the time.[15] This situation was connected to women's limited access to the two medical schools in the Old Kingdom, the one in Bucharest and the one in Iași. It wasn't until the beginning of the twentieth century that the number of female students in medicine became noticeable.[16] Still, records show that despite the low percentage of female graduates in the last decade of the nineteenth century, some managed to teach in secondary schools even prior to the 1900s.[17] After World War I, however, young generations from modest backgrounds began accessing university studies at an unprecedented pace. The number of women studying medicine increased compared to the years leading up to the conflict.[18] The same trends were visible among primary and secondary schools, which expanded considerably after the war, a direct consequence of a political commitment towards making education a universal right.

At the same time, Romania needed to rebuild its medical infrastructure and reorganize its sanitary administration, since the country had been devastated by the years of conflict and occupation. Consequently, medical services for middle-school and high-school students were created to fight social diseases; tuberculosis and venereal diseases were widespread due to poor health conditions caused by shortages of essential medical supplies and lack of information regarding prevention and contagion. Tuberculosis was a major health issue globally,[19] exacerbated by the war years. Documents show that for younger generations tuberculosis posed a major health threat and that the Ministry of Public Instruction wanted students to be subjected to periodical examinations to identify those sick or prone to developing the disease. Furthermore, the post-World War I years were marked by the spread of eugenics. Initially confined to specialists in medicine and biology, eugenics started reaching broader audiences, including politics, as it became increasingly clear that national states needed to design special social assistance policies to overcome the human losses caused by the war. As a result, protecting the 'biological reservoir of the nation' became a state policy in many countries in Central and Eastern Europe.[20] Healthy young generations depended entirely on the adults who became parents, and especially on women's health, since they were the ones bearing children. Therefore, the future of the nation relied on women being trained to protect themselves against social diseases and their knowledge of the hygienic care of children.[21] Education became the first line of defence, thus highlighting the importance of female doctors in girls' schools.

Politically, the first decade of the interwar period was dominated by the Liberal Party, which was responsible for drafting most of the laws and regulations that laid the foundation of Greater Romania during 1922–28. During that time, the Ministry of Public Instruction was coordinated by Constantin Angelescu, a doctor and professor at the Faculty of Medicine and a prominent figure of the liberal forces, who became the architect of the main education laws passed during the 1920s.

The first attempt to provide medical assistance in schools had begun in 1924, when the law on elementary schools was designed and introduced;[22] it included special articles regarding medical care for small children in primary schools. Their medical evaluation, however, was placed under the supervision of doctors leading the district to which these elementary schools belonged, since assigning a doctor for each institution surpassed the financial possibilities of the Ministry of Public Instruction. Therefore, a precedent already existed when, in 1928, the law on secondary schools was finally passed,[23] making medical assistance mandatory, while also stipulating that female doctors should be employed for girls' schools. Female doctors had been hired and taught in many secondary schools up to that point as well. However, this was not a systematic procedure, but rather an opportunity depending entirely on the budget of each institution. What the new law did was to enforce a tendency that had started to manifest in public schools from the final decade of the previous century and was made necessary by the postwar healthcare system and socio-political developments.

The major difference from the law on primary education, voted four years previously, was that the 1928 legislation concerned secondary schools, which at that time were defined by gender segregation; during secondary education mixed schools were no longer possible, female and male students being confined to separate institutions regardless of their profile. Therefore, the teaching personnel also tended to reflect this gender segregation which greatly influenced the way the medical service for such schools was organized. Even though the chapter on medical assistance was the result of a school policy aimed at the social body by targeting the young generations, resorting to women doctors was in fact the effect of a very conservative perception about male and female bodies[24] and the person best suited to evaluate and manage them. Starting with the 1928 law, female bodies were to be observed and touched exclusively by female doctors, a concern also discussed by Rai's and Edgerton-Tarpley's papers in this volume, with regards to India and China, respectively. Only where such physicians were not available could the principal of a girls' school hire a male doctor; however, he could not be granted the right to get tenure, as he was merely a provisional employee. Since the hygiene curriculum for girls' secondary schools included a component regarding pregnancy and child-rearing, one suspects that having a woman explain such notions was considered the most appropriate course of action to help students develop into responsible adults.

The Secondary Education Act

In 1928, the new law regarding public secondary schools was passed, which stated that each institution had to hire a doctor to monitor students' health and hygiene and evaluate the condition of the building and the boarding rooms. All doctors who wished to work in schools were obliged to prove they had graduated a school hygiene course, approved by a Hygiene Institute in Romania or abroad. In this way, doctors acquired a similar status to the other teachers, being appointed by the Ministry of Public Instruction, which monitored their activity and would eventually place them on the payroll. School doctors were responsible for implementing school hygiene and monitoring children's health; students had to be examined individually, to identify those who were sick or prone to diseases, treat them in schools or, if that was not possible, recommend clinics, hospitals or healing colonies at the seaside or the mountains. The results of individual medical exams had to be registered into a sanitary leaflet for each student for the entire duration of their studies.[25]

Perhaps the most rewarding task of all was teaching hygiene to middle- and high-school students. Following the nineteenth-century trend discussed above, hygiene became even more important after the war, being perceived as a central component of the culture of health that the biosciences promoted as indispensable to training responsible citizens. According to the theoretical high-school curriculum, hygiene was taught for one hour a week in the third grade and another hour a week in the seventh grade. The content for boys' schools was significantly different from that taught in girls' schools; even though they all learned general notions about germs, contagious diseases, proper nutrition or healthy behaviour for the body and the environment, during the final lessons boys learned about venereal diseases, while girls learned about the hygiene of the pregnancy and how to rear small children.[26] The gender separation of the academic content was directly linked to the future roles of male and female citizens, as hygiene played a significant role in the social project designed to prepare them for adulthood and citizenship.[27]

New law, old practices

The involvement of doctors in teaching institutions started before summer 1928, when the law on secondary education was finally passed. The previous year, Angelescu himself had ordered secondary-school principals all over the country to hire physicians and pay them out of their own incomes, provided by the School Committees.[28] The decision had been imposed on every high school with the important mention that no additional fees were to be collected from the students. This particular provision made school authorities open to collaboration with doctors who held public positions inside the medical administration, as they were the only ones willing to accept smaller pay, since they already had a main job. As it soon became obvious, the public personnel working for the sanitary administration were, in most cases, male, not female.

The collaboration between teaching institutions and doctors was needed prior to the war since medical examinations were part of the admission process for secondary schools. In fact, the study of hygiene had been part of primary and secondary curricula all over Europe. Still, this partnership became a necessity after the war, especially because of the long-lasting, debilitating effects it had on combatants and civilians alike; out of the second group, children were the most vulnerable category and attending to their medical needs became a priority to central authorities.

The same tendency to encourage and even legislate this partnership was also evident in the new provinces of the Romanian state. Transylvania was a good example: after the proclamation of the union, the province was governed by the Ruling Council (*Consiliul Dirigent*), until it was discharged by Alexandru Averescu's government in April 1920. Already in September 1919, Valeriu Braniște, Vasile Goldiș's successor at the Cults and Public

Instruction Portfolio (*Resortul Cultelor și Instrucțiunii Publice*), issued a provisional regulation regarding secondary schools, which revealed the importance of physicians. According to Article 27, the school doctor was expected to work closely with the head of the institution, monitoring the state of the building and removing possible health hazards for students. Therefore the contribution of such professionals to the well-being of students in secondary schools was being legislated right after the war and hygiene was retained as part of the curriculum, for both general high schools and technical secondary schools (*școala reală*).[29] However, the law did not make any references to the gender of such professionals; still, given the significant discrepancies between male and female graduates from medical schools in Eastern Europe, one can assume that in fact male doctors were the ones benefitting from such provisions. Therefore, there was a clear tendency for male doctors to monopolize such positions, as they were the ones already working in sanitary administration.[30]

Furthermore, to understand the context in which the law came into effect, one needs to examine two distinct events: the first was the political change that occurred during the fall of 1928, when the liberal regime came to an end. The Liberals were succeeded by the National Peasants' Party, a political force that emerged after Transylvanian politicians like Iuliu Maniu and Alexandru Vaida Voevod decided to join forces with the Agrarian Party from the Old Kingdom, placed under the authority of Ion Mihalache, a former teacher. It was the first time the party came to power; its leaders, among them Nicolae Costăchescu, appointed at the Ministry of Public Instruction, were the ones who enforced this law.

The second important event was the economic crisis that began in 1929. The economic situation deteriorated considerably during the time the National Peasants' Party governed the country, with devastating effects on the education budget. As a result, by 1931–32, the economy started spinning out of control and teachers were not being paid for months on end, with all additional benefits being cut. These difficult circumstances had a major impact on the way the law regarding the medical assistance of students in secondary schools was eventually implemented.

School hygiene courses and professional tensions

Immediately after the law came into effect, the Institute of Hygiene and Public Health in Bucharest, established in 1927,[31] started organizing classes for doctors who wanted to specialize in school hygiene.[32] Similar courses were organized by the Faculty of Medicine in Iași, as part of its Hygiene Laboratory,[33] and the Faculty of Medicine in Cluj,[34] in collaboration with the city's Social Hygiene Institute. Women were well represented in all these courses. Upon graduation, doctors applied for a position in a school, and the Ministry of Public Instruction appointed them as temporary hygiene teachers; after a probationary period of three years, they obtained tenure if they passed a special inspection.

This marked the beginning of a massive process of substitution, which followed two main directions: male doctors started applying for the positions of colleagues who did not hold a School Hygiene qualification but who had been working in schools, sometimes for a very long time. At the same time, female doctors started petitioning the Ministry to appoint them to positions in girls' schools occupied by male doctors (with or without the School Hygiene certificate). This was possible because the law was applied retroactively: no medical personnel, regardless of the number of years they had served as school doctors, were exempt from the obligation to attend a School Hygiene course. Those who had obtained their diploma earlier were allowed to choose the institutions where they wanted to work, as the number of such positions was limited.

Medical journals of the time attest to these developments. When doctor G. Banu launched the *Journal of Social Hygiene* in 1931, he explained in the inaugural issue that medicine was undergoing a major transformation from curative to preventive healthcare. Prevention and social hygiene were essential to the new era of medical interventionism and hygiene classes were designed to train and prepare doctors for such tasks.[35]

But how did one become a school doctor and, most importantly, how did women enter secondary schools as both physicians and hygiene teachers? Based on the available documents, schools hired female doctors either because they knew them or had already been working with them, or because they were appointed by the Ministry of Public Instruction, at their own request. Still, in the second case, the appointment was made upon the approval of the School Committee, as the new teaching staff member had to be accepted by the principal and their colleagues.

The real problem for women was that the Ministry of Public Instruction would not pay the salary for the school doctors unless they had tenure; until then they had to be paid by the School Committees, from their own incomes. Since schools were already struggling to cope with the numerous tasks the Ministry had placed upon them, it was more convenient to keep collaborating with male doctors who already held a position in the public health administration to avoid depending entirely on their school income. This situation occurred mostly in smaller, underfunded schools.

The entire legal and social context generated a massive correspondence between female doctors, who knew they were entitled to the medical positions in girls' schools, and the Ministry of Public Instruction. In demanding their right to occupy the available positions, female and male physicians were caught up in a spiral of accusations against those doctors who held them without a proper diploma or who were working in girls' schools. Surprisingly enough, the Ministry of Instruction did nothing to stop such denunciatory practices: on the contrary, it seemed to have encouraged them, to survey the situation in secondary schools in the country.[36]

On 9 September 1929, Alexandrina Ungureanu, a graduate of the School Hygiene course from Cluj, asked to be appointed as a doctor in one of the city's girls' secondary schools. 'Because no such positions are available at the moment', the petitioner wrote, 'I therefore ask you to appoint me in one of the many positions held by others, especially since I possess a School Hygiene diploma'.[37] Although she did not divulge any names, her letter was a direct accusation against those doctors who were simultaneously employed in several such positions. There is no doubt that before writing to the Ministry of Instruction, Ungureanu pondered the situation thoroughly.

Ungureanu took issue with the social and gender inequity created by the fact that some doctors were on the payroll of several public institutions, while others struggled to find a job at all. Her request was preceded by what might have been a letter from her husband, a doctor Valeriu Ungureanu, who voiced the same accusation that 'There are other colleagues who hold three or four or even more public positions and I only have one income that does not allow me to support myself in Cluj.'[38] Unlike Alexandrina Ungureanu, Valeriu Ungureanu already had a job. More surprising than the complaints themselves is the fact that the Ministry of Instruction took it upon itself to restore justice for such young doctors, by trying to accommodate them inside its administration network. Such social frustrations were very common at that time, since the number of students and young graduates was higher than ever before;[39] for those who did not have an urban social network to rely on, lack of a public-sector job became a major setback. On the other hand, the number of jobs that the state provided was limited, therefore at the end of the 1920s, younger graduates from the Faculty of Medicine started becoming resentful towards older professionals who had been working in different institutions prior to and after the war, not necessarily because they pursued such positions, but because they were needed. The Ministry of Instruction seemed to agree with their complaints, since from that moment on, due to economic difficulties, the holding of multiple public positions by a single individual was no longer permitted.

However, because doctors appointed to school positions did not have tenure for at least three years, their salary was paid by the School Committee, which meant that the amount was akin to a daily allowance. The Ministry of Instruction only paid their full salaries once doctors received tenure from the school. This was the reason some doctors juggled multiple jobs. In fact, some male doctors resigned because they found their pay incommensurate with the work they did in schools.

The Ministry of Instruction was thus trying to correct the consequences of a flawed law, which relied heavily on the collaboration of doctors who were also engaged elsewhere, as it lacked the funds to finance all specialists who obtained tenure. Another important reason for this situation was Constantin Angelescu's long-term vision regarding the creation of new schools and school services. In his opinion, what mattered most was passing the law and initiating the project; once this stage had been completed, financial means would be found more easily.[40] In fact, in the case of school medical services this way of thinking generated massive pressure on the School Committees' budget, as they were the ones covering the doctors' salaries until they received tenure.

Male versus female doctors

The new legislation generated inequity and frustration between male and female doctors. If women fared better in larger cities, since the number of secondary schools was higher, in smaller towns they faced numerous problems. Smaller towns meant fewer opportunities not only for women, but also for men; however, men were still better integrated into the public administration, as underdeveloped as it was, and they tried to protect their monopoly. One extraordinary example from Curtea de Argeș, a small town in the southern part of the country, shows that male doctors went to extreme lengths to preserve their privileges. In September 1929, shortly after being appointed doctor at the theological seminary boarding school by the Ministry of Instruction, George Ivancianu wrote to the school authorities asking for the same position at the town's girls' school. Since the authorities were reluctant to appoint a male doctor at a girls' school, the resolution clearly stating that he could not get tenure in such an institution, he asked the mayor of the town to issue a certificate testifying that no female doctors resided in the small urban community. The head doctor in this rural region happened to be a woman, married to the Commissioner of Argeș County, but lived in the town of Pitești. As the Ministry of Instruction was about to find out, Ivancianu had already been a doctor in the girls' middle school since January 1929, having been hired by the School Committee without its approval.[41] Apparently, the doctor wanted to cumulate as many school positions as possible to get better pay; however, in doing so he willingly denied female doctors their right to work in girls' schools, even though by law they were entitled to such positions.

What seems to have been an isolated case turned out to be a frequent dilemma that school authorities had to find a solution for. Even though the law stated that female physicians were to be assigned to the available positions in girls' schools if there were no suitable candidates, the principal of the institution could have accepted male doctors as a temporary solution. A similar case to the one in Curtea de Argeș is documented for another small town, this time from the province of Oltenia, Caracal. In November 1929, Petre Ioanițoiu wrote to the Ministry of Instruction, asking to be appointed doctor to the town's high school for girls. The head of the Elementary Schools Department (*Direcțiunea Învățământului Primar*) cautiously allowed him to fill the position, since the doctor formerly working there had resigned, but pointed out that this was a temporary measure. A few days later, the position was also requested by Xenia Ralef, a female doctor; interestingly enough, the reviser who received the document wrote that the same position had been requested first by Dr Ioanițoiu. His comment seems to imply that the order of the requests mattered when the Ministry assigned such positions, only because this criterion was clearly advantageous to the male doctor. In the end, school authorities in Bucharest followed the law, assigning the position to Dr Ralef.[42]

This case reveals a common misconception among local school authorities, from revisers to high-school principals, namely that male doctors were better suited for such positions, while women doctors were young and inexperienced. This is precisely what the principal of the Dej Girls' High School, a woman herself, argued in November 1929, when she pleaded that the Ministry of Instruction allow the male doctor already working in the school to continue. The document was a response to a job request made by Emilia Ciuta, a female doctor working at the public hospital in town who pursued the same position, but was dismissed by the principal, since she was perceived as not having a similar level of experience.[43]

Such attitudes were generated by the mistrust in female doctors' ability to work in schools, even though many women were already working in hospitals, maternities or rural healthcare institutions. However, the most important hurdle for principals must have been that of accommodating a new doctor and teacher they did not know; this change would have been even more difficult when the former male employee was of good conduct and had established himself as a good professional. Another factor that weighed heavily in this process was the pay; principals who had to pay doctors out of the School Committee funds with a daily allowance knew they needed to rely on a person having another job, or else they would not have been able to support themselves out of this little amount. Still, doctor Ciuta did have another job, proving that women also perceived school positions and the income generated by them as supplements, not salaries in their own right. Therefore, one cannot ignore the mistrust and even hostility that local school authorities tended to display towards such requests made by women, even though they were legally entitled to them. It seems their determination to have their right to work in girls' schools respected was bothering local authorities because it disrupted the way the system worked and led to the substitution of male doctors.

Despite such prejudices, women seemed aware of their legal rights and refused to accept that male doctors would continue to practise and teach in girls' schools. In September 1929, Dr Gudlea Ghelbruch wrote a petition to the Ministry of Instruction, asking to be appointed at the Carmen Sylva Girls' High School in Timișoara. At the time, she was working at the Social Security House in Timișoara, proving once again that women also tended to perceive such positions as a side-occupation. According to the documents she attached to her petition, she had attended the School Hygiene courses in Bucharest while working at the Social Security House because she wanted to teach and work in secondary schools.[44]

The person who processed the case wrote that the position had been filled by a male doctor, appointed 'years ago, when there were no female competitors'. He also explained that since the doctor did not have tenure, he was being paid at the end of each year, with a modest amount, implying that his main job was elsewhere. The way he insisted on the procedure of acquiring the position by publishing it as vacant in the Official Monitor must have triggered the suspicion of the Ministry of Instruction, since its resolution clearly stated that women doctors had to be appointed in girls' schools provided they met all the criteria of the law.[45] This type of attitude displayed by male inspectors shows that the job transfers from male to female school doctors created frictions between the two.

In the end, despite the obvious support from the Minister of Instruction, Dr Gudlea Gelbruch's petition led to the appointment of another colleague for the position available at the Carmen Sylva High School in Timișoara. Maria Cupcea Popovici was preferred because she was not an early career doctor and had a broader experience as a physician, teacher and school doctor.[46] However, it is not entirely clear if what mattered most was the ethnic origin, as Maria Cupcea was a Romanian born in Orhei County, Bessarabia, while Gudlea Gelbruch was a doctor of Jewish descent.

Female professionals at a crossroads: Masculinity, femininity and language barriers

Based on the cases discussed above, one might think that the substitution process for the positions of school doctor had only one direction: from male to female. In fact, documents show that the strict segregation of boys and girls into separate education institutions did not exclude the presence of female doctors in boys' schools. Even though such situations were exceptional, they present a more nuanced image of a complex social process. During the spring of 1929, the principal of the boys' middle school in Sulina, a small harbour town near the Danube Delta, faced a serious challenge when the Ministry of Instruction asked for the female doctor working there to be replaced by a male colleague. Doctor Virginia Stănescu, who oversaw the rural sanitary administration at the time, had also been willing to work as school doctor, hygiene teacher and French teacher. However, the Ministry of Instruction did not want to pay two different persons, insisting that one professional should cover both jobs.[47] The correspondence between the principal and the Ministry clearly shows that the authorities in Bucharest were increasingly concerned about their limited budget.

On the other hand, it becomes obvious that the Ministry of Instruction paid attention to the interactions between female doctors and male students. In fact, their concern was not directed towards the doctor, since she was already working with rural populations of different genders and ages, but towards the pre-adolescent boys, who were approaching puberty. The medical examinations required under the law, such as periodic body measurements and general examinations, were perceived as inappropriate when conducted by a professional of the opposite sex, more so if she was a woman. Among other things, this way of thinking was connected to the way secondary schools were designed. The law reflected the extreme shame surrounding the body and sexuality displayed by the general population,[48] in both urban and rural areas, on the one hand, and the ferociously prudish attitude of the political elite, on the other. It was in fact the very peak of a conservative school policy, the product of a political class that went out of its ways to dissociate reproduction from human anatomy for the sake of preserving the good morals of students, at least of those in theoretical schools.[49] However, in doing so, they offered new opportunities to women. At the same time, this was the outcome of an obvious attempt to modernize society by providing medical assistance and career guidance to the young generations, perceived as the most vulnerable and also bearers of a new way of life, healthy, rational and socially involved, a true preparation for adulthood and even more importantly, in the case of boys, for active citizenship.

Still, gender was not the only criterion for the exclusion of female and male doctors from secondary schools; language also played a role. Mastering Romanian became another standard by which medical professionals were being evaluated, mainly in provincial towns that prior to World War I had belonged to different states. For example, at the Principele Carol High School in Gura Humorului, a small town in the province of Bukovina, the local reviser considered appointing Dr Sonia Donetz, who had applied for the position. The main reason for this option was the fact that she was the only one among her colleagues in town who spoke Romanian fluently. The resolution issued by the Ministry of Instruction was however clear: no female doctor would be considered suited for a boys' school, the principal being urged to find a male alternative.[50] This case sheds light on a procedure that was highly discriminatory and against the best interests of the institution: a female doctor was being dismissed solely on the ground that it was preferable to leave students without any medical assistance than to have a woman examine and teach them, given the fact she was the only professional who knew the language.

In fact, the lengthy correspondence between the Ministry and local authorities could have been easily avoided by collaborating with the Ministry of Health to compile updated statistics of doctors, their education and place of residence. What central authorities must have regarded as a way to give local institutions the ability to decide for themselves was in fact a mere misconception: although they could appoint someone as school doctor, such institutions still needed the Ministry's consent.

Language turned out to be one of the most important criteria when settling a dispute between two female doctors aiming for the same position, when only one of them benefited from the support of the school's principal.[51] However, research conducted on documents from the Ministry of Instruction suggests that comments referring to the unsatisfactory degree of mastering the Romanian language constituted a common accusation among teaching staff and local school authorities. Its purpose was that of delegitimizing the opponent as a professional, since without the ability to speak Romanian the transfer of knowledge would have been impossible.

Despite these cases which reveal the numerous tensions generated by the mandatory gender segregation of medical staff working in secondary schools, not all substitutions were as difficult. On the contrary, in many schools, appointing a female doctor was the easiest decision the principal had to make; the board of the Girls' High School in Turnu Măgurele, which received two applications for the school doctor and hygiene teacher position, the first from a male doctor and the second from a female, chose the latter without any hesitation.[52]

Even more intriguing turned out to be the substitution of the male doctor working as a physician and hygiene teacher at the girls' school in the small town of Turda, Cluj County. In his resignation letter addressed to the Ministry of Instruction, he specifically mentioned the low monthly allowance paid by the School Committee.[53] He also recommended a female doctor as a replacement. Such attitudes show that not all men, be they doctors, principals, inspectors or revisers, feared and manifested hostility towards female doctors pursuing a position in girls' schools. More so since such disputes were clearly asymmetric, as boys' middle and high schools were in fact more numerous than the ones available for female students and, consequently, offered more positions for male doctors.[54]

Obtaining tenure

The 1928 secondary-school law was followed by the Regulation of the Secondary-Schools Teaching Personnel approved by the Parliament and published in the Official Monitor on 18 February 1929.[55] This second document regulated the way doctors obtained tenure; as Article 239 stated, all medical personnel already working in secondary schools, appointed and paid by the Ministry of Instruction, were able to get tenure provided they met the criteria of the law. Those who had served for at least three years could ask the Ministry to appoint them for special inspection; a reviser would visit the school, attend the classes, evaluate the medical office and inspect the registers and health leaflets for students before finally writing his report for the Board of General Revisers to give their resolution.

As it became obvious when analysing the Ministry of Instruction's archives, very few doctors (all men) were able to get tenure in 1929. It was not only a matter of such personnel not meeting every single criterion requested by the law, as many as there might have been; it was also a clear tendency among the central authorities. As one of the revisers appointed to evaluate and asses the request to get tenure for a female doctor put it, 'School doctors who do not have tenure do not enjoy any stability, they are being appointed and dismissed according to the Ministry's needs.'[56] This becomes clear when reading the documents: central authorities tended to avoid allowing too many doctors to get tenure at the same time because of budgetary reasons. Once a doctor got tenure, he would have been entitled to a full payment not from the School Committee, but from the Ministry itself, which represented an additional financial burden. Given the economic distress that followed the 1929 crisis, it is not entirely surprising that such practices appeared. Therefore, female doctors had to wait for the economy to recover, almost five years later, to benefit from the provisions of the law regarding tenure.

However, this longer period was not entirely the result of budgetary cuts or the way the Ministry tried to limit the number of medical professionals working in schools getting tenure. It was also because many of them had started their jobs as school doctors shortly before the law allowing them to work in secondary schools for girls was passed, in 1928. Consequently, they did not have the requested three provisional years, even though they possessed the mandatory School Hygiene diploma.

To illustrate the professional trajectory for women who decided to become school doctors, I have selected two cases that are representative for this career, made possible by the 1928 law. Medicine had been a highly respected profession since the nineteenth century, allowing those who practised it financial independence and bringing them social prestige. However, even though women were allowed to study in medical schools in Europe around the turn of the century,[57] not all of them ended up practising,[58] while others occupied subordinated positions inside the public health administration.[59] Although after the war the number of female students in medicine in Greater Romania increased and they began working in hospitals, maternities and other healthcare institutions, it was still male doctors who constituted the majority of medical personnel leading and working in the sanitary administration.[60] Therefore, when the 1928 law was passed, the recruitment of female doctors for the public health administration was still in its infancy. By then however there were two different generations of female medical graduates: women born at the end of the previous century, who had finished their studies prior to the war, and those born at the turn of the century, who had studied during the 1920s. The cases I have selected are representative of the two categories: the first doctor belongs to the first generation of female professionals, while the latter belongs to the second generation. Still, one can notice that both women had been appointed hygiene teachers and had also been granted tenure around the same year. As those moments were almost simultaneous, one can conclude that age or generational traits did not have a significant impact on the opportunities that the new law opened for female doctors. Their professional paths seemed to have depended entirely on the financial capacity of the public school administration to accommodate them.

Eliza Ștefănescu

Eliza Ștefănescu was born in 1882 and studied medicine in Bucharest, obtaining her medical diploma in 1907.[61] In 1929, she was appointed school doctor at the Elena Doamna Normal School in Bucharest, and was paid a daily wage, along with other female physicians recently designated as school doctors at different girls' high schools in the city.[62] However, she was able to pass the special inspection that allowed her to get tenure only in 1934. By then, she had transferred to the Central School Marica Brâncoveanu, one of the most prestigious and oldest high schools for girls in Bucharest.

According to the principal's recommendation, Eliza Ștefănescu had been appointed temporary school doctor in 1931; she was serving as both the boarding school's physician and hygiene teacher. Highly competent, with a spotless reputation and genuinely interested in the well-being of her students, she was willing to pay medical visits at school outside her schedule. Having a good working relationship with all her colleagues and being a respected doctor, she was praised by the principal, who highly recommended her for tenure. Similar appreciation expressed the inspector who assisted her seventh-grade hygiene course on alcoholism and smoking, valuing her ability to relate and explain the lesson to make it intelligible to her teenage students: 'I too, along with her students, have listened with great pleasure and interest to the recommendations the doctor has made ... to keep ourselves healthy.' Based on his report, there is no doubt that the inspector was truly impressed with doctor Ștefănescu's class activity and considered her worthy of getting tenure.[63]

According to the doctor's 1941 yearbook, during the war she retained her teaching position and was still working as a physician at the Central School. However, seven years later, her name was missing from the medical catalogue published in 1948, a clear sign that she had passed away, since even the retired doctors were included in the volume.[64]

Victoria Vasiliu

The second example is that of Victoria Vasiliu, born on 21 April 1896, in Botoșani, who graduated from the Faculty of Medicine in Bucharest in 1924.[65] After graduation, she passed the exam to become an intern at the Public Hospitals in Bucharest, being appointed school doctor at the Despina Doamna Normal School for Girls in Botoșani in 1928. To retain her position as a physician and a teacher, in May 1929 she graduated from the School Hygiene course of the Faculty of Medicine in Iași.

However, it was not until 1934 that Victoria Vasiliu requested to be evaluated for tenure at the boarding school where she had been serving since 1928. According to the principal's recommendation, she had shown genuine interest in students' health and well-being over the years, examining them individually each year, keeping their sanitary leaflets up to date and monitoring the boarding school's hygiene and housing conditions. Since Victoria Vasiliu had served for six years as school doctor, but only five as a hygiene teacher, the principal, Olga Savinescu, pleaded with the Ministry of Instruction to grant her tenure.

Still, for reasons that were not disclosed in the file, the evaluation did not take place until the last week of November 1936, when the designated inspector attended a seventh-grade medicine class on 'Bleeding and ways to stop it'. He evaluated the students as well prepared and appreciated the doctor's ability to use their answers to start the new lesson; she explained the procedure necessary to stop bleeding to the girls and showed them how to tie the knot, making them practise on each other's arms. The second class the inspector attended was a fifth-grade course on 'Eruptive fevers: Scarlet fever and measles'. Once again, the inspector appreciated her pedagogical skills and the clear and methodical approach of the lesson.

Her teaching was not the only target of evaluation. The inspector referred to the way she served as a doctor, taking care of the students and making sure contagious diseases were contained. He also added that she was highly regarded, in school and beyond, as she had served the high school with great dedication.[66] Twelve years later, according to the doctor's yearbook in 1948,[67] Victoria Vasiliu was still a school doctor, proving that such professional paths opened the possibility of a successful teaching career for female physicians.

Conclusions

Female professionals have been invisible in Romanian historiography. Despite the different dictionaries and encyclopaedias highlighting important women's trajectories, their presence in public sectors and their contributions have been generally overlooked. This study has tried to remedy this situation by focusing on female doctors working in schools during the 1920s and 1930s and their impact on the Romanian society. By analysing the legislation and institutional transformations generated by World War I, while also following the biographies of these female physicians, I have been able to show that the intersection of education and medicine opened new career paths for women.

School medicine in interwar Romania was an important development that helped authorities to expose young generations to medical discourse and ensure they would willingly subject themselves to medical examination. These developments were the outcome of changes in social assistance policy after the war. The human body was no longer considered private, since it was destined to be examined, measured and evaluated on a regular basis to be healthy, so that it could be productive and capable of reproduction. In the end, what was really at stake for school authorities was the need to determine students to obey the rules and be accountable for their own health and well-being, as a necessary condition for becoming responsible citizens.

At the same time, the obligation of secondary girls' schools to hire female doctors represented a major opportunity for women. Despite the initial tensions and hostility they faced in replacing men, they ended up occupying the positions they were entitled to. This process of substitution was made possible by their interaction with medical science. The opportunity to study medicine and their determination to obtain their diplomas did not lead to research-focused scientific careers, but as promoters of a new type of relation of high-school students with their own bodies, one based on sanitary principles. Even though women were not involved in the drafting of the medical assistance policy which promoted this culture of health, they served as dissemination and control agents, thus actively contributing to state efforts to modernize society. Their professional competence and dedication were of utmost importance in this role, as demonstrated above.

Notes

1 Y. Knibiehler and C. Fouquet, *La femme et les médecins: Analyse historique* [*The Woman and the Doctors: Historical Analysis*] (Paris: Hachette, 1983).
2 S. C. Martin and R. M. Arnold, 'Ruth M. Parker, gender and medical socialization', *Journal of Health and Social Behaviour*, 29: 4 (1988), 333–43.
3 C. G. Borst and K. W. Jones, 'As patients and healers: The history of women and medicine', *OAH Magazine of History*, 19: 5 (2005), 23–26.
4 D. R. Mandelbaum, 'Women in medicine', *Signs*, 4: 1 (1978), 136–45; M. W. Carpenter, *Health, Medicine and Society in Victorian England* (Santa Barbara, CA, Denver, CO and Oxford: Praeger, ABC Clio, 2009), pp. 166–75; T. Appel, 'Writing women into medical history in the 1930s', *Bulletin of the History of Medicine*, 88: 3 (2014), 457–92; K. Jensen, 'The "open way of opportunity": Colorado women physicians and World War I', *Western Historical Quarterly*, 27: 3 (1996), 327–48.
5 H. P. Freidenreich, 'Jewish women physicians in Central Europe in the early twentieth century', *Contemporary Jewry*, 17: 1 (1996), 79–105. N. M. Theriot, 'Women's voices in nineteenth-century medical discourse: A step toward deconstructing science', *Signs*, 19: 1 (1993), 1–31.
6 L. Trăușan-Matu, 'The doctor and the midwife: A study of two medical professions in the Romanian society of the 19th century (1831–1874)', in C. Bărbulescu and A. Ciupală (eds), *Medicine, Hygiene and Society from the Eighteenth to the Twentieth Centuries* (Cluj-Napoca: Mega, 2011), pp. 94–99. See also Edgerton-Tarpley's chapter in this volume.
7 N. Roman, *'Deznădăjduită muiere n-au fost ca mine': Femei, onoare și păcat în Valahia secolului al XIX-lea* ['*There Has Never Lived a Hopeless Woman Such as Myself': Women, Honour and Sin in Nineteenth-Century Wallachia*] (Bucharest: Humanitas, 2016), pp. 191–211.
8 Ministerul Instrucțiunii și Cultelor, *Anuarul oficial întocmit de Serviciul statelor personale și al statisticii la 15 octombrie 1910* [*The Ministry of Instruction and Cults, Official Yearbook of the Personnel and Statistics Department, 15 October 1910*] (Ploești: Institutul de Arte Grafice Progresul Societate Anonimă, 1910), pp. 376, 386–87.

9 A. Ciupală, *Bătălia lor: Femeile din România în Primul Război Mondial* [*Their Battle: Women in World War I Romania*] (Iași: Editura Polirom, 2017), pp. 159–71; M. Trancu Rainer, *File de jurnal: Mărturii* [*Leaves from a Journal: Confessions*] (Bucharest: Paul Editions, 2021), pp. 67–104.

10 Arhivele Naționale Istorice Centrale [National Archives of Romania, Central Branch, hereafter ANIC], Ministry of Public Health, Labour and Welfare, file 72/1923, pp. 227, 238.

11 The Ministry of Cults and Public Instruction has been the designated name since the establishment of the institution in 1859, when Alexandru Ioan Cuza was elected ruler (*domn*) of the Romanian Principalities. Cults or religious affairs were attached to it because they reflected the Church's involvement in education, prior to the establishment of the Romanian national state. However, after World War I, Instruction was separated from Cults, which from 1921 until 1930 became the Ministry of Cults and Arts. In 1930, it was merged again with the Ministry of Instruction. In November 1936, the Ministry of Instruction changed its name to the Ministry of Education, while Cults and Arts became an independent department with a different minister. S. Neagoe, *Istoria guvernelor României de la începuturi – 1859, până în zilele noastre – 2012*, ediția a III-a [*The History of Romanian Governments from the Beginning – 1859, Until Today – 2012*] (Bucharest: Machiavelli, 2013), pp. 95–139.

12 *Lege asupra învățământului secundar și superior (Sancționată prin înaltul decret regal Nr. 1.097 din 23 martie 1898, publicată în Monitorul Oficial Nr. 283 din 24 Martie 1898)* [Law on Secondary and Higher Education Sanctioned by Royal Decree no. 1097, March 1898, published in the Official Monitor No. 283, 24 March 1898] in *Antologia legilor învățământului din România* [*Anthology of Romanian Education Laws*] (Bucharest: Institutul de Științe ale Educației, 2004), p. 72; *Lege asupra învățământului primar și normal-primar (Sancționată prin înaltul decret regal Nr. 2.199 din 29 Aprilie 1896 și publicată în Monitorul Oficial nr. 24 din 30 Aprilie 1896)* [Law on Primary Education and Normal Education], in *Antologia legilor învățământului din România* [*Anthology of Romanian Education Laws*] (Bucharest: Institutul de Științe ale Educației, 2004), p. 67.

13 *Programe de studii pentru școlele secundare (licee, gimnazii și școle secundare de fete gradul I și II)* [Secondary Schools Curricula (High Schools, Middle Schools and Secondary Schools for Girls] (Bucharest: Imprimeria statului, 1899), pp. 187–97, 232–33.

14 P. Guillame, 'L'hygiène a l'école et par l'école' [Hygiene in school and through school], in P. Bourdelais and O. Faure (eds), *Les nouvelles pratiques de santé: Acteurs, objets, logiques sociales (XVIIIe-XXe siècles)* [*New Health Practices: Actors, Objects and Social Logic*] (Paris: Belin, 2005), pp. 213–26.

15 E. Grecu, *Azilul Elena Doamna și ajutorul domnesc dat orfanilor* [*The Elena Doamna Asylum and the Prince's Assistance to the Orphans*] (Bucharest: Editura Casa Școalelor, 1944), appendix 25, verso. This was the case with the Elena Doamna Asylum during the year 1882–83, when the students attending the normal school section were studying hygiene with a male doctor. In 1892,

at the Central School for Girls in Bucharest hygiene was being taught by a male doctor as well. Serviciul Municipiului București al Arhivelor Naționale, Fond Școala Centrală, dosar 190/1934 [Bucharest Municipal Service of the National Archives of Romania, Central School Collection, file 190/1934], unnumbered *Teaching personnel of the Central School for Girls in Bucharest.*

16 *Anuarul statistic al României* [*Romanian Statistical Yearbook*] (Bucharest: Imprimeria Statului, 1912), p. 465. This volume includes data for the years 1900–8, for the Faculty of Medicine at the Universities of Bucharest and Iași.

17 *Anuarul învățământului secundar din România pe anul 1924–1925 precedat de o expunere asupra dezvoltării învățământului secundar și a situației sale în anul școlar 1924–1925 întocmită de Const. Kirițescu* [*Romanian Secondary Education Yearbook 1924–1925 with an Introduction on the Development of Secondary Education and Its Situation during the School Year 1924–1925 by Const. Kirițescu*] (Bucharest: Tipografia Curții Regale F. Göbl Fii, 1925), pp. 239, 242. Teachers such as Ermina Kaminski, who graduated from the Medical School in Bucharest in 1889 and in 1925 was teaching natural sciences at the Regina Maria Secondary School for Girls, and Cornelia Kernbach, who graduated in Iași and was teaching hygiene at the Central School in Bucharest, from 1895 until 1924.

18 *Universitatea din București 1925–1926* [*The University of Bucharest 1925–1926*] (Bucharest: Institutul de Arte Grafice Răsăritul, 1927), pp. 206–9. In 1925–26, the number of female graduates at the Faculty of Medicine in Bucharest was 59, out of a total of 253 graduates. *Anuarul Universității din București pe anul școlar 1912–1913. Al XXI-lea anuar publicat de secretariatul Universității* [*The University of Bucharest Yearbook 1912–1913: The Twenty-First Yearbook Published by the Secretary of the University*] (Bucharest: Tipografia Profesională Dim. C. Ionescu, 1913), pp. 91–93. In 1913, out of the sixty-two graduates in Medicine, there were only fifteen female students.

19 A. Bonea, ' "Contagion by telephone": Print media and knowledge about infectious diseases in Britain, 1880s–1914', *Technology and Culture*, 62: 4 (2021), 1063–86.

20 M. Turda, *Eugenism și modernitate: Națiune, rasă și biopolitică în Europa (1870–1950)* [*Eugenics and Modernity: Nation, Race and Biopolitics in Europe (1870–1950)*] (Iași: Polirom, 2014), pp. 57–68.

21 C. Odeseanu, *Cartea femeii moderne: Ce trebuie să știe o femeie și chiar o fată* [*Modern Woman's Book: What Women and Even Girls Need to Know*] (Bucharest: Editura Cartea Românească, 1934), pp. 26–34.

22 *Lege pentru învățământul primar al statului (școli de copii mici, școli primare, școli și cursuri pentru adulți, școlile și clasele speciale pentru copii debili și anormali-educabili) și învățământul normal-primar* [*Law on Primary Public Education (Kindergartens, Primary Schools, Adult Schools, Schools and Special Classes for Mentally Impaired Children) and Normal Education*] (Bucharest: Cartea Românească, 1925), pp. 285–87.

23 *Legea învățământului secundar din 1928 precedată de expunere de motive* [*Law on Secondary Education in 1928 with an Explanatory Statement*] (Bucharest: Imprimeriile Statului, 1928), pp. 50–52.
24 B. Andrieu, 'Corps' [The Body], in B. Andrieu (ed.), *Le dictionnaire du corps en sciences humaines et sociales* [*Dictionary of the Body in Social and Human Sciences*] (Paris: CNRS Editions, 2006), pp. 103–104.
25 C. Zavarache, '"Numărul mare de copii … așteaptă o asistență medicală serioasă și la vreme făcută." Consecințele primei conflagrații mondiale asupra serviciului medical școlar din România' ["The Great Number of Children … Are in Need of Serious and Timely Medical Care." The Consequences of World War I on Medical Care in Schools in Romania], in C. Mihalache and N. Roman (eds), *Copilării trecute prin război: Povești de viață, politici sociale și reprezentări culturale în România anilor 1913–1923* [*Childhood in War: Life Stories, Social Policies and Cultural Representations in Romania, 1913–1923*] (Iași: Editura Universității Alexandru Ioan Cuza, 2020), pp. 192–93.
26 *Programele analitice ale învățământului secundar (Licee, gimnazii și clasele I-III ale școlilor normale) întocmite în conformitate cu Legea Învățământului secundar din 1928* [Secondary Schools Curricula (High Schools, Middle Schools and Ist to IIIrd Classes in Normal Schools) Written in Accordance with the Law on Secondary Education in 1928] (Bucharest: Imprimeriile Statului, 1929), pp. 12–13, 177–78.
27 P. K. Gilbert, *The Citizen's Body: Desire, Health and the Social in Victorian England* (Columbus: The Ohio State University Press, 2007), pp. 3–6.
28 ANIC, Fond Ministerul Cultelor și Instrucțiunii Publice [Ministry of Cults and Public Instruction Collection, hereafter MCIP Collection], file 308/1927, p. 280.
29 *Regulament provizor pentru școalele secundare (licee de băieți, de fete și gimnazii)* [Provisional Regulation for Secondary Schools (High Schools for Boys and Girls, Middle Schools)] (Sibiu: Tiparul Tipografiei Arhidiecezane, 1919), pp. 12–22, 39.
30 This was the case at the newly founded Domnița Anca Normal School for kindergarten teachers in Brașov, where during 1920–22 hygiene was taught by Dr Nicolae Căliman, who was also the school's physician and who was appointed hygiene inspector in Brașov in 1921. *Anuarul Școalei Normale de Conducătoare de grădini de copii din Brașov, pe anii școlari 1919, 1920, 1920–1921 și 1921–1922* [*The Normal School for Kindergarten Teachers in Brașov Yearbook, Years 1919, 1920, 1920–1921 and 1921–1922*] (Brașov: Tipografia Unirea, 1922), pp. 13, 15.
31 See 'Institutul de sănătate publică București' [The Public Health Institute in Bucharest], in N. Ursea, *Enciclopedie medicală românească: Secolul XX* [*The Romanian Medical Encyclopaedia: Twentieth Century*], vol. 4, (Bucharest: Universitatea de Medicină și Farmacie Carol Davila, 2001), pp. 929–32.
32 ANIC, MCIP Collection, file 329/1929, pp. 17–25.
33 ANIC, MCIP Collection, file 329/1929, pp. 73–79.
34 ANIC, MCIP Collection, file 329/1929, p. 173.

35 *Revista de Igienă Socială* [*Journal of Social Hygiene*], I: 1(January 1931), 61–63.
36 ANIC, MCIP Collection, file 260/1931, pp. 3–9.
37 ANIC, MCIP Collection, file 329/1929, pp. 229–30.
38 ANIC, MCIP Collection, file 329/1929, pp. 227–28.
39 D. Sdrobiș, *Limitele meritocrației într-o societate agrară: Șomaj intelectual și radicalizare politică a tineretului în România interbelică* [*The Limits of Meritocracy in an Agrarian Society: Intellectual Unemployment and Political Radicalisation in Interwar Romania*] (Iași: Polirom, 2015). The author has written extensively on the struggles university graduates faced in the context of the global economic crisis.
40 *Dezbaterile Adunării Deputaților, Ședința de sâmbătă 21 iunie 1924, Adunarea Deputaților Sesiunea prelungită 1923–1924* [The Debates of the Chamber of Deputies, The Meeting on Saturday, 21 June 1924, The Deputies Assembly, Prolonged Session 1923–1924] in *Monitorul Oficial*, no. 111, 30 July 1924, p. 3183.
41 ANIC, MCIP Collection, file 329/1929, pp. 165–72.
42 ANIC, MCIP Collection, file 329/1929, pp. 286, 291.
43 ANIC, MCIP Collection, file 329/1929, p. 257.
44 ANIC, MCIP Collection, file 329/1929, pp. 183–86.
45 ANIC, MCIP Collection, file 329/1929, p. 222.
46 ANIC, MCIP Collection, file 329/1929, pp. 236–50. Her CV was quite impressive. She had been working as an eye doctor, had experience as a school doctor and teacher and had worked in gynaecology offices. She had also published different sanitary articles for broader audiences.
47 ANIC, MCIP Collection, 329/1929, pp. 83–85.
48 C. Odeseanu, *Cartea femeii moderne: Ce trebuie să știe o femeie și chiar o fată* [*Modern Woman's Book: What Women and Even Girls Need to Know*] (Bucharest: Editura Cartea Românească, 1934), pp. 19–21. She quoted numerous examples of female patients who had given birth multiple times but were still reluctant to let themselves be examined by a gynaecologist; also, the author highlighted the erroneous information even educated patients had regarding their reproductive anatomy and physiology.
49 I. Chrisopol, *Curs de igienă pentru școlile profesionale și cele de ucenici din atelierele și depourile căilor ferate* [*Hygiene Manual for Professional Schools and Apprentice Schools Associated with Railway Workshops and Depots*] (Bucharest: Tipografia Cultura, 1925), p. 61. Students who were trained in professional institutions were given free access to this type of information, precisely because of the way they were regarded: coming from rural regions or peripheral neighbourhoods, they were seen as prone to early, unsafe sexual encounters and therefore they needed to be warned about the dangers of venereal diseases.
50 ANIC, MCIP Collection, file 329/1929, p. 133 front and back.
51 ANIC, MCIP Collection, file 329/1929, p. 192.
52 ANIC, MCIP Collection, file 329/1929, pp. 284, 266.

53 ANIC, MCIP Collection, file 329/1929, p. 217.
54 *Anuarul Învățământului secundar din România pe anul 1924–1925, Precedat de o expunere asupra desvoltării învățământului secundar și a situației sale în anul școlar 1924–1925 întocmită de Constantin Kirițescu* [Romanian Secondary Education Yearbook 1924–1925, With an Introduction on the Development of Secondary Education and Its Situation during the School Year 1924–1925 by Const. Kirițescu] (Bucharest: Tipografia Curții Regale F. Göbl Fiii, 1925), XL–XLIV. In 1924–25 there were 116 boys' high schools and 75 gymnasiums throughout the entire country; at the same time, there were only 63 high schools and secondary schools and just 25 middle schools for girls.
55 *Regulamentul personalului didactic al școalelor secundare* [Regulations for Secondary Schools Teaching Personnel], *Monitorul Oficial*, no. 39, 18 February 1929, p. 1429.
56 ANIC, MCIP Collection, file 329/1929, pp. 20–21, 121–28, 180.
57 H. P. Freidenreich, 'Jewish women physicians in Central Europe in the early twentieth century', in *Contemporary Jewry*, 17: 1 (1996), 80.
58 Cornelia Brădiceanu, wife of the important Romanian poet and philosopher Lucian Blaga, had studied medicine in Vienna and Cluj; however, after marriage she only practised dentistry for a while, before devoting herself to Blaga's work. D. Blaga, *Tatăl meu, Lucian Blaga* [My Father, Lucian Blaga] (Bucharest: Editura Humanitas, 2015), pp. 63–64.
59 Looking at the doctors who were publishing articles on their work as surgeons at the Maternity Institute, in the *Revista de obstetrica, ginecologie și puericultură* in 1925, one can easily notice that they were predominantly male, women being a minority. *Revista de obstetrica, ginecologie și puericultură* [Journal of Obstetrics, Gynaecology and Puericulture], 1 (January–February 1925), see table of contents.
60 Archives of the Public Health Ministry, file 5/1927 Decorations. The file includes lists and royal decrees regarding the persons who had contributed to the successful implementation of sanitary legislation and were recommended to be decorated with the Cross of Sanitary Merit. Most of them were male doctors, civilian or military, sanitary agents, accountants and pharmacists, local administrative officials, Romanian but also foreign. Women were also on the list but as social workers, members of the philanthropic societies or spouses of male professionals being decorated, midwives; only a minority were actual doctors.
61 *Anuarul General al medicilor din România 1941 după datele recensământului din 15 martie 1941* [The 1941 Doctors in Romania General Yearbook Compiled after the Data Collected during the 15 March 1941 Census] (Bucharest: Tipografia Universul, 1941), p. 262.
62 *Anuarul General al medicilor din România 1941* [The 1941 Doctors in Romania General Yearbook], p. 287.
63 Serviciul Municipiului București al Arhivelor Naționale, Fond Școala Centrală, dosar 190/1934, nenumerotat [Bucharest Municipal Service of the National Archives of Romania, Central School Collection], file 190/1934, unnumbered.

64 *Anuarul medicilor 1948 cu un index al medicilor primari* [*The 1948 Doctors' Yearbook with an Index of All Primary Medical Personnel*] (Bucharest: Imprimeria Centrală, 1948), p. 506.
65 *Anuarul general al medicilor din România 1941 după datele recensământului din 15 martie 1941* [*The 1941 Romanian Doctors' General Yearbook Compiled after the Data Collected during the 15 March 1941 Census*] (Bucharest: Tipografia Universul, 1941), p. 46.
66 ANIC, MCIP Collection, file 330/1934, pp. 248–54.
67 *Anuarul medicilor 1948* [*The 1948 Doctors' Yearbook*], p. 64.

7

Women and the practice of Western medicine in late Republican China: Evidence from Sichuan

Jean Corbi

In a text published in 1948 under the title 'Liu Yunbo, female physician', the Chinese poet and essayist Zhu Ziqing provides a portrait of the young woman, whom he probably met in Chengdu during the war against Japan (1937–45):

> Liu Yunbo is a female obstetrician and gynaecologist in Chengdu. She has been practising medicine in Chengdu for ten years. She opened the Hongji Hospital by herself. During the war of resistance, she also served as the director of obstetrics and gynaecology at the Chengdu Central Military Academy Hospital and as the director of obstetrics and gynaecology at Chengdu Municipal Hospital. After the victory, the Military Academy Hospital was demobilized to Nanjing. As she could not be in both places she reluctantly had to stop working in this hospital. Last year, she also served as the president of the Chengdu Superior Medical Vocational School. I wrote this résumé and saw that she was a busy person.[1]

Liu Yunbo was born in the district of Suining, in central Sichuan, in 1905. She completed her secondary education in a Christian school in Shanghai and moved to Germany in 1921 to study medicine. As with other Chinese women who studied Western medicine or science abroad, her proximity to foreign missionaries in China facilitated international mobility. In 1935, she earned a doctoral degree from the University of Jena. She then returned to China and briefly practised in Shanghai before the war started. After the Japanese invasion of China and the fall of Nanjing in December 1937, the nationalist government of Chang Kai-shek retreated to Chongqing, in the province of Sichuan, where it remained until the end of the war. The occupation of the eastern part of the country by Japanese troops led to the relocation of many government-related institutions, followed by many in the educated elite. Thus, after 1937, like Zhu Ziqing – who was a teacher at the Tsinghua University in Beijing – Liu Yunbo left the Japanese-occupied territory and settled in south-western China. In Chengdu, the capital of

Sichuan, she founded her hospital while assuming responsibilities in many other medical institutions in the city. Some of these institutions played a vital role in the development of Western medicine in the province.[2] For example, she headed the Chengdu Superior Medical Vocational School, which was the first non-missionary school in the province to train nurses and midwives.

The Chengdu Superior Medical Vocational School was part of a broader strategy of promoting modern childbirth and childcare that became central during the war against Japan, as Nicole Elizabeth Barnes shows.[3] It was part of an even broader intellectual trend that linked the promotion of Western science and Western medicine to the strengthening of the nation, defined in biological and racial terms.[4] From this perspective, developing modern, scientific obstetrics and midwifery and educating mothers in hygiene was indispensable to the strengthening of the Chinese nation.[5] This wartime strategy sought to continue the public health efforts initiated during the 1927–37 period: the school, in a certain way, built on the experience of other institutions, such as the First National Midwifery School, set up in 1929 in Peking by another woman, Yang Chongrui.[6] Yang had graduated from Peking Union Medical College (hereafter, PUMC) and had also studied at Johns Hopkins University. She participated in the foundation of the National Midwifery Board and worked to improve childbirth practices in China. During the war she followed the PUMC, moving from Beijing to Sichuan. Although mainly committed to war orphan relief, she published several texts on maternal and child healthcare, including midwifery teaching materials.[7]

Recreating a public health system in Sichuan like the one built between 1927 and 1937 in the east was obviously a challenge. The constraints of war limited the resources available for such a plan. Moreover, the health facilities in Sichuan were much less advanced than in the east of China. According to the main architect of wartime public health policy in Sichuan, C. C. Chen, 'no progress in health matters had been made [between 1911 and 1939]'[8]. A major difficulty was the lack of Western-style physicians, medical workers trained in modern techniques and modern hospitals, making people such as Liu Yunbo even more valuable. Thus, the knowledge, skills and experience of highly trained women physicians such as Liu Yunbo and Yang Chongrui were undoubtedly very important for the modern medical services the government sought to set up in Sichuan during the war.

These women thus fully participated in the propagation of 'science' in China. During the Republican period, the term 'science' became increasingly important, even omnipresent, in Chinese society.[9] It encapsulated the promise of modernity and the possibility for China to rise to the rank of 'advanced' nation.[10] Yet this discourse on science and modernity most often placed women among those who needed to be educated in science and very rarely among the educators themselves. Science for women generally derived from the gendered role they were assigned to, as mothers or homemakers, focusing on health, hygiene and domestic science.[11] In the field of medicine, women were more easily depicted as midwives who needed to be trained in scientific techniques. But the important role women physicians played in educating modern midwives – which this chapter will evidence – has been far less visible.

Despite this, were it not for Zhu Ziqing's text, Liu Yunbo's activities during the war against Japan would have likely remained invisible, which is still the case for many women physicians in early twentieth-century China, in Sichuan and elsewhere. Indeed, the brilliant career of some women about whom we have relatively precise information should not prevent us from considering female physicians as a group that was by no means marginal within the medical profession.[12] Moreover, while most of the studies cited consider female doctors separately from their male counterparts, our approach in this chapter is fundamentally comparative, enabling us to highlight certain gender-related specificities in study and career paths. This comparative dimension seems essential to understanding and explaining the lesser visibility of Chinese women doctors. Finally, we intend to show that the feminization of the medical profession was not limited to the eastern part of the country, on which most of the historiography focused, but it also applied to the case of inland provinces such as Sichuan. Of course, such a regional focus implies that our findings may not necessarily be generalized to the rest of the country, but it allows the historian to reconstruct a picture of the profession in its full diversity.

The problem of the invisibility of some physicians is not specific to women, although it is more acute for them. Documentation is usually rich for a few famous physicians, due to their positions or publications, while the majority remain anonymous.[13] Female physicians are no exception, especially since they seemed less likely to rise to prominence. What makes this imbalance particularly problematic for them is that it obscures their position and contribution to the expansion of Western medicine in China. We know that women acquired more social visibility in the final decades of the Qing dynasty (1644–1912) and during the Republican period (1912–49).[14] They entered public spaces, as Wang Di demonstrates for Chengdu: 'In the early twentieth century, women in Chengdu were increasingly showing up in places that were traditionally for men only, such as teahouses and theatres.'[15] They also gained access to educated professions becoming teachers or doctors, which embodied the figure of this 'new woman'. Medicine thus opened new career opportunities for Chinese women. We may wonder in return how the presence of these women affected the history of modern medicine in China. What percentage of the medical profession did women represent? What position did women occupy among physicians? Were there specific, gendered career profiles or specialties in the medical field?

To answer these questions, we will draw on the vast documentation produced by the public health administration in the 1930s and 1940s, especially the Sichuan Provincial Health Administration. In 1929, soon after the creation of the first Ministry of Health, the Nanjing government legislated that all practitioners of Western medicine had to be registered.[16] Although similar regulations had been implemented locally before 1929, it was the first time that all Western-trained physicians had to register to be allowed to practice.[17] This registration process generated an unprecedented mass of documents providing rather detailed information on virtually every Western-style physician in the country.[18] These new documents provide a more comprehensive view on the profession of 'Western-style physician' compared to previous studies that have tended to focus on some individuals or specific institutions.[19] This is particularly true for the provinces of inland China, some of which have remained poorly documented.[20]

Women as Western-style physicians

The proportion of women among Chinese physicians was far from negligible, especially in the case of Western medicine, which attracted proportionally more women than Chinese medicine. The link between missionary and philanthropic works, the circulation of Western scientific knowledge and the education of women resulted in the rapid feminization of the profession in the province, so much so that on the eve of World War II, the share of women among Sichuan's Western-style physicians was comparable to that of the United States.

Overall, among both Chinese-style and Western-style physicians, women made up a minority of Sichuanese practitioners. Out of more than 2,300 physicians whose registration files were archived by the provincial administration between 1937 and 1947, only 60 (2.6 per cent) were women (Table 7.1). However, there is a stark difference concerning the proportion of women physicians practising Western and Chinese-style medicine. Chinese-style physicians constitute a large majority of the total number of physicians registered but of these only 1.5 per cent are women, compared to 11 per cent for Western-style physicians.

This disparity indicates that Western medicine provided more professional opportunities for women than Chinese medicine. This was partly due to the situation of women within Chinese medicine. State-sanctioned registration of Chinese-style physicians favoured orthodox, literati medicine, based on the knowledge of the Chinese medical classics. In this regard, female healers

Table 7.1 Physicians registered in Sichuan, 1937–47

Type of medicine	Western style (*xiyi*)	Chinese style (*zhonyi*)	Total
Women	31 (9%)	29 (1.5%)	60 (2.6%)
Men	308 (89%)	1,884 (96%)	2,192 (94.9%)
Unknown gender	8 (2%)	49 (2.5%)	57 (2.5%)
Total	347 (100%)	1,962 (100%)	2,309 (100%)

Source: SPA, 民113 (Provincial Department of Hygiene). Read: Out of 353 Western-style physicians, 9 per cent (31) were women. Fisher test result (for 'Women' and 'Men') < 0.01 (significant).

suffered from long-term marginalization from literati physicians.[21] The modes of transmission of Chinese medicine – master to apprentice, father to son, uncle to nephew and so on – tended to exclude women from literati medicine, although there were some exceptions. This does not mean that there were no women healers. They were rather involved in more informal, popular healing techniques, especially midwifery, and were heavily criticized by literati physicians.[22] As demonstrated by Angela Ki-che Leung, from the Song dynasty (960–1279) to the Qing dynasty (1644–1911), increasingly stricter sex segregation made female healers indispensable to treat wives, concubines or daughters; but the more indispensable they became, the more distrusted they were. As a consequence, it is likely that most of these female healers did not meet the criteria for official registration as Chinese-style physicians.

On the other hand, Western medical education may have been more welcoming for women from an educated social background. In the early twentieth century, many of the first universities and medical schools were affiliated to protestant missions, which made significant efforts to educate women.[23] This was partly due to the presence of women among Western missionaries themselves, in Sichuan as in the rest of the country. A branch of the Women's Missionary Society was active in the province from the end of the nineteenth century. One of its members, Retta Gifford, established a hospital for women and children in the city of Chengdu in 1896.[24] She then participated in the organization of the West China Union University (WCUU, *Huaxi xiehe daxue*) which was founded in 1914 and later became one of the major institutions for higher education in the province until the foundation of the People's Republic of China. According to Karen Minden, 579 students graduated from the WCUU College of Medicine and Dentistry between 1920 and 1949.[25] Among the thirty-one women physicians registered in the province between 1937 and 1947, twenty-two were born in

Sichuan, and nine had graduated from the WCUU College of Medicine. Higher-education institutions supported by missionary organizations, such as the WCUU, or foreign philanthropists, such as the Peking Union Medical College sponsored by the Rockefeller Foundation, recruited a fair share of their students from missionary schools. These schools pioneered education for girls in China: in Sichuan, in 1919, the 109 Canadian Methodist Mission lower-primary schools taught 2,911 boys and 1,245 girls.[26] According to Paul Bailey, in 1923, girls represented only 5.56 per cent of students in lower-primary schools.[27] Thus, the link with missionary education, among other factors, may explain why Western medicine was relatively more accessible to Sichuanese women than Chinese medicine.

It is difficult to know if the proportion of women practitioners obtained through the analysis of registration files is representative of the overall situation in Sichuan at the time. The figure of one tenth of women physicians (Table 7.1) is, however, consistent with numbers drawn from other sources. One of these sources is a series of 'surveys of hospitals and clinics' conducted by the provincial administration in 1938–39 in ninety-three of its districts (*xian*).[28] Hospitals were usually bigger and included accommodation to allow patients to stay for at least several days. On the contrary, clinics were generally designed to treat patients over a shorter period, usually within a day. While clinics rarely had more than one physician – the director of the clinic – and a maximum of two, most hospitals had at least two physicians. Out of these 93 districts, 49 declared no hospital or clinic, and 41 declared at least one establishment, for a total of 219 establishments.

The majority (158) of these institutions only had one physician – their director – which explains why the number of directors was much

Table 7.2 Physicians practising in the hospitals and clinics surveyed in 1939

Position in the hospital/clinic	[Employee] Physician (*yishi*)	Director (*yuan/suozhang*)	Total no. of physicians
Women	28 (33%)	9 (4%)	37 (12%)
Men	57 (67%)	202 (96%)	259 (88%)
Total	85 (100%)	211 (100%)	296 (100%)

Source: SPA, 民113-01-0023, 113-01-0296, 113-01-0511, 113-01-1730. Read: Out of 86 physicians employed in the hospitals and clinics surveyed in 1939, 33 per cent (twenty-eight) of them were women. Fisher test < 0.01 (significant).

For four hospitals and clinics, the gender of the director is unknown. Another four establishments do not have any physician, their director being a nurse or midwife. Two of these are headed by women (both midwives), and the two others are headed by men (both nurses).

higher than that of employee physicians. Forty-two establishments had one employee physician and sixteen had more than one. The total number of physicians practising in the hospitals and clinics surveyed is close to a tenth. Although the registration files and the survey of hospitals and clinics together do not encompass all the physicians practising in the province, they do provide useful and consistent statistical insights into the proportion of women physicians in the province during the wartime period.

The number of female physicians registered in Sichuan increased significantly between the early 1930s and the wartime period (Table 7.3). While there are only 5 women out of 139 physicians registered in Sichuan from 1929 to 1934, the situation between 1937 and 1947 became similar to the national gender distribution. This statistical gap may be explained by two factors. First, the influx of physicians from the rest of the country, especially the eastern provinces and cities, brought the gender distribution of Sichuanese physicians closer to the national level. Some came from other provinces: out of thirty-four female physicians, twelve were born outside of Sichuan. All came from provinces that were occupied by Japan in 1938, whether entirely (Jiangsu, Zhejiang, Liaoning, Shandong, Anhui, Hebei) or in part (Guangdong, Fujian). It is often impossible to know precisely when they moved to Sichuan. When this information is available, it indicates that they moved at the beginning of the war. Other female physicians were born in Sichuan but had left the province during their studies and graduated from institutions located outside the province or in foreign countries: this was the case for seven of the twenty-two women born in Sichuan. Again, they most likely went back to Sichuan during the war. All, except Liu Yunbo, studied in Shanghai.

For example, Deng Xiaoli was among the first Sichuanese women to earn a medical degree outside the province (according to the 1937–47 registration files). She graduated from Nanyang University in Shanghai in 1935.[29] During the war, she took the position of intern in the Red Cross Hospital

Table 7.3 Physicians registered between 1929 and 1934

Age group	20–29	30–39	40–49	50–59	60–69	Total
Women	254	234	89	25	3	605 (Sichuan: 5)
Men	1,524	3,020	1,071	224	26	5,865 (Sichuan: 134)
Total	1,778	3,254	1,160	249	29	6,470 (Sichuan: 139)

Source: *Internal Affairs Statistical Yearbook (Neizheng nianjian)* (Shanghai: Shangwu yinshuguan, 1936), G134–35.

in Shanghai and settled in Chongqing in 1937, where she worked at the municipal hospital. Second, the training of female physicians began later in Sichuan. In Shanghai, the entry of women into medical education began with the May Fourth Movement which played a crucial role in legitimizing the presence of women in the public sphere.[30] The same happened in Sichuan a decade later. The first woman trained at the WCUU, Yue Yicheng, entered the university in 1924 and graduated in 1932.[31] Among the physicians registered during the wartime period, she was the first to be trained in Sichuan, while many others graduated in the 1930s. This trend appears clearly in the WCUU archives. Among the alumni of the WCUU Medical College, Yue Yicheng was the only woman to join before 1935. From 1935 to 1940, 27 women joined the network and 38 did so between 1941 and 1945, representing more than a quarter of the total 232 new alumni over the 1935–45 period.[32]

Thus, by the late 1930s and early 1940s, the percentage of registered women physicians in Sichuan had become comparable to that of the eastern provinces, such as Jiangsu or Zhejiang, a few years before. It was higher than the percentage of women physicians in the United States during the same period (4.4 per cent of women physicians in 1930) and matched or even exceeded that of US cities such as Boston (8.7 per cent in 1930), Chicago (7.5 per cent), Los Angeles (9.5 per cent) or New York (5.5 per cent).[33] Of course, in absolute numbers, female doctors were more numerous in each of these cities than in the whole province of Sichuan. Moreover, since statistical data in China are less precise that in the US, we must not overstate this comparison. Above all, this shows that the feminization of Western medicine in China was already significant and increasing, while in the United States it has been decreasing since 1870.

Gendered medical careers

At the same time, some medical professions witnessed a rapid feminization during the first half of the twentieth century. While the first nurses trained by missionaries were primarily men, the number of female nurses increased quickly in the 1920s and 1930s.[34] On the other hand, modern midwifery had been, since the birth of the profession, mainly women's business.[35] Both midwifery and nursing were thus far more accessible for women. Because of this, Tina Johnson considers that 'Chinese women entered the medical field, often as low-level technicians.'[36] While it is true that women were more oriented towards paramedical or technical professions, this should not lead to the assumption that women physicians were trained in a different, shorter or less prestigious way than men.

A closer look at the profile of women within the profession of Western-style physicians shows that the differences between the medical education of men and women were relatively small. At a time when the profession of 'Western-style physician' was in considerable flux and no standard medical curriculum existed in China, women physicians, as a group, were in no way less educated than men. However, a medical gender gap did grow once they started practising, as women seemed to occupy overall lower professional positions than men. It is possible to draw a comparison between female and male physicians registered in Sichuan using the registration files.[37] Women were slightly younger than male physicians – the average birth date for women is 1908 and 1905 for men. This gap illustrates the later entry of women into the profession. While twelve male physicians were born before 1900, only one woman was born before that date.

The medical curriculum for women is quite similar to that of male physicians. Women even studied medicine a little longer than their male counterparts. The registration files show that most of the physicians declaring four years of medical studies or less were men (Table 7.4). On the contrary, the difference between men and women regarding the duration of medical studies becomes less significant when considering longer durations, for example, six years (Table 7.5). In this case, there is no longer any significant difference between women and men. This means that the key factor in this discrepancy is that men were more likely to graduate from a short-duration medical curriculum. Indeed, men had access to some shorter medical curricula from which women were apparently excluded.

The institutional organization of medical education in China is still the subject of much debate, including in government circles. Medical education was divided into many different schools, each with its own study programme and different admission requirements. To standardize the training of physicians, the Ministry of Education created a Commission on Medical

Table 7.4 Male physicians more likely to graduate from a short-term medical course

Duration of medical studies	Four years and less	More than four years	Unknown	Total
Women	2	21	8	31
Men	15	22	7	44
Total	17	43	15	78

Source: Sichuan Provincial Archives registration files (1937–47). Fisher test (first two columns) < 0.01(significant)

Table 7.5 Women as likely as men to graduate from a longer medical course

Duration of medical studies	Less than six years	Six years and more	Unknown	Total
Women	14	9	8	31
Men	17	20	7	44
Total	31	29	15	78

Source: Sichuan Provincial Archives registration files (1937–47). Fisher test (first two columns) > 0.1 (not significant).

Education in 1929 with the support of the Ministry of Health.[38] The Commission established a two-tier system for training physicians. Medical colleges and universities aimed to train the first-rate physicians through a six-year curriculum. 'Technical medical schools' (*zhuanmen xuexiao*) 'were designed to produce a large number of doctors in a shorter period of time' over four years.[39] This organization was a compromise between two visions on physician education: giving Chinese physicians advanced scientific training similar to that provided by European and American universities or training as quickly as possible the medical workforce the country desperately needed. Of course, the reality did not entirely match the administration's design and not all schools consistently met the standards set by the Ministry. Some graduates from technical schools would declare more or less than four years of study. Besides, there was a minority of physicians (four) trained in 'military medical schools' (*junyi xuexiao*) whose curriculum seem even shorter, with a maximum of three years. All these four physicians were men and it is highly likely that these institutions were mostly masculine.[40] In other words, despite the Chinese government's attempts to standardize physician training, registration files show that medical education remained relatively disparate and messy.

Women physicians were well integrated into transnational scientific and medical education networks. Apparently, many of them studied in middle schools founded by foreign missionaries in China, just like Liu Yunbo, who studied in a Christian school in Shanghai. They often studied medicine in medical schools or universities administered by missionaries or transnational organizations such as the Rockefeller Foundation. Although most of these schools were run by Americans, when Chinese physicians travelled abroad to study medicine, they went to Germany and Japan, women and men alike.[41] Japan was geographically close, and, as an Asian country, it had managed to rise to the rank of modern nations. Germany, on the other hand, enjoyed a great scientific reputation – at the time, it had produced more Nobel Prize-winners in scientific fields than any other nation – which might explain its appeal despite the weaker German academic networks in China itself.[42] Women and men seemed equally likely to study at a university in China, whether in their home province or outside. According to the data available for Sichuan, there was no prestigious type of study exclusive to men or women, at least in the 1930s and 1940s. This is also true regarding the possibility of studying at a foreign university: women were as likely as men to travel across the country, or even the globe, to pursue their medical studies.

While there was no specifically female profile regarding medical training, important differences emerged between men and women once they had graduated and began to practise medicine. First, it seems that women tended to occupy lower positions than male physicians. The case of Chen Weixi even constitutes an example of a graduate of a medical school taking a position that did not match their credentials. Chen graduated from the Sichuan Province Medical Technical School in 1934 after studying medicine for six years. She took a position as an obstetrician at the local hospital in Chongqing (*Chongqing defang yiyuan*).[43] The following year, she entered the Neijiang's detoxification hospital (*Neijiang jieyan yiyuan*) as an assistant physician. Surprisingly, she then assumed the position of midwife in the public health centre at Fushun (*Fushun weisheng yuan*). It seems very unusual for a physician registered in Western medicine to take a different position than that for which she was trained. It is therefore difficult to generalize based on this single case.

We can state with much more certainty that it was harder for women to reach positions of higher responsibility in medical institutions. Out of the forty-six male physicians sampled, fifteen assumed some sort of management position at least once in their career, and for eight of them it was the first position they held.[44] Six out of thirty-one women held management positions, but it was never the first position they had in their career. For all except one, it was in the field of gynaecology and obstetrics (*fuchan ke*). The analysis of surveys of hospitals and clinics provides further evidence supporting the idea of a 'medical glass ceiling' (Table 7.2). Although women practising in the establishments of the province represented 12 per cent of the total number of physicians, only 4 per cent (nine) were directors of their clinic or hospital. In these institutions run by women, there was not a single male doctor, except in one hospital. Two of these nine hospitals were affiliated to missionary organizations, including the one that employed a male physician, and at least one specialized in gynaecology and obstetrics. Therefore, in the rare cases where women oversaw a medical institution, its medical staff consisted almost exclusively of women. This survey also suggests that women were less likely than men to open their own clinic. The registration files confirm this point. They make a distinction between 'position held' (*zhiye*) and 'independent practice' (*kaiye*). Out of forty-four male physicians, more than half (twenty-seven) set up an independent practice at least once in their career. For women, less than a third (ten out of thirty-one) did so.

Hence, the main differences attributable to gender regarding the profile of physicians registered in Sichuan does not seem to stem from the medical curriculum they followed but emerged after they graduated. Although in the hospitals of the province women physicians had equivalent – or even better – medical education than their male counterparts, it was usually a man who was in charge.

The central role of women in the development of gynaecology, obstetrics and midwifery

The fact that institutions or hospital departments managed by women often specialized in gynaecology and obstetrics is consistent with a more general trend among women physicians, who usually specialized in this field. In the beginning of the twentieth century, gender segregation made childbirth and midwifery an exclusively female sphere, and medicine for women was often in the hands of female doctors. Of course, there were exceptions, such as Dr Viéron, a French doctor based in Chongqing, who in the late 1920s treated mostly women. But according to him, this uncommon situation showed that he was particularly trusted by his patients.[45] In the résumé provided by physicians applying for registration, it was common for women to mention a medical specialty. On the contrary, most medical positions that male physicians declared in their registration résumé do not mention any medical specialty. In other words, in the 'positions held' rubric, most male physicians simply declared 'physician' (*yishi*), whereas several women wrote, for instance, 'obstetrician and gynaecologist'.[46]

Some women physicians practised exclusively in this medical domain. For example, Jiang Liangying graduated from WCUU in 1937 and took up a position as 'physician of obstetrics and gynecology' (*chanfu ke yishi*) in one of the WCUU hospitals – likely the hospital for women and children founded by Retta Gifford in 1896 – and in the Three University Joint Hospital (*San daxue lianhe yiyuan*) for four years.[47] Then, she became director of the Second Centre for the Protection of Infants (*Dier baoying shiwu suo*) for two years. Zheng Qiyin, after graduating from Tokyo Women's Medical School (*Dongjing nüzi yixue zhuanmen xuexiao*) in 1919, went back to Zhejiang, where she was born.[48] There, she practised in the department of gynaecology and obstetrics at the Zhejiang Hospital (*Zhejiang bingyuan*). She then assumed the position of 'chief obstetrician' in the provinces of Fujian, Guizhou and eventually in Chengdu. This mention of obstetrics and gynaecology is specific to women: it is never mentioned in any male physicians' résumé.

Less frequently, women's résumés featured paediatrics, and when they did, it was usually infant paediatrics. For example, Tian Yingzhao, born in Hebei Province, graduated from the PUMC in 1925 and took the position of physician in departments of obstetrics and gynaecology (*chanfu ke*) and paediatrics (*xiaoer ke*).[49] Zhang Jingfen, born in Yunnan and trained in Japan, practised at the Kunhua Hospital (Yunnan) and in the Guangxi Provincial Hospital, each time in the paediatrics department.[50] However, unlike gynaecology and obstetrics, one male physician did specialize in paediatrics.

Women physicians also played a decisive role in training other women in modern midwifery who were greatly needed at the time. We mentioned Liu Yunbo's role in this matter, but she was far from being alone. Zheng Qiyin, cited above, was director of the Zhejiang School of Midwifery for five years. Li Shuyuan, born in Hebei and a graduate of the provincial medical school of Hebei, worked in the Superior School of Midwifery in Tianjin before the war.[51] Zhang Fengtong, a native of Liaoning, worked in the Superior School of Midwifery of Nanchang for one year, and then in the Chengdu Superior Medical Vocational School – of which Liu Yunbo became the director after the war.[52] According to a report on the work of the Chengdu Superior Medical Vocational School submitted to the provincial administration in 1941, Jiang Liangying, mentioned above, also taught obstetrics to the third-year students.[53] Zhou Jixian, a graduate of Shanghai Dongnan Medical School who also worked in the Sichuan Provincial Health Administration, taught gynaecology and paediatrics to second-year students, while Zhang Fengtong taught them obstetrics. The school curriculum was split between classes for nurses and midwives, starting from the second year of study. The lessons taught by Jiang, Zhou and Zhang represented all the theoretical courses taken by the midwifery students in the second and third years. Wang Guoying, the head midwife of the institution, taught most of the practical courses. In short, in the Chengdu Superior Medical Vocational School – an important institution in the provincial government's effort to train a modern medical workforce – the training of midwives was essentially conducted by women.

This evidence supports the idea that gynaecology and obstetrics were medical fields reserved almost exclusively for women. Tina Johnson has shown that the advancement of Western medicine in China strengthened the place of women in the management of childbirth by constituting a 'new and respected female midwifery profession'.[54] Despite calls from some PUMC figures to apply the US model in China, where the doctor, usually a man, played a larger role in childbirth, the female midwives retained a central role. The information provided by the Sichuan archives enables us to take this idea further: not only did women play an essential role in the supervision of childbirth as midwives, but when doctors were also involved, they were most likely women too.

Conclusion

Archival material from Sichuan helps to highlight the growing importance of women in medicine, which might otherwise be obscured by the simple analysis of a few prominent women physicians. The number of women doctors in Sichuan at the end of the Republic was far from insignificant: almost one in ten doctors was a woman, bringing Sichuan closer to the national average in 1936 and placing the province on a par with American cities at the same time. Moreover, Western medicine may have been more accessible to women than Chinese medicine, where the share of women remained very small. This suggests that scientific education provided women with new career opportunities. Female Chinese physicians played important roles in hospitals, especially in obstetrics and gynaecology departments, and in teaching facilities, including less-visible but no-less-important schools of midwifery. In these various positions, they participated in the circulation of Western medical and scientific knowledge in China, especially for midwives, whose education in hygiene and modern childbirth techniques was considered essential to the health of the nation.

Overall, however, women were less likely than men to reach management positions, both in hospitals and medical schools, which explains why they do not appear in sources that pay less attention to practitioners lower in the hierarchy. The fact that women were more present in certain medical specialties, such as gynaecology and obstetrics, may also have contributed to their lower visibility in the medical profession as a whole. While gynaecology and obstetrics were indeed very feminized, general medicine, on the other hand, seemed to have been a man's world. It should not be forgotten however, that women were also present, albeit in a very small minority, in the other medical fields. For example, Deng Xiaoli, born in Sichuan and trained at the Shanghai Dongnan Medical College, specialized in X-ray (*guangke*) and internal medicine (*neike*).[55] Another graduate of Dongnan Medical College, Guo Shiduan, after working at the Dongnan College Hospital, joined the military and worked in several military hospitals during the war.[56]

From a methodological point of view, the use of registration files, surveys of hospitals and, generally speaking, the abundance of archival material produced by the health administration in the 1930s and 1940s makes it possible to take a look at a much more comprehensive number of physicians, prominent or anonymous, men or women alike. It is difficult to know whether the results presented here for the case of Sichuan are representative of the rest of China. This might not be the case since the situation in Sichuan is very specific during the Republican period. After two decades during which the province had been on the margins of national politics, it became the centre of free China during the war against Japan. As a result, the province saw a sharp increase in the number of doctors and the resources allocated to public health. The presence of the central government in Chongqing also made the documentation regarding physicians' registration easily available. Nevertheless, registration records undoubtedly existed in other provinces and it would therefore be interesting to find points of comparison in a survey that goes beyond the scope of this chapter.

Notes

1 Z. Zhu, *Collection of Zhu Ziqing's Prose* [*Zhu zi qing san wen ji*] (Nanjing: Nanjing chuban she, 2018), pp. 9–12. The 'war of resistance' refers to the Sino-Japanese War of 1937–45.
2 In this chapter, the term 'Western medicine' or 'Western-style physicians' refers to scientific biomedicine, in opposition to 'Chinese medicine' or 'Chinese-style physicians', referring to indigenous medical knowledge and techniques. In using those terms, I tried to follow as closely as possible the words employed in the sources I draw upon. During the first half of the century, the term *xiyi* (*xi* meaning 'West', *yi* meaning 'medicine' or 'physician') and its counterpart *zhongyi* (*Zhong* meaning 'Chinese') became increasingly common, although alternative designations were still in use. These terms acquired a legal meaning in 1936, when two regulations set up an official status for both 'Chinese-style physicians' (*zhong yishi*, *yishi* meaning 'physician') and 'Western-style physicians' (*xi yishi*). However, these designations must not obscure the heterogeneity of both categories or presume their mutual exclusivity. For a discussion of the progressive emergence and mutual construction of both categories, see B. Andrews, *The Making of Modern Chinese Medicine, 1850–1960* (Vancouver: University of British Columbia Press, 2014); S. H.-L. Lei, *Neither Donkey nor Horse: Medicine in the Struggle Over China's Modernity* (Chicago, IL: University of Chicago Press, 2014).
3 N. E. Barnes, 'Protecting the national body: Gender, public health and medicine in southwest China during the War of Resistance against Japan, 1937–1945' (PhD dissertation, University of California, Irvine, 2012), pp. 276–91.
4 F. Dikötter, *The Discourse of Race in Modern China* (Oxford: Oxford University Press, 2015), chapters 4–5.

5 J. Judge, *Republican Lens: Gender, Visuality, and Experience in the Early Chinese Periodical Press* (Oakland: University of California Press, 2015), pp. 117–18.
6 T. P. Johnson, 'Yang Chongrui and the first National Midwifery School: Childbirth reform in early twentieth-century China', *Asian Medicine*, 4: 2 (2008), 280–302.
7 Barnes, 'Protecting the national body', p. 284.
8 C. C. Chen (Chen Zhiqian) was born in Sichuan at the beginning of the twentieth century. In 1911, he left the province and later joined the PUMC. He went on to graduate from Harvard in 1931. Between 1931 and 1937, he assumed the role of director of the Ding County Health Department, where he experimented with the development of a public health system as part of the Rural Reconstruction Movement. During the war, he founded the Sichuan Provincial Health Administration in 1939 and became its director until 1946.
9 D. W. Y. Kwok, *Scientism in Chinese Thought, 1900–1950* (New Haven, CT: Yale University Press, 1965); H. Wang, 'The fate of "Mr. Science" in China: The concept of science and its application in modern Chinese thought', *Positions: Asia Critique*, 3: 1 (1995), 1–68.
10 R. Rogaski, *Hygienic Modernity: Meanings of Health and Disease in Treaty-Port China* (Berkeley: University of California Press, 2004), chapter 4.
11 N. Nahimas, 'Making science popular: Readers, nation, and the universe in Chinese popular science periodicals, 1933–1952' (PhD dissertation, York University, Toronto, 2022), p. 72.
12 Over the past decade, there has been an increasing number of studies on female Chinese physicians. These include works on prominent physicians, such as C. A. Shemo, *The Chinese Medical Ministries of Kang Cheng and Shi Meiyu, 1872–1937: On a Cross-Cultural Frontier of Gender, Race, and Nation* (Bethlehem, PA: Lehigh University Press, 2011); S.-T. Lin, 'The female hand: The making of western medicine for women in China, 1880s–1920s' (PhD dissertation, Columbia University, New York, 2015); C. A. Shemo, ' "Wants learn cut, finish people": American missionary medical education for Chinese women and cultural imperialism in the missionary enterprise, 1890s–1920', *The Chinese Historical Review*, 20: 1 (2013), 54–69. Some works in Chinese took a more general perspective, trying to encompass the whole group of female physicians in a specific area. However, they generally focus on eastern regions, especially Shanghai, for which there are more sources; L. Zhao, 'Qingmo Minchu Zhongguo Nü Xiyi Yanjiu' [Study of Chinese female doctors of Western medicine in late Qing dynasty and early Republic of China] (master's dissertation, Hunan Normal University, 2013); S. Xu, 'Minguo shiji Shanghai nü xiyi yanjiu, 1919–1937' [Study on female doctors of Western medicine in Shanghai during the Republican period, 1919–1937] (master's dissertation, Jinan University, 2020).
13 Until the beginning of the twentieth century, sources generally mention certain doctors because they published books or were well regarded by the literate elite. Women very seldomly appeared in those sources. F. Bretelle-Establet,

La Santé En Chine Du Sud, 1898–1928 (Paris: CNRS, 2002), chapter 3; F. Bretelle-Establet, 'Chinese biographies of experts in medicine: What uses can we make of them?' *East Asian Science, Technology and Society*, 3: 4 (2009), 421–51.

14 J. Judge, *The Precious Raft of History: The Past, the West, and the Woman Question in China* (Stanford, CA: Stanford University Press, 2008), chapter 2; X. Shi, *At Home in the World: Women and Charity in Late Qing and Early Republican China* (New York: Columbia University Press, 2018), p. 4.

15 D. Wang, *Street Culture in Chengdu: Public Space, Urban Commoners, and Local Politics, 1870–1930* (Stanford, CA: Stanford University Press, 2013), p. 180.

16 K.-C. Yip, *Health and National Reconstruction in Nationalist China: The Development of Modern Health Services, 1928–1937* (Ann Arbor, MI: Association for Asian Studies, 1995), pp. 58–59; Y. Hu, 'Minguo shiqi yisheng zhi zhenxun he pinghe' [The screening and evaluation of doctors in the Republican period], *Zhejiang xue kan*, 5 (2008), 88–94.

17 Chinese-style physicians started to register only in 1936.

18 However, in practice, the registration files do not provide a complete overview of the population of physicians. First, it is impossible to know if every Western-style physician did register. For those who had been trained in medical schools or universities in China or abroad, there was no reason not to register. But it is very likely that many still practised Western medicine without due qualifications. Second, for various reasons, there are some gaps in the available archival records. In the case of Sichuan, registration files before 1937 are kept in municipal archives and in Sichuan provincial archives (SPA) from 1937 on. Because of restrictive archival policies, some of these files are not yet accessible – for example, in the Chengdu municipal archives, at least when I did this research. For these reasons, the present research is based on the registration files kept in the Sichuan provincial archives, produced between 1937 and 1947.

19 M. B. Bullock, *An American Transplant: the Rockefeller Foundation and Peking Union Medical College* (Berkeley: University of California Press, 1980); K. Minden, *Bamboo Stone: The Evolution of a Chinese Medical Elite* (Toronto: University of Toronto Press, 1994).

20 From 1912 to 1937, the province had been divided into 'garrison areas' controlled by rival warlords. On this, see R. A. Kapp, *Szechwan and the Chinese Republic: Provincial Militarism and Central Power, 1911–1938* (New Haven, CT: Yale University Press, 1973). Independence from central government, fragmentation of the province and the replacement of civil administration by the military, more than the absence of medical policies, contributed to a lack of documentation on these issues.

21 C. Furth, *A Flourishing Yin: Gender in China's Medical History, 960–1665* (Berkeley: University of California Press, 1999), p. 305; A. K. C. Leung, 'Women practicing medicine in pre-modern China', in H. T. Zurndorfer (ed.), *Chinese Women in the Imperial Past: New Perspectives* (Leiden: Brill, 1999),

pp. 101–34. Like in India, childbirth was dominated by female practitioners. See Rai in this volume.
22 Andrews, *The Making of Modern Chinese Medicine, 1850–1960*, p. 38.
23 A. R. Drucker, 'The role of the YWCA in the development of the Chinese women's movement, 1890–1927', *Social Service Review*, 53: 3 (1979), 421–40; P.-L. Kwok, *Chinese Women and Christianity, 1860–1927* (Atlanta, GA: Scholars Press, 1992).
24 'Report of the Chengdu Hospitals Board', 1931, WCUU, 279–4410, p. 25.
25 K. Minden, *Bamboo Stone: The Evolution of a Chinese Medical Elite* (Toronto: University of Toronto Press, 1994), p. 106.
26 *Our West China Mission* (Toronto: Missionary Society of the Methodist Church, 1920), p. 456.
27 P. J. Bailey, *Gender and Education in China: Gender Discourses and Women's Schooling in the Early Twentieth Century*, Routledge Contemporary China Series 15 (New York: Routledge, 2007), p. 85.
28 We used all the surveys accessible in the Sichuan Provincial Archives: SPA, 113-01-0023 (three districts), 113-01-0296 (fifteen districts), 113-01-0511 (sixty-four districts), 113-01-1730 (eleven districts). These surveys do not include the Chengdu and Chongqing urban areas. The districts that include part of these two towns in their administrative border only reported establishments located outside of Chengdu and Chongqing municipal jurisdiction.
29 'Zhuce lülibiao: Deng Xiaoli' [Registration résumé: Deng Xiaoli], 1942, SPA, 113-01-2304.
30 Xu, 'Study on Female Doctors of Western Medicine in Shanghai during the Republican Period, 1919–1937'.
31 'Zhuce lülibiao: Yue Yicheng' [Registration résumé: Yue Yicheng], 1942, SPA, 113-01-2304.
32 'Number of the alumni of the West China Union University', November 1945, WCUU Archives, 11-279-4404. Apparently not all WCUU graduates joined the alumni association.
33 M. R. Walsh, *'Doctors Wanted, No Women Need Apply': Sexual Barriers in the Medical Profession, 1835–1975* (New Haven, CT: Yale University Press, 1977), pp. 185–86.
34 The nursing school established in 1915 by the Canadian Methodist Church in Chengdu at first only recruited men, *Our West China Mission*, p. 394. According to Michelle Renshaw, this was a general trend in the whole country: M. Renshaw, 'Accommodating the Chinese: The American Hospital in China, 1880–1920' (PhD dissertation, University of Adelaide, 2003), p. 254. In the 1920s and after, the number of female nurses grew in missionary and Chinese hospitals. This trend appears clearly in the surveys of hospitals and clinics from 1938 to 1939, where a third of practising nurses were women.
35 T. P. Johnson, *Childbirth in Republican China: Delivering Modernity* (Lanham, MD: Lexington Books, 2011). Modern midwifery, based on biomedical knowledge and techniques, progressively replaced old-style midwives, who were also exclusively women.

36 Johnson, *Childbirth in Republican China*, p. 80.
37 This analysis includes the registration files of all thirty-four women physicians in the province, and a random sample of forty-four male Western-style physicians.
38 Yip, *Health and National Reconstruction in Nationalist China*, pp. 134–39.
39 We do not know precisely how these technical schools worked. Between 1929 and 1931, the Commission for Medical Education and an expert sent by the League of Nations formulated recommendations regarding the organization of these schools; K. Faber, *Report in Medical Schools in China* (Geneva: League of Nations Health Organization, 1931). However, it seems that these recommendations were not systematically applied.
40 There was one very specific situation in which a woman graduated from such a military institution. As she explains in her resume, '[I] studied at the Sichuan province medical technical school. Because the technical school was transformed into a course for medical officers, [I] graduated from the course for medical officers.' The very fact that she had to provide a thorough explanation of her situation may indicate that it was rather unconventional. 'Zhuce lülibiao: Peng Zurong' [Registration résumé: Peng Zurong], 1941, SPA, 113-01-2301.
41 Out of a total of seventy-eight physicians, four studied in Germany (two women, two men) and three in Japan (one woman, two men). These were the only two destinations mentioned by physicians.
42 P. S. Richards, 'The movement of scientific knowledge from and to Germany under national socialism', *Minerva*, 28: 4 (1990), 401–25.
43 'Zhuce lülibiao: Chen Weixi' [Registration résumé: Chen Weixi], 1942, SPA, 113-01-2303.
44 This includes the positions of 'director' (*yuanzhang*), 'team leader' (*duizhang*) or 'chief physician' (*zhuren yishi*).
45 Diplomatic archives of Nantes, 513PO/A, n°196, rapport de Viéron, 25 January 1928.
46 This was the case for seven out of twenty-three who mentioned at least one position (eleven did not mention any); two mentioned another specialty (surgery and X-ray medicine).
47 'Zhuce lülibiao: Jiang Liangying' [Registration résumé: Jiang Liangying], 1943, SPA, 113-01-0327.
48 'Zhuce lülibiao: Zheng Qiyin' [Registration résumé: Zheng Qiyin], 1943, SPA, 113-01-0327.
49 'Zhuce lülibiao: Tian Yingzhao' [Registration résumé: Tian Yingzhao], 1943, SPA, 113-01-2304.
50 'Suining yiyuan xianyou zhiyuan mingce' [Suining Hospital staff chart], 1946, SPA, 113-01-2187.
51 'Zhuce lülibiao: Li Shuyuan' [Registration résumé: Li Shuyuan], 1941, SPA, 113-01-0331.
52 'Zhuce lülibiao: Zhang Fengtong' [Registration résumé: Zhang Fengtong], 1943, SPA, 113-01-0327.

53 'Shichuan shengli Chengdu gaoji yishi zhiye xuexiao geke jiaoyuan xingming biao' [List of faculty members of the Chengdu Superior Medical Vocational School], 1941, APS, 113-01-1215.
54 Johnson, *Childbirth in Republican China*, p. 74.
55 'Zhuce lülibiao: Deng Xiaoli' [Registration résumé: Deng Xiaoli], 1943, SPA, 113-01-2304.
56 'Zhuce lülibiao: Guo Shiduan' [Registration résumé: Guo Shiduan], 1940, SPA, 113-01-0331.

8

Agency and coercion: Fighting 'women's illnesses' with grassroots science and medicine during the Great Famine in China, 1958–1962

Kathryn Edgerton-Tarpley

Introduction

During China's Great Leap Famine, when approximately 36 million people starved to death between 1958 and 1962,[1] a female medical student from Wuhan Medical College was sent to the countryside for six months 'to take part in a campaign to improve rural health before the spring ploughing'. This young woman, who was interviewed in early 1962 by the Hong Kong-based journal *Current Scene* after she left mainland China, was part of a group of 400 medical students from her college who were sent to Xiangyang County in central China's Hubei Province to 'exterminate the prevalent diseases in that area', among them oedema and uterine prolapse.[2] *Current Scene* did not name the young interviewee. Similarly, the largely female medical personnel discussed in county-level reports about attempts to treat gynaecological ailments during the Great Famine were almost never named. The identity of these women health workers is thus to some extent 'hidden'. Nevertheless, the first-hand account provided by the *Current Scene* interviewee, when read in concert with national-level reports from China's Ministry of Public Health and local reports compiled by cadres in one particularly hard-hit county in eastern China, demonstrates that women health workers and researchers played a key role in fostering Maoist 'grassroots science' during China's famine-era campaign to treat 'women's illnesses' (*funü bing*).[3]

The *Current Scene* interviewee imparted a vivid description of what practising grassroots medicine in rural China during the famine entailed. 'I was placed in charge of a disease prevention station at the Tso-lien-chiao People's Commune', she recalled, 'with four assistants under me'. She noted that the food situation was getting worse when she arrived in October 1960, and there was no hospital in the area. Her station focused on using a combination of Chinese and Western medicine to treat uterine prolapse, one of the two so-called 'women's illnesses' that received a great deal of attention during the Great Famine of 1958–62. The health station was housed in a clay-walled building that had formerly been the ancestral hall of a local family. 'When I took over I covered the brick floor with dry grass and wheat stalks', she recalled. 'Then I used to encourage the patients to bring their bedding and stay in the station when necessary, even if it meant bringing their babies with them.' The interviewee used Chinese herbal medicine to treat a total of eighty uterine prolapse patients during her six-month posting in the countryside. 'I usually had between 10 and 20 patients staying in the station at one time', she said. 'The most I ever had was 43.'[4]

Given that this young woman was only nineteen years old when she was sent to serve in the countryside and her four years of training in biomedicine had been repeatedly interrupted by political campaigns, she achieved a striking degree of agency and authority at her station. The disease prevention station provided treatment for people in six surrounding villages, she recalled. Commune officials assigned two cooks to prepare food and clean and wash patient clothing at the station, and her four assistants brought patients to the station for her to diagnose and treat. There was no physician in her team due to the paucity of qualified doctors, so the interviewee was authorized to prescribe treatment when no 'roving doctor' was in her area. 'During all those months I ran my station as a one-man team', she told her *Current Scene* interviewer. 'I organized everything, I diagnosed, and I acted as nurse.'[5]

This evocative interview raises a series of questions discussed in this essay. First, how and why did young female health workers like this nineteen-year-old medical student come to play such a prominent role in the treatment of 'women's illnesses' during China's Great Leap Forward campaign (1958–62, hereafter GLF) and the famine it gave rise to, and what can this tell us about the role of women in Mao-era science and medicine? Second, why did so many rural women suffer from uterine prolapse and amenorrhea during the Great Famine, and why did the Chinese Communist Party (CCP) put so much emphasis on treating these non-fatal maladies during a time of mass starvation? Finally, how did Mao-era 'mass science' shape the Traditional Chinese Medicine and biomedicine remedies used by the predominantly female healthcare workers dispatched to treat uterine prolapse and amenorrhea in 1958–62, and how were these treatments received by rural women? I find that while medical interventions did little to address the root cause – starvation and intensive labour – of the gynaecological ailments discussed, the exhaustive search for cures provides a vivid snapshot of both Maoist 'grassroots science' and Chinese Communist expectations of women. Women medical practitioners and researchers played an important role during the Great Famine, I argue, and in some ways the decision of the CCP leadership to put female health workers in charge of treating 'women's illnesses' opened a space within which women had a notable degree of agency. At the same time, similar to Saurav Kumar Rai's findings in this volume regarding the reinforcement of patriarchy found in India's early twentieth-century Ayurvedic medical texts,[6] the CCP's focus on treating amenorrhea and uterine prolapse also reflected the Party's continued foregrounding of women's roles in reproduction and production. Moreover, the Maoist state expected female medical practitioners to coerce other women into accepting treatment for 'women's illnesses' to advance the Party's goals.

Grassroots science and women doctors in China's Great Leap Forward

The turn to Maoist mass science during the GLF

The case of the nineteen-year-old female medical student sent to a rural health station to provide rudimentary treatment for women suffering from uterine prolapse reflects several important aspects of the GLF, a utopian campaign that aspired to radically accelerate China's industrialization by unleashing the productive energies of China's mobilized masses, but ultimately led to a massive famine. Disturbed by criticism of the Party, in mid-1957 China's leadership unleashed a virulent Anti-Rightist Campaign in which an estimated 400,000–700,000 intellectuals, among them leading scientists, engineers and physicians, were denounced. Many of those branded as 'rightists' lost their positions and were sent down to the countryside to learn from the peasantry.[7] By the end of 1957, writes Roderick Macfarquhar, there was 'a growing mood of disenchantment' with the Soviet model of state-led economic development that the CCP had followed since coming to power in 1949, and 'a feeling among many Chinese leaders that new methods were needed if China were to break out of its economic backwardness'.[8] In response, in 1958 the CCP launched the Great Leap, which aimed to achieve an economic breakthrough based on local resources, mass mobilization and 'the transformative enthusiasm and strength of the peasantry'.[9] Within several months, smaller agricultural collectives throughout the country were combined into giant 'People's Communes' on which private plots were banned, people were expected to eat in communal dining halls and peasants were organized into quasi-military production brigades that demanded intensive labour in agriculture or industry.[10]

In part because much of the impetus for the GLF came from 'Mao's disavowal of the Soviet Union and her "advanced experience"', the campaign had significant repercussions for all fields of Chinese science.[11] Embracing a 'voluntarist faith in the power of the masses', Mao and his allies sought to form a science that advanced the Maoist principle of 'anti-imperialist self-reliance' and was 'produced by the broad masses for the fulfilment of socialist revolutionary goals'. They thus championed *tu* ('native, Chinese, local, rustic, mass, crude') over *yang* ('foreign, Western, elite, professional, ivory-tower'), in many cases trumpeting the value of *tu* 'to the point of denigrating *yang*'.[12] The resulting Maoist grassroots science encouraged rural leaders to conduct field investigations and process their data to find solutions for pragmatic problems',[13] among them the upsurge of amenorrhea and uterine prolapse cases across rural China. The turn to mass or 'grassroots' science during the GLF years also had a profound impact on medicine and public health in China. As demonstrated by the remedies employed by the health workers assigned to treat 'women's illnesses', the GLF contributed greatly to 'the popularization of Chinese medicine', and gave Chinese medicine a more prominent role in public health campaigns.[14] Beginning in 1956 and continuing through the GLF, what has become known in English as Traditional Chinese Medicine (TCM) was institutionalized as a standardized, government-created medicine, in part by preparations to establish four Academies of TCM.[15] In October 1958, Mao proclaimed that 'China's medicine and pharmacology is a great treasure-house', and directed the Ministry of Health to arrange for doctors trained in biomedicine to study Chinese medicine, with the aim of integrating Chinese and Western medicines. China's national newspapers were immediately 'ablaze with these words', writes Kim Taylor, leading to the promotion of TCM across the country as well as a pronounced emphasis on combining Chinese and Western medicine. Particularly in rural China, this valorization of TCM, combined with the GLF emphasis on self-reliance and *tu* science, led to the widespread promotion of acupuncture, Chinese herbal medicine and folk or 'home-grown remedies' during the 1958–62 period.[16]

The privileging of grassroots over professional science during the GLF also shaped the medical personnel, such as the *Current Scene* interviewee, who took part in the mass public health campaigns launched to eliminate 'women's illnesses' and other diseases. In her work on China's lengthy campaign to eradicate schistosomiasis, Miriam Gross finds that as the GLF unfolded in 1958, the Party was busy 'dismembering professional science in the wake of the 1957 Anti-Rightist Campaign', and was at the same time expanding 'grassroots scientifically oriented campaigns'.[17] Grassroots science 'initially often lacked standard scientific criteria, such as reproducibility or control groups', continues Gross, but participating in public health campaigns introduced local cadres and educated rural youth to basic technical and medical skills as well as statistics, forms and records, and simple experiments.[18] Moreover, Gross observes that while sending leading physicians and scientists to the countryside to learn from the masses during the GLF 'devastated professional science', it proved helpful to grass-roots public health campaigns. The GLF and the early years of the Cultural Revolution (1966–71) 'were the only times that the urban medical establishment was emptied out and dispatched to the far hinterland, staying long enough to really make a difference', she explains. In time, educated youth like the *Current Scene* interviewee were able to take over low-level technical and medical work, allowing the limited pool of sent-down doctors to focus on treatment and training. In sum, 'the unintended consequence of banishing doctors was a dissemination of urban medical knowledge to the countryside'.[19]

The role of women health workers during the Great Leap Famine

The experiences the young medical student interviewed by *Current Scene* had while overseeing a rural health station reflect the emphasis that Mao-era grassroots science placed on self-reliance and the use of local resources. Because of the lack of doctors in the area she was assigned to, the interviewee prescribed treatments on her own, and even described running her station as 'a one-man team'. She also drew heavily on local sources, such as housing patients in a building that had previously been a local family's ancestral hall and relying on local staff to cook and clean for the station.[20] The fact that a nineteen-year-old woman was put in charge of a disease prevention station also reflects a second important aspect of the GLF, the championing of women who served as 'icons of modernity' by entering fields traditionally dominated by men.[21] In the decades before and after the 1949 revolution, female doctors were depicted as 'progressive, scientifically minded, ambitious, and professional women with patriotic ideals'. As such, they represented 'a powerful symbol that fully articulated the ideals of China's modernization, and the country's aspiration of becoming a new and strong nation'.[22] After the CCP victory, 'women's health was articulated as a state priority' and the new Party-state sought to 'bring scientific knowledge and practice to the countryside', often via midwives and other female health workers. Beginning in the early 1950s, the CCP worked to retrain rural midwives in new-style midwifery that emphasized the sterilization of delivery instruments,[23] and from 1969 on stipulated that each production brigade should have at least one female 'barefoot doctor'.[24] Positive portrayals of female physicians may well have spurred the *Current Scene* interviewee's desire to become a doctor. She was among the first generation of urban medical graduates trained in biomedicine who were 'sent down' to serve in rural areas as part of the Mao-era effort to address glaring rural–urban inequities in healthcare.[25] The interviewee was from an elite family that quickly fell afoul of the new government – her father, a county magistrate, was arrested by the Communists in 1949 and executed in 1951, leaving her and her mother to survive on monthly remittances sent by an aunt in Hong Kong. Yet in 1957, she passed an exam that enabled her to enter the Wuhan Medical College in Hankou, a major city in central China's Hubei Province.[26]

When the nineteen-year-old interviewee was sent to rural Hubei for six months, she was part of a regular Mao-era practice of sending medical teams comprised primarily of Traditional Chinese Medicine practitioners, with a small number of biomedical physicians, nurses and medical students, into rural communes to treat common diseases. The medical personnel in such teams often had rudimentary training, notes Sun Qi, and medicines were always in short supply.[27] The experience of the *Current Scene* interviewee during her four years at the Wuhan Medical College provides a striking example of the type of interrupted training many health workers received during the GLF. When she entered college in September 1957, all academic activities were suspended due to the Anti-Rightist Campaign, and the following summer the college was closed for two months so that medical students, doctors and nurses could take part in a campaign to smelt iron and steel. 'We built over 20 native-style furnaces in the college grounds', she recalled.

In fall 1958, medical students were asked to criticize older experienced professors as part of the 'reforming-the-teaching-methods' campaign, and then in summer 1959 each student had to do two months of manual labour to understand the perspective of labourers. The interviewee was assigned to the college farm, where she 'spread manure and hoed the fields'. Lessons missed during these campaigns were 'simply skipped'.[28] The food situation worsened in late 1959, and in February 1960, the medical college was closed yet again when she and her classmates were sent to northern Hubei to help with a railway construction project. 'We lost a month's studies but in one way it was a most welcome change for us because we had so much better food', she told *Current Scene*. 'We students could eat our fill on the construction site.' In contrast, she observed peasants in their host village consuming meals consisting of 'only a handful of steamed rice and a few chunks of "cotton-seed cake'". The interviewee and her classmates started clinical practice at the college hospital after returning from their railway construction work and were then assigned to serve in rural Hubei for six months from October 1960 to March 1961. Due to all the interruptions, after reaching Hong Kong the interviewee expressed little confidence in her medical training. 'I wonder if I'll ever become a fully-fledged doctor', she lamented. 'At the moment I'm spending all my time trying to catch up on the work I missed.'[29]

There were practical as well as ideological reasons why the state-sponsored campaign to prevent and treat 'women's illnesses' during the famine depended heavily on female healthcare workers like the young interviewee. Xiaoping Fang finds that 'more women urban medical personnel with lower designations were sent down [to the countryside] than men'.[30] Moreover, female medical personnel were often assigned to treat gynaecological ailments because many rural women were uncomfortable having such conditions treated by a male doctor. The *Current Scene* interviewee explained as follows: 'I was sent to that area particularly because there were a lot of cases of prolapsed uterus and country women, as you know, are very conservative and flatly refuse to be treated for ailments of that sort by a man.'[31] Her statement alludes to the segregation of the sexes that became increasingly strict from the Song period (960–1279) and lasted well into the twentieth century in rural areas.[32] In imperial China male physicians were trained to use diagnostic methods that required limited physical contact with female patients. Even in the 1930s, finds Fang, doctors promoting public health programmes experienced difficulties vaccinating girls and women who were ashamed to expose their arms to a male vaccinator, and into the 1960s, some village women were too embarrassed to 'let a male doctor press a stethoscope through their clothes and onto their chests'. Training women doctors was thus an important aspect of the CCP's rural healthcare programme.[33]

During the famine years, All-China Women's Federation (ACWF) personnel in each commune worked with health stations to organize the work of treating 'women's illnesses'.[34] For instance, an October 1959 report from Xinmin Commune in Wuwei, a county in eastern China's Anhui Province, specified that the commune party committee had made female brigade leaders responsible for the administrative management and ideological education work related to the campaign to treat prolapse. Women cadres were assigned to visit patients and treat them according to the level of their prolapse.[35] Another report from Wuwei County specifies that because uterine prolapse was seriously affecting women's health and labour capacity in rural areas, in July 1960, the health division of the Wuwei County People's Committee organized a training course for preventing and treating the condition. Each of Wuwei's twenty-seven communes sent a female health worker to attend the training. When health workers returned to their commune hospital or station, they shared the treatment methods they had learned with other medical personnel and in some cases launched pilot treatment plans.[36] The heralding of both grassroots science and women doctors during the GLF meant that female medical personnel, many of them with only rudimentary training, played a key role in the Party's famine-era campaign to treat 'women's illnesses'.

The medicalization of starvation: Prioritizing production and reproduction during the Famine

The sharp increase in gynaecological ailments suffered by rural women is rarely discussed in detail in existing scholarship on the GLF of 1958–62 but was an issue that received considerable attention in both central-government and county-level reports during the famine years. What factors led to the upsurge of cases and why did the government respond with such alarm? The rising number of cases of uterine prolapse and amenorrhea (the cessation of menstruation) was directly related to GLF policies and the famine they gave rise to. The drive to harness women's labour productivity was a defining aspect of the GLF. The Chinese Communists came to power promising to liberate Chinese women, and women in rural China became 'both objects and agents of revolutionary change' after 1949.[37] The CCP viewed mobilizing women to conduct agricultural and industrial labour outside the home, rather than focusing on the traditional women's work of spinning and weaving at home, as key to women's liberation. Both the CCP and the ACWF 'argued that women would only become free from the fetters of the feudal past when they participated in productive work outside the home'.[38] Women's labour became all the more crucial during the 1958–62 period due to the urgent need for labourers for GLF projects. Beginning in 1958, rural women were given increasing responsibility for planting and harvesting crops, while most able-bodied men were sent to work on massive irrigation, mining and steel-smelting projects.[39] According to estimates compiled by Marina Thorborg, the percentage of Chinese women working in agriculture jumped from a range of 30–50 per cent in the years before 1955 to 80–95 per cent in 1958–59.[40]

While CCP and ACWF leaders emphasized women's labour outside the home, they also expected women to continue to handle childcare and other domestic responsibilities.[41] Pre-1949 scientific discourses that 'tied women, and especially mothers, to national projects of construction' continued to resonate in the 1950s, notes Kimberley Ens Manning, when dominant discourses 'made reproduction of the next generation a woman's "natural duty"'.[42] A volume on Chinese medicine treatments for gynaecological maladies published in 1958 by Pu Fuzhou, an esteemed doctor of traditional Chinese medicine, provides an example of Mao-era rhetoric on the importance of women's roles. 'In old China, women were oppressed and despised', wrote Pu. Because they were often ignored, many had died of diseases. Although women's political status and health had improved since Liberation (1949), it was important to continue the trend. 'Since women make up almost fifty percent of the population', he noted, 'their health is crucial for the development of socialist construction. We, scientific and medical staff, should regard the treatment of women's illnesses as a political task and make it a priority.'[43]

Contrary to the Party's bright hopes for the GLF, policies connected to the campaign led to a severe famine that resulted in the death of an estimated 36 million people and a shortfall of 40 million births, making it the most lethal famine in Chinese and world history.[44] The rapid establishment of People's Communes caused a sharp decline in agricultural production, and decisions made by Mao in the summer of 1959 made the situation worse. When Defence Minister Peng Dehuai voiced serious critiques of the GLF at the Lushan Conference in July 1959, Mao responded by launching a renewed Anti-Rightist Campaign that silenced all dissent. In the following months, local cadres submitted increasingly exaggerated estimates of their grain yields to avoid being branded as rightists. The state, which based its procurement requirements on these wildly inflated estimates, then requisitioned dangerous amounts of grain in its effort to provision China's cities and fund rapid industrialization. The state also rejected international aid and even exported more grain abroad to earn foreign exchange currency. Natural disasters and the worsening Sino-Soviet split exacerbated the crisis. The harvest of 1959 was a miserable failure. Short of food, communal mess halls repeatedly cut rations and, in some cases, completely ran out of grain.[45] In the winter and spring of 1959–60, widespread starvation ensued in multiple provinces across China, accompanied by a sharp increase in cases of amenorrhea, uterine prolapse and oedema.

During the Great Famine, extreme political pressure made it increasingly dangerous for observers at any level to directly acknowledge that famine conditions were occurring. Especially after the deepening of the Anti-Rightist Campaign in fall 1959, 'famine as a topic became taboo not just for the media, but also within the internal information system of the party'.[46] Prevented by political realities from acknowledging the existence of mass starvation, but unable to let disturbing demographic trends go wholly unreported, during the Great Leap disaster county and provincial officials across China employed a pronounced medicalization of starvation. They attributed so-called 'unnatural deaths' in their locales to oedema (*fuzhong bing*) and blamed falling birth rates and a decrease in women's labour productivity on 'women's illnesses' (*funü bing*), in particular amenorrhea (*bijing*) and uterine prolapse (*zigong tuochui*). The connection between malnutrition and amenorrhea, or the absence of three or more successive menstrual periods in a woman of childbearing age who previously had periods,[47] was not fully understood in the 1950s and 1960s. In January 1961, China's Ministry of Health posited that the increase in amenorrhea cases was due to a wide variety of factors, including insufficient nutrition and improper labour arrangements, poor sanitary protection for women during menstruation, pregnancy, childbirth and lactation, weakened constitutions, mental factors, other diseases or sudden changes in lifestyle.[48] Research conducted at Harvard a decade after the GLF by Rose Epstein Frisch, a pioneer in discovering the biological mechanisms of fertility in women, demonstrated that insufficient body fat, whether from malnutrition or very intensive physical activity, leads to amenorrhea and decreased fertility in otherwise healthy women.[49] 'Menstrual function stops in girls and women ages 16 years and over when body weight falls below the minimum fatness level', explains Frisch, 'and is restored following a gain in body weight sufficient to restore fatness above the minimum level'.[50]

Uterine prolapse occurs when pelvic floor muscles and ligaments become so stretched and weakened that they no longer provide sufficient support for the uterus and the uterus slips down into, or in severe cases drops completely outside of, the vagina. It most often affects postmenopausal women who have had one or more vaginal deliveries and is exacerbated by conducting heavy physical labour.[51] In today's US, uterine prolapse is often addressed by having a woman insert a vaginal pessary, or a plastic or rubber ring, into the vagina to support the prolapsed tissue. Alternatively, a surgeon can address the issue by repairing weakened pelvic floor tissues with a graft of synthetic material, or in some cases by removing the uterus.[52] As was the case with amenorrhea, in 1961, China's Ministry of Health put forth multiple factors believed to cause uterine prolapse, including injury during childbirth, returning to physical work too soon after giving birth, excessive labour intensity, illness that led to increased abdominal pressure, lack of attention to menstrual periods and issues related to hygiene during pregnancy.[53] The young female medical student who treated uterine prolapse cases at her rural health station in 1960–61 gave a blunter assessment of reasons for the rising incidence of prolapsed uterus: 'It was usually caused by the woman carrying too heavy a load or having to work far too hard in the fields', she told *Current Scene* in 1962. 'There has been a marked increase in prolapses since the end of 1958 when women were liberated from domestic chores to work in the fields as the equals of men. Women are not physically fitted to do such heavy work, particularly immediately after childbirth.' The interviewee also pointed out the gap between official policy and practice during the Great Leap. 'Officially a woman was supposed to take forty days leave before and after childbirth', she noted, 'but many of them were sent back to the fields too soon. Malnutrition also aggravates this condition.'[54]

As famine conditions worsened in many parts of China in late 1959 but could not be acknowledged directly, 'women's illnesses' became a pronounced focal point in reports from the national, provincial, county and even commune levels of the Chinese bureaucracy. It was the fact that amenorrhea made it impossible for young women to bear children while uterine prolapse curtailed women's ability to conduct intensive physical labour that made these ailments so alarming to the CCP. A lengthy report on China's disease situation compiled by the Ministry of Health in January 1961 highlights both the national scope of the crisis and the central government's marked focus on women's production and reproduction. This report stated that based on local surveys, uterine prolapse was affecting about 2 per cent of the full-time and part-time female workforce in northern China and about 10 per cent of the female labour force in southern regions. Amenorrhea was even more widespread: by early 1961, it was affecting roughly 30 per cent of all women of childbearing age and over half of them in the most severe disaster areas. Due to the harm caused by these diseases, noted the report, China's birth rate had declined.[55] Anhui Province in eastern China was one of China's most severely affected areas during the famine – Cao Shuji posits that the province suffered 6.33 million excess deaths during the disaster, giving it an unnatural death rate of 18.37 per cent, and estimates derived from provincial figures indicate that Anhui had over 2.4 million missing births from 1958 to 1961.[56] This dramatic shortfall in births is reflected in the number of Anhui women suffering from prolapse or amenorrhea as famine conditions persisted into 1961. According to a report compiled by the Anhui Provincial Health Bureau Party Group, in late March 1961, there were 324,000 patients in Anhui with uterine prolapse, and over 948,000 women suffering from amenorrhea.[57] Reports compiled by cadres in Wuwei County, located in south-eastern Anhui, provide local-level numbers from one of China's hardest-hit counties. 'Wuwei was one of the most notorious counties during the Great Famine because of the death of about a quarter of the approximately one million local population', note historians Cao Shuji and Yang Bin.[58] In fall 1959, the Wuwei County party committee reported that all the communes in Wuwei had a significant number of women suffering from uterine prolapse and amenorrhea. Of the 320 female labourers in the Fenghuang Brigade in Shushan Commune, for instance, 51 women, or 15.93 per cent of them, had uterine prolapse, while 245 of the women, or 76.56 per cent, had amenorrhea. This had 'negatively influenced production'.[59] Spurred by such reports, efforts to address 'women's illnesses' became an important aspect of Chinese responses to the Great Famine. Just as the Ayurvedic medical texts analysed by Rai in this volume supported Indian patriarchy by discussing the female body only in connection with diseases of the uterus and genital area, so the CCP's hyper-focus on treating amenorrhea and uterine prolapse during the GLF intimated the Party's view of women as first and foremost reproducers and producers.[60]

Treating women's illnesses with a combination of Chinese and Western medicine

Female medical personnel like the *Current Scene* interviewee and commune health workers in Wuwei County drew on Maoist grassroots science and medicine to treat gynaecological ailments, though their efforts were not always welcomed by rural women. The medical student interviewed by *Current Scene* gave few details about how she treated uterine prolapse cases at her rural health station. 'In normal circumstances one would perform a surgical operation to relieve the condition', she explained, 'but as this was impossible in the countryside, where there were no surgical instruments at all, I had to fall back on a patent herb medicine'.[61] Influenced by the GLF-era emphasis on integrating Chinese and Western medicine, county-level reports from different parts of China championed the importance of drawing on both TCM and biomedicine treatments.[62] In practical terms, China had only 51,000 Western-trained doctors in the 1950s, so it made sense to draw on the roughly 360,000 Chinese medicine practitioners working in rural areas, especially since they used inexpensive locally grown herbs rather than costly drugs and equipment often imported from abroad.[63] Moreover, doctors of Chinese medicine had a long history of employing Chinese herbal medicine and acupuncture to address 'dysfunctions of Blood and womb'.[64] During the Famine practitioners of grassroots medicine made significant use of this heritage, in combination with biomedicine, in their attempts to treat women suffering from amenorrhea or uterine prolapse.

In another example of the prominence of women in the campaigns to promote both TCM and the importance of treating 'women's illnesses', a female doctor of Chinese medicine, Zeng Jingguang (1918–2010), played a leading role in developing a foundational Mao-era textbook on TCM gynaecology. Zeng, a native of Sichuan, began her study of Chinese medicine in 1937 at one of the new colleges established during the Republican era (1912–49). She practised Chinese medicine throughout the war years and in 1957 was assigned to the Gynaecology Teaching and Research Group at the Chengdu College of Chinese Medicine.[65] In 1959, representatives from the five main Academies of TCM that had been recently established in Beijing, Nanjing, Shanghai, Guangzhou and Chengdu met in Chengdu to begin the work of developing a national set of textbooks that could provide uniform, comprehensive instruction in TCM.[66] Zeng Jingguang subsequently led her Chengdu research group to compile the national textbook on TCM gynaecology, titled *Lecture Notes for Chinese Medicine Gynaecology* (*Zhongyi fukexue jiangyi*). The textbook, which appeared in 1960, discusses treatments for both amenorrhea and uterine prolapse, but for reasons of space this chapter examines only the latter. Zeng and her Chengdu research group posited that uterine prolapse was caused either by a deficiency of *qi* (vital force) due to a weak physique, excessive weight-bearing or sex, undue strain in childbirth and returning to physical labour too soon after childbirth, or by 'dampness' due to spleen deficiency.[67] Treatment for uterine prolapse focused on herbal medicine believed to nourish and lift, combined with acupuncture and moxibustion, rest, and abstention from both sex and bearing heavy loads. To boost a women's *qi* so that her body could lift the uterus back into place, the textbook authors recommended a decoction called *Bu zhong yi qi tang*. This prescription, which was first recorded in a medicinal work in 1249 and remains on the market today, is composed of a hot water extract made from eight medicinal herbs. Among them, astragalus, ginseng, atractylodes, liquorice and angelica were also used to treat amenorrhea in late-imperial China. To treat uterine prolapse caused by dampness, the textbook advocated a gentian decoction for purging the liver. This prescription again included ingredients, such as liquorice, angelica and rehmannia root, commonly used in Ming-Qing prescriptions for gynaecological ailments.[68]

Many of the herbal medicines recommended in the national TCM gynaecology textbook compiled by Zeng and her colleagues also appear in Wuwei County documents concerning the treatment of 'women's illnesses'. Perhaps because Wuwei suffered so severely during the famine, its archival records from the disaster years are particularly rich.[69] Wuwei reports give fuller overviews of remedies for uterine prolapse than for amenorrhea, most likely because it was more feasible to treat prolapse without providing patients with additional food that health workers had no way to procure. In 1959 and 1960, the combined biomedicine-and-TCM treatment discussed most frequently in Wuwei reports was a so-called 'alcohol block therapy' (*jiujing fengbi liaofa*). This method required doctors to inject anhydrous alcohol and procaine into the connective tissue near a woman's cervix to get the uterus to contract back into the vaginal orifice, and eventually back to its normal location. After each of three treatments, explained a commune-level report from September 1960, the patient was instructed to lie on her back with gauze in her vagina for twenty-four hours and to drink TCM herbs with water.[70]

A compilation of material from a professional conference on maternal and child health work held in mid-1960 in Wuwei's Chaohu District introduced TCM treatments for multiple gynaecological ailments, among them uterine prolapse and amenorrhea, and for common childhood diseases. The clinical details provided demonstrate that the compilation was written by and for medical personnel rather than party cadres. The unnamed authors described their own experiences carrying out the treatment methods introduced and displayed a willingness to change course based on patient feedback. The conference compilation included four treatments for uterine prolapse.

The first and most involved treatment method called for employing a combination of Chinese herbal medicine and smoke fumigation (*yan xun*) to cure prolapse. On the first day of this treatment, patients were to take the *Bu zhong yi qi tang* decoction, the ancient formula also prescribed in Zeng's 1960 TCM gynaecology textbook.[71] The smoke fumigation part of the treatment began on day two. After removing the hornets and pupa from a hornet's nest, stated the compilation, medical personnel would grind the nest into powder, pour the powder into a white paper bag, put the small mouth of a cylindrical cardboard tube into the bag of powder and aim the larger mouth of the tube at the prolapsed part of the patient's uterus. Health workers would then light the paper bag and use the smoke to fumigate the uterus through the tube. 'We changed the original sitting fumigation to lying down fumigation and set up simple beds for patients in hospitals and health stations', explained the unnamed medical personnel who wrote the account.[72]

After trying this treatment on nine women, the presumably female medical personnel in charge of the fumigation treatment took the time to seek feedback from their patients. A patient named Chen Lizhen told them: 'When fumigating longer, [I] feel the uterus contracting and I feel very comfortable after fumigating.' Ms Chen asked to have the treatment a few more times, for a longer period of time. 'We thus changed from fumigating once a day to twice a day and changed from 15 minutes each time to 30 minutes', explained the health workers, displaying considerable flexibility. The revised treatment lasted for five days and included ten smoke fumigations and five doses of medicine. Most patients were cured by that point, claimed the report, though a few needed additional days of treatment. After completing their course of smoke fumigation, women were given two or three doses of herbal medicine to stabilize the position of the contracted uterus and instructed to avoid heavy work for at least twenty days. 'All women with uterine prolapse can be cured', concluded the overview. 'It costs only about five yuan per person, and is welcomed by the masses.'[73] Like the educated rural youth who took part in the anti-schistosomiasis campaigns discussed by Miriam Gross, local health workers who tried different treatments for uterine prolapse also gained familiarity with 'scientific tools' such as forms and records, evaluation of treatment work and simple experiments.[74] Moreover, just as Clara Park and other American mothers discussed in Marga Vicedo's chapter in this volume attended scientific meetings on autism and then shared the knowledge they gained with other parents of autistic children, so the primarily female health workers who attended professional conferences and training courses on how to treat gynaecological ailments returned to their communes to implement what they had learned and train other commune members.[75]

While Wuwei County reports claimed that using 'alcohol block therapy' or a combination of herbal medicine and smoke fumigation cured uterine prolapse patients, the *Current Scene* interviewee did not have much success using Chinese herbal medicine. 'I'm sorry to say that there was not a single complete recovery out of all the eighty cases I treated', she stated bluntly. 'Quite apart from the need for a surgical operation', she continued, 'patients need plenty of nourishment and rest and I couldn't give them either'. Although this interviewee fled to Hong Kong soon after completing her six months in the countryside, she spoke positively of her time at the health station. She had enjoyed her work there, she explained, because 'my patients were so grateful and so very kind to me'. Given that the herbal remedies she provided did not cure their condition, it is important to ask what precisely the rural women treated by this nineteen-year-old medical school student were grateful for. It seems likely that the opportunity to gain a brief respite from the heavy labour required during the GLF by staying in the health station for a few days, combined with the fact that patients in the station received a daily ration of eighteen ounces of mixed grain at a time when ordinary villagers 'were only getting eight to nine ounces', were key factors that gave the health station and the young woman who staffed it some appeal.[76]

The food situation in rural China began to stabilize in mid-1961, due to the government's belated decision to decrease grain requisition and import food from abroad.[77] In Anhui the birth-rate recovered somewhat from its nadir in 1960, but the province still had a shortfall of over 704,000 births in 1961.[78] Women's reproduction and labour productivity remained a key concern for the Party, so the focus on preventing and treating 'women's illnesses' continued into 1962. For amenorrhea, taking herbal medicines such as motherwort remained the primary treatment, but it was not until women received sufficient food that the caseload declined significantly.[79] As for uterine prolapse, by the spring of 1961, it became clear to central and provincial-level health bureaus that the curative effect of treating uterine prolapse with 'alcohol block therapy', herbal medicines and fumigation was only temporary for most women. Beginning in April 1961, provincial authorities in Anhui began promoting the distribution and use of pessaries (*zigong tuo*) to prevent recurrence. This was a new strategy not mentioned in reports from 1959 to 1960, the *Current Scene* interview or Zeng Jingguang's 1960 TCM gynaecology textbook.[80] Perhaps because pessaries, in contrast to alcohol block therapy and fumigation treatment, did not give emaciated and exhausted women the opportunity to lie down and rest for a few days, the use of pessaries to treat uterine prolapse does not appear to have been welcomed by rural women. A Wuwei County report from August 1961, for instance, stated that pessaries had been distributed to women suffering from prolapse of the uterus, 'but they are not properly used because many patients still do not understand how to use them'.[81] A provincial-level report from Anhui reiterated that point in October 1961, stating, 'Some women know how to put them [pessaries] in, but not how to take them out. Other women fear using them, and in some places pessaries are even being used as toys or clothing hangers.'[82]

Effective treatment for gynaecological maladies continued to elude Wuwei County into 1962, months after the worst period of the famine had ended. According to a report compiled by the Wuwei County Health Department in February 1962, as the spring planting season approached the county had cured only 11.7 per cent of its existing uterine prolapse patients and just 1.8 per cent of amenorrhea patients.[83] In some cases women repeatedly delayed getting treatment because they had many children and were very busy with housework, explained the health department report. Moreover, at times cadres neglected the treatment of women's disease in favour of treating oedema (which affected men as well as women), and some medical personnel feared treating gynaecological ailments because they believed that patients with such maladies were especially quick to relapse. 'The treatment of women's illnesses develops slowly', acknowledged the report. The health department placed the burden for addressing such problems largely on female cadres and medical personnel. Cadres from the Women's Federation must encourage uterine prolapse patients to get treatment and arrange housing and bedding for them during their treatment, stipulated the report.[84]

In Anhui efforts to persuade women with uterine prolapse to wear a pessary intensified in spring 1962, when party committees met to discuss the urgency of treating gynaecological ailments that were having a negative influence on women's labour productivity during the spring planting season. In response, in April, Wuwei County authorities assigned 402 people, among them 226 female cadres, 139 medical personnel and 37 staff, to examine patients suffering from women's illnesses and distribute pessaries to them.[85] Reports concerning the spring 1962 campaign hint at a significant amount of opposition to pessaries on the part of rural women in Wuwei. A 23 April report from the county's Wucheng Commune, for example, states that while 243 pessaries had been distributed to patients, only 160 women were wearing them and just 87 of them knew how to use them correctly. 'Some patients refuse to be treated, using old age or embarrassment as an excuse', continued the authors. Similarly, in Hongqiao Commune twenty-five women had refused treatment. It was thus crucial for female cadres to teach patients how to insert and remove pessaries, stated the report, and how to apply medicine to them and wash them.[86] By June 1962, the food situation in Wuwei had improved enough that the number of patients suffering from uterine prolapse and amenorrhea finally decreased significantly.

The resistance to wearing pessaries did not follow suit. 'Most women say they have already recovered by using a pessary and they refuse to be checked again', stated a June report from Wucheng Commune. 'Women cadres need to check whether patients are wearing their pessaries', continued the report, and medical workers needed to 'patiently persuade patients to accept regular checks'.[87] In short, as famine conditions ended in 1962, female cadres and health workers were increasingly called on to coerce uterine prolapse patients to wear pessaries, less for their own well-being than to ensure that 'women's illnesses' no longer impeded women's labour productivity.

Conclusion

This essay has demonstrated that during China's GLF and famine, the pronounced medicalization of starvation made necessary by the Anti-Rightist Campaign and GLF led cadres to blame declining birth-rates and low women's labour productivity on 'women's illnesses' rather than malnutrition and overwork. Medical researchers and health workers influenced by Mao-era grassroots science employed a wide variety of TCM and biomedicine treatments to curtail the alarming increase in cases of amenorrhea and uterine prolapse during the disastrous 1958–62 period. Women medical practitioners played a newly prominent role during the Great Leap disaster, I argue, both by providing the brunt of the hands-on treatment for 'women's illnesses' and by publishing influential texts on TCM therapies for these ailments. In some ways, the CCP's decision to put female health workers in charge of treating 'women's illnesses' created space for women like the *Current Scene* interviewee and Zeng Jingguang to display an impressive degree of agency. Yet the degree of emphasis Party leaders at all levels placed on treating these non-fatal maladies during a time of mass starvation reveals the CCP's continued positioning of women as above all reproducers and labourers. Moreover, the Party's expectation that women cadres, doctors and health workers would coerce rural women into wearing pessaries in spite of their disinclination to do so showcases the Mao-era state's willingness to override women's desires and agency in order to advance the Party's goals. The explicit use of women trained in grassroots science and medicine to persuade and monitor other women also foreshadows in sobering ways the heavy-handed role of women who served as 'population police' with the advent of China's one-child policy two decades later.[88] In sum, the medical interventions employed during the Great Leap Famine did little to address the starvation and overwork that caused amenorrhea and uterine prolapse, but they illustrate the multifaceted role of women in Mao-era mass science. The medical personnel and interventions discussed highlight as well the development of a heroic narrative of 'curing women' that celebrated women doctors and affirmed the liberative posture of the CCP, while obscuring the profound difficulties that new labour demands coupled with old expectations placed upon Chinese women in the 1950s and 1960s.

Notes

1 J. Yang, *Tombstone: The Great Chinese Famine, 1958–1962*, trans. S. Mosher and J. Guo (New York: Farrar, Straus and Giroux, 2012), p. 430.
2 'Interview: "Politics Must Always Take Command…"', *Current Scene: Developments in Mainland China*, 1: 26 (5 March 1962), 8–9. *Current Scene* was published in Hong Kong from 1961 to 1978. In addition to scholarly essays and English translations of excerpts from mainland newspapers, early volumes of the journal include multiple interviews with Chinese refugees who fled to Hong Kong during the famine years. The journal has a pronounced anti-Communist bent, but many interviewees discuss positive as well as negative experiences they had in Maoist China.
3 For an example of women's hidden contributions to science during a similar timeframe but in a US setting, see Vicedo's chapter in this volume.
4 'Interview: "Politics Must Always Take Command…"'
5 'Interview: "Politics Must Always Take Command…"'
6 See Rai's chapter in this volume.
7 K. Taylor, *Chinese Medicine in Early Communist China, 1945–63: A Medicine of Revolution* (New York: Routledge, 2004), p. 113; D. Pietz, *The Yellow River: The Problem of Water in Modern China* (Cambridge, MA: Harvard University Press, 2015), p. 187.
8 R. Macfarquhar, *The Origins of the Cultural Revolution*, vol. 2: *The Great Leap Forward, 1958–1960* (New York: Columbia University Press, 1983), pp. 1–2.
9 Pietz, *The Yellow River*, p. 195; J.-L. Domenach, *The Origins of the Great Leap Forward: The Case of One Chinese Province*, trans. A. M. Berrett (Boulder, CO: Westview Press, 1995), pp. 139–41.
10 X. Meng, N. Qian and P. Yared, 'The institutional causes of China's Great Famine, 1959–61', *Review of Economic Studies*, 82 (2015), 1568–1611.
11 A. Ghosh, *Making It Count: Statistics and Statecraft in the Early People's Republic of China* (Princeton, NJ: Princeton University Press, 2020), pp. 16, 28, 249–50.
12 S. Schmalzer, *Red Revolution, Green Revolution: Scientific Farming in Socialist China* (Chicago, IL: University of Chicago Press, 2016), pp. 5, 28, 34, 37.
13 M. Gross, *Farewell to the God of Plague: Chairman Mao's Campaign to Deworm China* (Berkeley: University of California Press, 2016), pp. 11, 204.
14 Taylor, *Chinese Medicine in Early Communist China*, pp. 110, 117.

15 Taylor, *Chinese Medicine*, pp. 82–84, 103. The use of the term 'Traditional Chinese Medicine' or TCM is foreign language specific. In Chinese the term is simply 'Chinese medicine' (*zhongguo yixue* or *zhongyi*).
16 Taylor, *Chinese Medicine*, pp. 117–23.
17 Gross, *Farewell to the God of Plague*, pp. 203–4.
18 Gross, *Farewell to the God of Plague*, pp. 11, 211, 235.
19 Gross, *Farewell to the God of Plague*, pp. 34, 40, 235.
20 'Interview: "Politics Must Always Take Command…"'
21 R. King, 'Romancing the Leap: Euphoria in the moment before disaster', in K. E. Manning and F. Wemheuer (eds), *Eating Bitterness: New Perspectives on China's Great Leap Forward and Famine* (Vancouver: University of British Columbia Press, 2011), pp. 57–58.
22 A. K. C. Leung, 'Dignity of the nation, gender equality, or charity for all? Options for the first modern Chinese women doctors', in S. Y. S. Chien and J. Fitzgerald (eds), *The Dignity of Nations: Equality, Competition, and Honor in East Asian Nationalisms* (Hong Kong: Hong Kong University Press, 2006), pp. 73, 77–78. For more on women doctors in late-imperial and Republican China, see A. K. C. Leung, 'Women practicing medicine', in H. Zurndorfer (ed.), *Chinese Women in the Imperial Past: New Perspectives* (Leiden: Brill Academic Publishers, 1999); B. Andrews, *The Making of Modern Chinese Medicine, 1850–1960* (Honolulu: University of Hawai'i Press, 2014); N. E. Barnes, *Intimate Communities: Wartime Healthcare and the Birth of Modern China, 1937–1945* (Berkeley: University of California Press, 2018).
23 G. Hershatter, *The Gender of Memory: Rural Women and China's Collective Past* (Berkeley: University of California Press, 2014), pp. 156–57.
24 X. Fang, *Barefoot Doctors and Western Medicine in China* (Rochester, NY: University of Rochester Press, 2012), p. 52.
25 Fang, *Barefoot Doctors*, pp. 31, 37–41, 47.
26 'Interview: "Politics Must Always Take Command…"', pp. 1–2.
27 Q. Sun, "Reluctant 'Maocare': The shaping and practice of the rural healthcare system in the Wenzhou Region, 1949–1978' (PhD dissertation, University of Hong Kong, 2017), pp. 104–7, 116–17. See also a *Current Scene* interview with a male refugee doctor who was assigned to a rural health station for six months after graduating from medical college. 'Interview: "Years of Constant Caution…"', *Current Scene*, 3: 14 (3 October 1961), 1–2.
28 'Interview: "Politics Must Always Take Command…"', pp. 3–5.
29 'Interview: "Politics Must Always Take Command…"', pp. 3–5.
30 Fang, *Barefoot Doctors*, p. 40.
31 'Interview: "Politics Must Always Take Command…"', p. 8. For similar findings regarding women's hesitance to discuss their health problems with male doctors in early twentieth-century India, see Rai's chapter in this volume.
32 Leung, 'Women practicing medicine', p. 101; Fang, *Barefoot Doctors*, p. 51.
33 Fang, *Barefoot Doctors*, pp. 51–53.
34 Shijian Commune Health Station, 'Guanyu zhiliao zigong xiachui bing gong-zuo jihua' [Working plan for treating uterine prolapse], 25 December 1960,

pp. 1–3, File 1-55-1-1960-028, Anhui Province, Wuwei County Archives, in History Department Library, Shanghai Jiao Tong University. Hereafter cited as WCA.
35 Wuwei County Maternal and Child Health Station, 'Xinmin gongshe diaobo zhiliao zigong tuochui de gongzuo zongjie' [Work summary of the allocation and treatment of uterine prolapse in Xinmin Commune], 15 October 1959, pp. 1–6, File 1-55-1959-022, WCA.
36 Health Division of the Wuwei County People's Committee, 'Guanyu zigong tuochui shidian zhiliao zongjie de baogao' [Report on a summary of the pilot treatment for uterine prolapse], no date (1960), p. 1, File 1-55-1-1960-002, WCA. For a US example of women, in this case mothers of autistic children, who attended scientific meetings and shared what they learned with other parents, see Vicedo's chapter in this volume.
37 Hershatter, *Gender of Memory*, pp. 4–5.
38 K. E. Manning, 'Making a Great Leap Forward? The politics of women's liberation in Maoist China', *Gender & History*, 18: 3 (2006), 579.
39 King, 'Romancing the Leap', pp. 57, 61–62.
40 M. Thorborg, 'Chinese employment policy in 1949–78 with special emphasis on women in rural production', with tables, in *Chinese Economy Post-Mao: A Compendium of Papers Submitted to the Joint Economic Committee Congress of the United States*, vol 1: *Policy and Performance* (Washington, DC: US Government Printing Office, 1978), p. 582.
41 Hershatter, *Gender of Memory*, pp. 182–87.
42 Manning, 'Making a Great Leap Forward?' p. 578.
43 F. Pu, *Zhongyi dui jizhong funubing de zhiliao fa* [*Chinese Medicine Treatment Methods for Several Women's Illnesses*] (Beijing: Popular Science Press, 1958), pp. 1–2.
44 J. Yang, *Tombstone*, pp. 394–430. There continue to be widely differing estimates of the total number of famine deaths.
45 A. Hu, *The Great Leap Forward: 1957–1965*, trans. G. Hu and V. C. W. Hui (Singapore: Enrich Professional Publishing, 2014), pp. 114–18; R. Thaxton Jr., *Catastrophe and Contention in Rural China: Mao's Great Leap Forward Famine and the Origins of Righteous Resistance in Da Fo Village* (New York: Cambridge University Press, 2008), chapters 4–5; D. Yang, *Calamity and Reform in China: State, Rural Society, and Institutional Change Since the Great Leap Famine* (Stanford, CA: Stanford University Press, 1996), p. 66.
46 F. Wemheuer, *Famine Politics in Maoist China and The Soviet Union* (New Haven, CT: Yale University Press, 2014), pp. 99–101.
47 'Amenorrhea', 9 February 2023, www.mayoclinic.org/diseases-conditions/amenorrhea/symptoms-causes/syc-20369299 (accessed 14 May 2021).
48 Ministry of Health, 'Guanyu fuzhongbing deng liaofa jianjie' [Introduction to treatment for fuzhongbing and other illnesses], 7 January 1961, p. 8, File B242-1-1318-49, Shanghai Municipal Archives. Hereafter cited as SMA.
49 R. E. Frisch, *Female Fertility and the Body Fat Connection* (Chicago, IL: University of Chicago Press, 2002), pp. 10–17.

50 R. E. Frisch, 'Demographic implications of the biological determinants of female fecundity', *Social Biology*, 2: 1 (1975), 17–22.
51 'Uterine prolapse', 8 September 2022, www.mayoclinic.org/diseases-conditions/uterine-prolapse/symptoms-causes/syc-20353458 (accessed 29 July 2019).
52 A.-L. W. M. Coolen et al., 'Primary treatment of pelvic organ prolapse: Pessary use versus prolapse surgery', *International Urogynecology Journal*, 29: 1 (2018), 99–107.
53 Ministry of Health, 'Introduction to treatment', 7 January 1961, pp. 11–12, File B242-1-1318-49, SMA.
54 'Interview: "Politics Must Always Take Command…"', p. 8. From 1958 on, explains Hershatter, women were entitled to a month of rest in late pregnancy and a month for postpartum recovery. The post-pregnancy rest echoed the traditional practice of 'sitting the month' (*zuo yuezi*) or taking a month's bed rest. During the GLF, however, the pressure to contribute to agricultural production meant that many women were unable to take these leaves. Hershatter, *The Gender of Memory*, pp. 159–61, 174.
55 Ministry of Health Party Group, 'Guanyu fangzhi dangqian zhuyao jibing de baogao' [Report on prevention and treatment of current major diseases], January 20, 1961, in the Editorial Board of *The Chinese Great Leap Forward and Great Famine Database* (chief editor: Y. Song; Editorial Board: J. Guo, S. Ding, Y. Zhou, G. Dong, Z. Shen, Z. Zhou, X. Yu; foreword by J. Hu), *Database of Chinese Great Leap Forward and Great Famine 1958–1962* (Hong Kong: Fairbank Center for Chinese Studies at Harvard University; Universities Service Centre at the Chinese University of Hong Kong, 2013). Hereafter cited as GLF database.
56 J. Yang, *Tombstone*, pp. 395, 411–12, 416. Yang cautions that figures on unnatural deaths drawn from provincial statistics 'must be considered an underestimate'.
57 Anhui Provincial Health Bureau Party Group, 'Sheng weisheng ting dangzu guanyu dangqian jibing fangzhi qingkuang he jinhou yijian de baogao' [Report by the Provincial Health Bureau Party Group on the current state of disease prevention and treatment, with suggestions for future work], 16 April 1961, GLF database.
58 B. Yang and S. Cao, 'Cadres, grain, and sexual abuse in Wuwei County, Mao's China', *Journal of Women's History*, 28: 2 (2016), 35. See also S. Cao and B. Yang, 'Grain, local politics, and the making of Mao's Famine in Wuwei, 1958–1961', *Modern Asian Studies*, 49: 6 (2015), 1675–703. According to the Wuwei Gazetteer, Wuwei's population decreased by 245,000 during the famine, resulting in a death rate of 25.8 per cent.
59 Wuwei County Committee, 'Guanyu zigong tuo chui, bijing jiancha ji fangzhi fangfa de baogao' [A report on suggestions for treatment and prevention methods for uterine prolapse and amenorrhea], undated, likely from August 1959, pp. 25–33, File 1-55-1-1959-022, WCA.
60 Rai's chapter in this volume.
61 'Interview: "Politics Must Always Take Command…"', p. 9.

62 Zunyi Prefectural Party Committee, 'Zhonggong Zunyi diwei guanyu dangqian nongcun jibing qingkuang de baogao' [Report on the current state of epidemic diseases in rural areas], 14 July 1959, GLF database; Guangshan County Party Committee, 'Zhonggong Guangshan xian wei guanyu liaoyangyuan siwang qingkuang baogao – jibing qingkuang' [Report on deaths in convalescent hospitals – the state of diseases], exact date unknown, probably second half of 1960, GLF database.
63 Gross, *Farewell to the God of Plague*, pp. 27–28; Taylor, *Chinese Medicine*, p. 109.
64 Y.-L. Wu, *Reproducing Women: Medicine, Metaphor, and Childbirth in Late Imperial China* (Berkeley: University of California Press, 2010), pp. 25, 84–85. See also C. Furth, *A Flourishing Yin: Gender in China's Medical History, 960–1665* (Berkeley: University of California Press, 1999), pp. 170–73.
65 H. Zhu (ed.), *Zeng Jingguang* (Beijing: Zhongguo Zhong yiyao chubanshe, 2018), pp. 2–4.
66 Taylor, *Chinese Medicine*, pp. 127–31; Gynecology Teaching and Research Group at the Chengdu College of Chinese Medicine, *Zhongyi xueyuan shiyong jiaocai: Zhongyi fuke xue jiangyi* [Trial Textbooks for Chinese Medicine Colleges: Lecture Notes in Chinese Gynecology] (Beijing: Renmin weisheng chubanshe, 1960).
67 *Zhongyi fuke xue jiangyi* (1960), p. 85.
68 *Zhongyi fuke xue jiangyi* (1960), pp. 36, 41, 85–86; X.-F. Zheng , J.-S. Tian, P. Liu, J. Xing and X.-M. Qin, 'Analysis of the restorative effect of *Bu-zhong-yi-qi-tang* in the spleen-qi deficiency rat model using ^{1}H-NMR-based metabonomics', *Journal of Ethnopharmacology*, 151: 2 (2014), 912–20. For examples of medicinal herbs commonly prescribed to treat gynaecological ailments in late-imperial China, see Furth, *A Flourishing Yin*, pp. 85–87, 164–65, 204, 238–40; Wu, *Reproducing Women*, pp. 2, 176, 210–12, 215.
69 Cao and Yang, 'Grain, local politics, and the making of Mao's famine in Wuwei', pp. 1675–703.
70 Youfang Health Center of Hongmiao Commune, 'Guanyu shi zhi zigong xiachui bing' (xiaojie) [Summary of trial treatment for uterine prolapse], 26 September 1960, pp. 1–4, File 1-55-1-1960-029, WCA. Many additional Wuwei County reports written from fall 1959 through early 1961 recommended the 'alcohol block' treatment.
71 'Chaohu zhuanqu fuyou weisheng zhuanye huiyi ziliao' [Material from the professional Maternal and Child Health Conference in Chaohu special district], undated (1960), pp. 1–14, File 1-55-1-1960-002, WCA.
72 'Chaohu zhuanqu fuyou weisheng zhuanye huiyi ziliao' (1960), pp. 5–6.
73 'Chaohu zhuanqu' (1960), p. 6. In 1960, paying five yuan and missing multiple days of work was a substantial cost for commune members.
74 Gross, *Farewell to the God of Plague*, pp. 11, 211, 235.
75 'Chaohu zhuanqu' (1960), pp. 1–14; Vicedo's chapter in this volume.
76 'Interview: "Politics Must Always Take Command…"', p. 9.

77 Wemheuer, *Famine Politics*, pp. 142–47. The first grain shipment from abroad arrived in February 1961.
78 Yang, *Tombstone*, pp. 411–12.
79 Anhui Province Health Bureau Party Group, 'Guanyu dangqian jibing fangzhi qingkuang he jinhou yijian de baogao' [Report on the current state of disease prevention and treatment, with suggestions for the future], 16 April 1961, GLF database.
80 Report by the Anhui Provincial Health Bureau, the Provincial Women's Federation, and the Provincial Science and Technology Commission, 'Guanyu funubing fangzhi qingkuang he jinhou yijian de baogao' [On preventing and curing women's diseases, with suggestions for the future], 30 October 1961, GLF database.
81 Wuwei County Committee Office for Eliminating Pests and Diseases, 'Wuwei xian Yaogou gongshe Wuzhou dadui mie luo xiaoguo kaohe yu sibing qingkuang de diaocha baogao' [Investigative report on the effect of snail elimination and the situation of the four diseases in Wuzhou Brigade, Yaogou Commune, Wuwei County], 23 August 1961, p. 61, File 1-56-1-1961-009, WCA.
82 Report by the Anhui Provincial Health Bureau, 30 October 1961, GLF database.
83 Wuwei County Health Department, 'Fangzhi funubing de gongzuo yijian (chugao)' [Suggestions for the prevention and treatment of Women's Diseases (first draft)], February 1962, File 1-55-1-1962-045, WCA.
84 'Fangzhi funubing', February 1962, File 1-55-1-1962-045, WCA.
85 Wuwei County Committee Leadership Group, Office for Eliminating Pests and Diseases, 'Guanyu dangqian fuke bing fangzhi qingkuang jianbao' [Briefing on the current situation of gynecological disease prevention and treatment], April 16, 1962, File 1-55-1-1962-045, WCA.
86 'Wucheng qu zhiliao fukebing zongjie huibao' [Summary report on the treatment of gynecological diseases in Wuchen District], 23 April 1962, File 1-55-1-1962-045, WCA.
87 Health Center of Wucheng People's Commune, Wuwei County, '1962 nian shang bannian fang bing zhi bing gongzuo zongjie' [Summary of disease prevention and treatment work in the first half of 1962], 25 June 1962, File 1-55-1-1962-045, WCA.
88 M. Fong, *One Child: The Story of China's Most Radical Experiment* (Boston, MA: Houghton Mifflin, 2016), pp. 65–84.

IV

Intimate knowledge and in/visible domesticities: Science, medicine and the home

9

The curious case of Yashoda Devi, a woman Ayurvedic practitioner in colonial India

Saurav Kumar Rai

On 2 April 1870, the acclaimed *British Medical Journal* (hereafter *BMJ*)[1] published a lead article under the title 'Lady surgeons'.[2] Apart from debating the suitability of women for the medical profession, the article also pondered whether their mental faculties equalled those of men. In trying to answer these questions, it emphasized a judicious division of labour as one of the chief gains of civilization and progress, first and foremost among them being the distribution of suitable tasks in life between the two sexes. The anonymous writer also stated that lady-doctors were 'traitress[es] to [their own] sex'.[3] The article thus exposed entrenched notions about the scientific acumen of women, especially those pursuing the medical profession.[4] In fact, studies show that gender inequality in this field continues to be prevalent in terms of lower pay, inappropriate conduct and paltry leadership roles.[5]

In this context, if we look at the field of 'indigenous' medicine in India in the late nineteenth and early twentieth centuries, we notice that it was almost exclusively dominated by male practitioners. This chapter explores an exception to that rule, namely the case of Yashoda Devi.[6] Devi, as she was popularly referred to, was a famous Ayurvedic practitioner from early twentieth-century United Provinces, in colonial North India, who wrote and edited several volumes and journals on Ayurveda. She was based in Allahabad, the capital city of the United Provinces of Agra and Oudh. She was the founder of Stri Shiksha Pustakalaya, a publishing house, and editor of the only Ayurvedic journal devoted exclusively to women and children, *Stri Chikitsak (Female Practitioner)*. Her proactive presence in a field hegemonized by male practitioners was revolutionary in itself. However, as this chapter shows, patriarchy was often rearticulated and reinforced even in Yashoda Devi's writings. Such considerations notwithstanding, it is difficult to locate her in any static frame. Although reinforcing patriarchal norms on some occasions, through her focus on relatively taboo subjects like women's sexual pleasure and non-consensual sex, Yashoda Devi also offered a critique of male sexual behaviour and occasionally gender hierarchies. In other words, her writings are fragmented and kaleidoscopic, a topic discussed in detail in what follows.

This chapter thus seeks to address one of the fundamental objectives of the volume, namely, to promote new agendas for the study of women in the history of science by moving beyond the popular 'heroine' model to investigate the many hidden figures who contributed to processes of knowledge-making. Although Yashoda Devi has been a relatively visible figure, she is still a marginal one in the broader discourse on Ayurvedic revivalism. This can be discerned from the fact that her name hardly appeared in the professional circles of contemporary Ayurvedic practitioners. There was widespread indifference towards her writings, coupled with hostile antagonism towards her from Ayurveda's normative authority.[7] Even in later historiographical analyses the role of women practitioners in the making of 'modern' Ayurveda is hardly documented, with the narrative being heavily skewed in favour of male practitioners. As Jean Corbi argues in this volume, it is not that women healers were absent; rather, they were involved in more informal healing practices. Yashoda Devi certainly traversed this binary of 'formal' and 'informal', thereby offering an interesting case study for women and science in twentieth century.

Western medical hegemony and Ayurvedic revivalism

Colonialism and the spread of Western medicine were inextricably connected. In many ways, colonialism provided a broader horizon to Western medicine[8] to test its efficacy and carry out new research which was otherwise not possible in the metropole.[9] In this context, one can see a clear colonial bias in favour of Western medicine in India at least from the second half of the nineteenth century on. The 'indigenous' ways of healing were simultaneously regarded as grossly 'unscientific' and often devalued to the extent of quackery.[10] The frequent colonial criticism of Indian system(s) of healing like Ayurveda and Unani as 'primitive', 'prehistoric', 'stagnant' and 'non-scientific' methods of treatment, amounting to no more than 'quackery', disgruntled the reformist nationalist elite of the time. The revival of a 'truly' Indian 'indigenous' system of healing, worthier than Western medicine, became crucial to their very identity.[11]

The opportunity for such revivalism was created largely by the failure of colonial public health infrastructure,[12] which was centred primarily around Western medicine, during the plague epidemic of the 1890s. This was the decade when a growing political consciousness, along with the active pursuits of 'indigenous' medical practitioners, led to the rise and growth of what may be broadly termed 'medical nationalism'. One important tenet of this medical nationalism was the revival of 'indigenous' healing systems as well as practices of Indian medicine like Ayurveda, Unani, Yoga and Siddha. Ayurvedic revivalism took the lead, with various organizations coming up in the early part of the twentieth century which significantly transformed the very texture and form of Ayurvedic healing.[13]

Ayurveda and its utility were debated and discussed in public forums and attempts were made to obtain for it the 'due' place and the necessary patronage it 'rightfully deserved'.[14] Print, organization and mobilization were the three major ways in which attempts were made to establish the sanctity of Ayurveda not only over Western medicine but over all other forms of 'indigenous' healing systems. This led to the creation of a substantial Ayurvedic discourse which did not pertain only to the medical aspects of the healing system. Rather, a careful reading of the Ayurvedic journals, books, tracts, pamphlets and so on of this period, an analysis of the Ayurvedic debates in the public sphere and an exploration of the activities of organizations such as All India Vaidya Sammelan[15] and its provincial units throw light on the socio-cultural and religious processes of the time. The kind of social culture which early twentieth-century mainstream Ayurvedic discourse, at least in North India, exhibited was highly casteist, communal and gender- and class-biased in its content as well as intent.[16] Let us discuss the gendered prejudices of that Ayurvedic discourse in detail to contextualize Yashoda Devi's work.

Gender predilections in late-colonial Ayurvedic discourse

In 1912, Rai Pooran Chand, who was associated with the All India Vaidyak Vidyapith and was one of the members of the commission appointed by the All India Vaidya Sammelan to enquire into plague, cholera, malaria and other epidemic diseases in India and their remedies, delivered a speech at the annual session of the Sammelan held at Kanpur. He described an incident when he had gone to examine a female patient on the request of a reputed client. When he visited her for the first time, she was unconscious and unveiled. He examined her pulse and prescribed some medicine. By the time Rai Pooran Chand visited her a second time, she had regained consciousness. When he was examining her pulse this time, she veiled herself in modesty. As Rai Pooran Chand put it, 'That day the patient showed me the pulse on her own. She veiled herself in modesty, although twelve hours before (when she was unconscious) despite the attempts of the family members she had shown no such signs of modesty.'[17] In this description, we find an indirect reinforcement of the *purdah* or veil by a medical practitioner. In his speech, he very cleverly highlighted the importance of *purdah* insofar as a conscious person was concerned. Thus, as the woman patient regained her consciousness, she found it obligatory to veil herself during check-up.

The excerpt from Rai Pooran Chand's speech is a very subtle example of gendered predilections as manifested in early twentieth-century mainstream Ayurvedic discourse. The Ayurvedic practitioners of the period under discussion often reinforced patriarchy, even if not always consciously. While discussing diseases related to women, texts like *Arogya Darpan* (1898) often assumed a didactic tone, advising women how to conduct themselves in the social sphere. Here, it is noticeable that a very popular genre of literary production of early twentieth-century North India was didactic literature. This didactic literature attempted to enshrine upper-caste/middle-class norms, values and concerns within society. Of course, control of women's sexuality and mobility was one of the foremost concerns of such literature.[18]

Furthermore, if we look at the Ayurvedic tracts on childcare (referred to as *Santati Shastra*) of the period under discussion, we find that these tracts also reflected gender biases. While the healthy upbringing of a male child was extremely significant, the female child's upbringing was virtually a non-issue for most of the Ayurvedic childcare tracts produced during this period.[19] It is instructive of the social culture of the Ayurvedic discourse of the period that childcare was intimately associated with parenting. In the case of a girl child what was significant was only her proper social upbringing, which is evident from numerous didactic texts on this matter. Female health became a significant issue only once they reached adolescence. This was largely because they were supposed to play significant social and familial roles – particularly reproductive roles – in their adulthood, which was not possible if they remained ill.

As Shalini Shah argues, even seminal Ayurvedic texts like *Charaka Samhita*, *Vagabhatta Samhita* and *Madhavanidana* marginalize, if not exclude, women in their discussion of diseases. The only discussions of the female body in these texts are in connection with the diseases of the uterus (*garbhavyapta*) and genital area (*yonivyapta*) largely because, as Shah argues, it was only the uterus and genital area of women which were of some significance to a patriarchal male society and related health discourse. Thus, even the discussion on *garbhavyapta* and *yonivyapta* was not generated by concern about women's health per se, but by the fact that a healthy uterus was crucial for a healthy childbirth. In all other cases, men's health and virility took prevalence over the health of women in these texts, a situation which reduced them to the role of 'sex women'.[20]

Furthermore, one of the main areas of interest of the Ayurvedic texts of this period was to look at the possibilities of the birth of a male child. This was quite natural in a patriarchal society where female infanticide and dowry were prevalent.[21] In addition, a male child was required to perform various religious roles, including funeral rites.²² Many texts on this topic often contained a subsection called '*Manmani* or *Manchahi Santan*' ('Desired child'). One can find texts devoted entirely to this theme and one can cite here *Manchahi Santan* (Allahabad, 1928), written by Rishilal Agarwal, as an example. *Vaidya Priya*, an Ayurvedic text written in verse form, came up with bizarre suggestions advocating the use of pigeon's excrement, cannabis seeds and so on to secure a male child.²³

In fact, many pseudo-scientific theories regarding ways and appropriate timings of sexual intercourse were propounded by several Ayurvedic tracts to realize the patriarchal desire of obtaining a male child. Even well-established Ayurvedic journals like *Dhanvantari*²⁴ could not escape this dominant template of the late-colonial Ayurvedic discourse. For example, one of the articles published in *Dhanvantari* (September 1934) by Vaidya Govind Prasad Varshaneya of Moradabad proposed a unique theory according to which the birth of a male child could be achieved through the fertilization of an ovum from the right ovary of a woman by the sperm from the right testis of a man. By contrast, the fertilization of sperm and an ovum produced by the left testis and ovary respectively was believed to result in the birth of a female child. Govind Prasad Varshaneya went on to suggest ways in which the activities of the left testis or ovary could be suppressed, for example, by tying the left testis during copulation.²⁵

The late-nineteenth- and early twentieth-century Ayurvedic discourse on midwifery was equally gender-biased. Despite being engaged in a head-to-head fight against Western medicine, the Ayurvedic practitioners of the time cherished the idea of the *sarkari* or state-sponsored training of the traditional *dais* (midwives). Besides caste, class and communal biases, gender was also involved in this reformist urge to professionalize traditional *dais*. Prior to the twentieth century, the issue of childbirth in India was a female-dominated sphere. Men were excluded completely from *sutika griha* or *antur ghar*²⁶ and they hardly had any say in birth rituals and activities. However, the professionalization of midwifery was supposed to open this sphere even for males, as it did in Europe. Queen Victoria had herself been delivered by a male surgeon.²⁷ In fact, controlling women's reproductive health was central to securing a middle-class life in the metropolitan world as well as in colonial India. Exclusive women's spheres were looked upon with a lot of suspicion and disdain by the male members of the reformist middle class and they attacked these spheres which were associated directly with women under some pretext or the other. Sometimes the language of this attack was communal (such as in the case of attack on women's practice of visiting the *pirs*)²⁸ and sometimes casteist (as in the case of the *dai*), but a hidden gendered agenda of exercising control and regulating these relatively 'semi-autonomous' spheres was always present.

Equally relevant is the contemporaneous Ayurvedic discourse on *Brahmacharya* or celibacy.[29] This was an issue significant for the Ayurvedic practitioners, not only from the viewpoint of health, but also in the broader interests of the 'community', 'society' and the 'nation'. Hence, one can find repeated references to *Brahmacharya* and its significance in the Ayurvedic tracts of the period under discussion. Celibacy and sexual self-control came to be viewed as the most important way to discipline one's own body and to make oneself 'healthy' and 'masculine' so that one could serve in the interest of the broader public and the 'nation'.[30] In this regard, the Ayurvedic texts severely criticized masturbation and anal sex or sodomy. In fact, these two were often regarded as 'horrendous' and 'immoral' forms of sex.[31] Many of the Ayurvedic tracts reproduced popular prejudices regarding the overt sexual appetite of women and sexual corruption of girls belonging to lower castes and class. The author of *Arogya Darpan* stated that 'it has often been seen that if a gorgeous, charming and extremely beautiful girl is born in the lower clan or in the poor's quarter, then she becomes sexually corrupt (*vyabhicharini*)'.[32] Similarly, Chatursen Shastri blamed the low-caste women who frequented the homes of respected families as servants for corrupting young girls, widows and daughters-in-law sexually.[33]

Ayurveda and the world of Yashoda Devi

It was in this gendered atmosphere of Ayurvedic revivalism that Yashoda Devi treaded her path as a female Ayurvedic practitioner. According to the available sources, Yashoda Devi received her training in Ayurvedic healing from her father, who was himself an Ayurvedic practitioner. In this regard, she was very much different from the institutionally trained *vaids*[34] of the era. She devoted her attention and energy especially to the treatment of women. For this, she considered Satyabhama Bai, a female doctor from Nepal, her role model.[35] She set up her clinic in Allahabad, named Stri Aushadhalaya, which became quite popular among women. She also established a publishing house, Stri Shiksha Pustakalaya, and was the editor of magazines devoted to women and child health such as *Stri Chikitsak, Kanya Sarvasva* and *Stridharma Shikshak*. At the same time, she authored many books and booklets related to women's health and daily routines (see Figure 9.1).

It is noticeable that Yashoda Devi's oeuvre, besides being medical, also reflected concerns which aimed to improve the conduct of women and make them ideal housewives. In other words, her writings can be classified within the genre of didactic literature, besides being Ayurvedic. Yashoda Devi also claimed a larger status than that of a mere medical practitioner

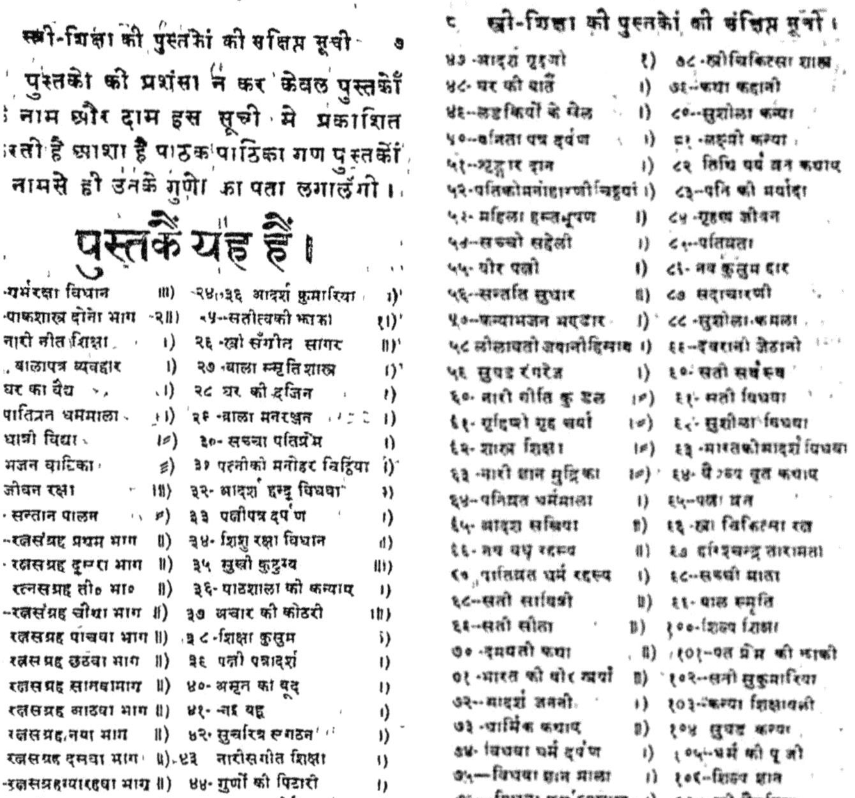

स्त्री-शिक्षा की पुस्तकों की संक्षिप्त सूची ७

पुस्तकों की प्रशंसा न कर केवल पुस्तकों के नाम और दाम इस सूची में प्रकाशित करती हैं आशा है पाठक पाठिका गण पुस्तकों के नामसे ही उनके गुणों का पता लगालेंगे।

पुस्तकें यह हैं।

गर्भरक्षा विधान ।।।)
पाकशास्त्र दोनों भाग २।।)
नारी नीत शिक्षा ।)
बालापत्र व्यवहार ।)
घर का वैद्य ।)
पातिव्रत धर्ममाला ।)
धात्री विद्या ।=)
भजन वाटिका =)
जीवन रक्षा ।।।)
सन्तान पालन =)
रत्नसंग्रह प्रथम भाग ।।)
रत्नसंग्रह दूसरा भाग ।।)
रत्नसंग्रह ती० भा० ।।)
रत्नसंग्रह चौथा भाग ।।)
रत्नसंग्रह पांचवा भाग ।।)
रत्नसंग्रह छठवां भाग ।।)
रत्नसंग्रह सातवां भाग ।।)
रत्नसंग्रह आठवां भाग ।।)
रत्नसंग्रह नवां भाग ।।)
रत्नसंग्रह दसवां भाग ।।)
रत्नसंग्रह ग्यारहवां भाग ।।)
रत्नसंग्रह बारहवां भाग ।।)
रत्नसंग्रह तेरहवां भाग ।।)
२४।३६ आदर्श कुमारियां ।)
२५--सतीत्वकी झांकी १।)
२६-स्त्री संगीत सागर ।।)
२७-बाला स्मृति शास्त्र ।)
२८ घर की दुलहिन ।)
२९-बाला मनरञ्जन ।)
३०-सच्चा पतिप्रेम ।)
३१ पत्नीको मनोहर चिट्ठियां ।)
३२-आदर्श हिन्दू विधवा ।)
३३ पतीपत्र दर्पण ।)
३४-शिशु रक्षा विधान ।)
३५ सुखी कुटुम्ब ।।।)
३६-पाठशाला की कन्याएं ।)
३७ अचार की कोठरी ।।।)
३८-शिक्षा कुसुम ।)
३९ पती पत्नादर्श ।)
४०-अमृत का बूंद ।)
४१-नई बहू ।)
४२-सुचरित्र संगठन ।)
४३ नारीसंगीत शिक्षा ।)
४४-गुणों की पिटारी ।)
४५ आदर्श-रसोइया ।)
४६-सती भूषण ।)

८ स्त्री-शिक्षा की पुस्तकों की संक्षिप्त सूची।

४७-आदर्श गृहिणी १)
४८-घर की बातें ।)
४९--लड़कियों के खेल ।)
५०--वनिता पत्र दर्पण ।)
५१--शृङ्गार ज्ञान ।)
५२-पतिकोमनोहारणी चिट्ठियां ।)
५३-महिला हस्तभूषण ।)
५४--सच्ची सहेली ।)
५५-वीर पत्नी ।)
५६--सन्तति सुधार ।।)
५७--कन्याभजन भण्डार ।)
५८ लीलावती भवानीहिम्मत ।)
५९ सुघड़ रंगरेज ।)
६०-नारी नीति कुण्डल ।=)
६१-सुहिणी गृह चर्या ।=)
६२-शास्त्र शिक्षा ।=)
६३-नारी ज्ञान मुद्रिका ।=)
६४--पतिव्रत धर्ममाला ।)
६५-आदर्श सखिया ।।)
६६-नव वधू रहस्य ।।)
६७-पातिव्रत धर्म रहस्य ।)
६८--सती सावित्री ।।)
६९--सती सीता ।।)
७०-दमयन्ती कथा ।।)
७१-भारत की वीर स्त्रियां ।।)
७२--आदर्श जननी ।)
७३-धार्मिक कथाएं ।।)
७४-विधवा धर्म दर्पण ।)
७५--विधवा ज्ञान माला ।)
७६--विधवा कर्त्तव्यशास्त्र ।)
७७-विधवा धर्म भूषण ।)
७८-स्त्रीचिकित्सा शास्त्र
७९--कथा कहानी
८०--सुशीला कन्या
८१-लक्ष्मी कन्या
८२ तिथि पर्व व्रत कथाएं
८३--पति की मर्यादा
८४-गृहस्थ जीवन
८५--पतिव्रता
८६-नव कुसुम हार
८७ सदाचारणी
८८-सुशीला कमला
८९--देवरानी जेठानी
९०-सती सर्वस्व
९१-सती विधवा
९२-सुशीला विधवा
९३-भारतकी आदर्श विधवा
९४-पैंसठ व्रत कथाएं
९५--पत्नी व्रत
९६-स्त्री चिकित्सा रत्न
९७ हरिश्चन्द्र तारामती
९८--सच्ची माता
९९-बाल स्मृति
१००-शिल्प शिक्षा
१०१-पति प्रेम की झांकी
१०२--सती सुकुमारियां
१०३-कन्या शिक्षावली
१०४ सुघड़ कन्या
१०५--धर्म की पूंजी
१०६--शिल्प ज्ञान
१०७-स्त्री वैद्यविद्या
१०८-भारतका नारीइतिहास

पता:-यशोदादेवी पे० बक्स न० ४ कर्नलगंज प्र०

Figure 9.1 List of 108 books published by Yashoda Devi (illustration from Yashoda Devi, *Samsar ka Nari Itihas (Bharat ka Nari Itihas)*, vol. 2, Allahabad, 1922)

who specialized in female and child healthcare. She described herself as *Stridharma Sikshak*.[36] This self-assumed title reflected a coalescing of social and health concerns in the Ayurvedic discourse of the time, whereby a medical practitioner was also assuming the title and role of a social mentor.

In fact, the range of books published by Yashoda Devi clearly reflects her didactic concerns. Apart from publishing books on Ayurveda, female health and childcare, she also produced publications like *Nari Niti Shiksha* (*Teachings on Ethics for Women*, 1910), *Sachcha Pati Prem* (*True Wifely Love*, 1910), *Sugharh Grihani* (*Expert Housewife*, 1924), *Grihani Kartavya Shastra Arogya Shastra athva Pak Shastra* (*Manual for Housewife for a Healthy Family or on Cookery*, 1924), *Adarsh Pati-Patni aur Santati Sudhar*

(*Ideal Husband-Wife and Eugenics*, 1924), *Pati-bhakti ki Shakti arthat Pati ki Maryada* (*Power of Devotion to Husband or Husband's Honour*, 1925), *Kanya Kartavya* (*Duties of Unmarried Girls*, 1925), *Pativrat Dharma Mala* (*Duties of a Married Woman*, 1926), *Nari Dharma Shastra Grih Prabandh Shiksha* (*Education on Home Management for Women*, 1931) and so on. These books claimed to reveal the secrets of a happy household life to women.

As Gyan Prakash argues, such prescriptive texts were in line with the typical middle-class imagery of Hindu wives who had to preserve the health of their families and thereby of the 'Hindu' nation through 'scientific' management within the household.[37] Devi believed that the root cause of many a women's health problems was connected to their ignorance and troubled married life. That is why she emphasized female education, in particular the moral and ethical education of women. She constantly underlined the necessity of *Grihastha Shiksha* (education related to domestic affairs) to promote the healthy life of the entire family. If in the case of males the *Brahmacharya ashram*[38] was the focal point of attention in Ayurvedic discourse, in the case of females it was the *Grihastha ashram* which received emphasis. In fact, in one of Yashoda Devi's writings, *Grihani Kartavya Shastra Arogyashastra athva Pakshastra* (1924), cooking became intrinsic to the proper healthcare of family, the responsibility of which fell upon housewives.[39] It is noticeable that for Devi, cooking based on rules of health and hygiene was a 'scientific' activity and not merely a matter of taste. She even seized upon an opportunity to portray the fictional near-death of a Hindu male caused by the neglect of the rules of nutrition by his wife.[40] According to her, the ideal housewife rose early, took charge of the household and prepared food in accordance with the laws of nature and the season. This approach provided scientific legitimation to patriarchal norms prescribed for Hindu housewives.

Similarly, if we look at the childcare tracts written by Yashoda Devi we notice that they embodied the gendered predilections of the late-colonial Ayurvedic discourse. For instance, while talking of *Santan Palan* (the rearing of progeny), Devi focuses entirely on the healthy upbringing of the male child, not of the female child. It should be noted that *santan* (progeny) is a gender-neutral term. But in her text *Santan Palan* (*Rearing of Progeny*, n.d.) she uses either the term *ladka* or *bachcha* (boy) and nowhere *ladki* or *bachchi* (girl).[41] A similar case can be seen in *Shishu Raksha Vidhan* (*Manual for the Protection of Children*, 1912). *Shishu* (baby) is also a gender-neutral term but in the text she uses exclusively the term *balak* (boy).[42] Furthermore, one of her texts, *Dampati Arogyata Jeevanshastra* (*The Science of a Healthy Conjugal Life*, 1927), contains a section on *Manmani* or *Manchahi Santan*, or the 'desired child'.[43]

The dominant Ayurvedic theme of sexual self-control or the discourse on *Brahmacharya*, as discussed in the preceding section, also featured in the writings of Yashoda Devi. Stressing the necessity for conservation of semen through the observation of strict celibate behaviour, Yashoda Devi wrote that, 'Semen is like the king matter of the body, a man who does not conserve it and indulges in its wasteful outflow, remains afflicted with various diseases throughout his life and never gets marital happiness as he wants. Hence, man should conserve the semen diligently.'[44] In fact, she considered that the loss of semen was responsible for the onset of constipation.[45] Similarly, attacking excessive sexual desire, masturbation and wet dreams (*swapnadosh*), she believed that the large percentage of tuberculosis patients were guilty of uncontrolled sexual behaviours, followed by those who masturbated and finally those 'suffering' from wet dreams.[46] Furthermore, she argued that masturbation, anal sex, excessive uncontrolled sex and visits to prostitutes were responsible for male impotency as well.[47] Thus, in her writings, Yashoda Devi regarded the conservation of semen as the root of good health, a happy married life and the production of healthy offspring. These ideas strengthened upper-caste, middle-class Hindu notions of *Brahmacharya* that were in vogue during the period under discussion. To be sure, this was a very common discourse in eighteenth- and nineteenth-century Europe as well. Unease about masturbation and consequent loss of semen began in the early eighteenth century, when a book titled *Onania; or The Heinous Sin of Self-Pollution (1707–17)* appeared anonymously in Holland. By the middle of the century, noted Swiss physician Samuel-Auguste Tissot gave scientific validity to this 'vice' through his treatise *Onanism: A Treatise upon the Disorders Produced by Masturbation* (1760). Tissot argued that loss of semen led to general debility, consumption, deterioration of eyesight, nervous disorders and so on.[48]

Yashoda Devi also advocated controlling the reading habits of the common people. Here it is worth noting that according to the Ayurvedic concept of *Ashta-maithun* (eight kinds of sex),[49] which supposedly led to the loss of *Brahmacharya*, *guhya bhasan* or secret talk was one of them. *Guhya bhasan* referred to chatting secretly with women or indulging in lustful talk with them, as well as reading sleazy novels, stories, plays and so on. Complaining about the reading habits of lay readers, especially women, Yashoda Devi lamented the fact that she failed to meet even the cost of putting a stall at the *Magh Mela*,[50] while at the same time stalls selling sleazy novels and songs recorded a robust sale.[51] Furthermore, she complained that literate women were so fond of such 'dirty' literature that one could easily find one or two of them in their trunk.[52] She also claimed that she used to receive several letters daily demanding such sleazy literature.[53] This bewailing tone clearly reflected her concerns about what was regarded as the 'rightful conduct' of women in general.

All this is not to undermine the 'unique' position that Yashoda Devi held in the field of a healing system (Ayurveda) which was dominated almost exclusively by male practitioners. It is obvious that women could freely discuss their medical conditions with a woman practitioner in ways they could not do with a male counterpart.[54] In this regard, the very presence of Yashoda Devi in an exclusive male sphere had its own significance as she must have had a comforting impact on her clientele. She also published several self-healing manuals for women whose original purpose was to make Ayurvedic treatment accessible to women even in the confines of the home. The language of these manuals was plain and the remedies suggested quite handy.[55] Given the kind of restrictions placed on the mobility of women in the patriarchal society of the period, these self-help and other such booklets pertaining exclusively to the daily familial concerns of women had a considerable readership.[56]

Along with this, Yashoda Devi used to provide medical advice to women through the post as well. In fact, she was so popular, writes Charu Gupta, that letters addressed just 'Devi, Allahabad' usually reached her.[57] Free postal assistance and medical advice to patients were offered by many of these practitioners as well as Ayurvedic pharmacies. They also arranged for the pre-paid delivery of medicines through the post. For instance, Dhanvantari Aushadhalaya of Aligarh used to provide guidance to those facing health-related problems and dispatched medicines using the postal services. Certainly, it helped practitioners like Yashoda Devi to reach a wider clientele.

Yashoda Devi not only challenged but also occasionally inverted the prevailing male-centric discourse on couple relationships through her discussion of women's sexual pleasure, condemnation of sex against one's will or deliberation on the role of males behind non-procreative marriages. Her discussion of sexual intercourse was remarkable in itself. As Charu Gupta rightly observed, Yashoda Devi's writing was 'ambiguous and fractured at various levels' and it is difficult 'to enframe her within any fixed category'. According to Gupta, her writings simultaneously reiterated gender hierarchies and questioned them.[58]

Some of Devi's books, such as *Dampati Arogyata Jeevanshashtra arthat Ratishastra Santatishastra* (*The Science of Healthy Conjugal Life or the Science of Sexual Intercourse and Procreation*, 1927), *Dampatya Prem aur Ratikriya ka Gupt Rahasya* (*Conjugal Love and Secrets of Sexual Intercourse*, 1933), *Nari Sharir Vigyan Stri Chikitsa Sagar: Sambhog Vigyan* (*Women's Physiology and Medical Treatment: Science of Intercourse*, 1938) and *Vivah Vigyan Kamshastra* (*Science of Marriage and Sex*, n.d.) discussed female sexual pleasure in detail and successfully brought this subject within the ambit of the ongoing Ayurvedic discourse and medical science. Yashoda Devi condemned those males who were unable to provide sexual pleasure to their wives.[59] Invoking *Dharmashastras*,[60] she argued that it was the duty of a husband to satisfy his wife after *ritusnan*,[61] failing which he was guilty of a sin.[62] She further wrote that the very term *sambhog*[63] literally meant equal enjoyment, both by man and woman, during sexual intercourse. This alone was the real marital pleasure, and not the fact that the wife was suffering from diseases and sorrow while her husband kept satisfying his sexual appetite by indulging in forced intercourse.[64] In this connection, Yashoda Devi severely condemned non-consensual sex. According to her, husbands who did not respect the sexual consent of their respective wives were worse than animals.[65] In fact, she praised here Western culture, where there was a special emphasis on consent and the sexual pleasure of women.[66]

At the same time, it is worth mentioning that in a society where women were often held responsible for the failure of giving birth to a child (especially a male child) after marriage, Yashoda Devi dared to write that men were not only equally responsible for non-procreation, but in most of the cases they were actually more responsible than women.[67] As a corollary, she lamented the social ostracization of females in this regard, stating that her clinical experiences evidently showed that women mostly faced the brunt of deficiency of their respective husbands in this matter.[68] Further, being a moralist, Yashoda Devi advocated identical sexual morality both for husband and wife, thereby vehemently criticizing sexual conduct outside marriage. Such adulterous relationships, argued Devi, not only bred several diseases including infertility, but people who engaged in them met devastating ends (see Figure 9.2).

Equally interesting are her views on child marriage, which was quite common during the period under discussion. She described child marriages as 'mismatched marriages' with dire consequences. Quoting from Ayurvedic scriptures, she wrote that until the woman was sixteen and the man twenty, their marriage should not be solemnized, otherwise it may cause serious diseases.[69] Furthermore, according to her, the child born from the union of a husband and wife younger than the prescribed age remained either sick or short-lived. It was quite possible, according to her, that such 'mismatched' couples could not produce children. She also described the excessive age gap between husband and wife as detrimental to a happy married life. According to her, there should be no difference between the age of husband and wife beyond the prescribed limit.[70] These views regarding the proper age of marriage were certainly very progressive and reformist.

Finally, Yashoda Devi also prepared a multivolume encyclopaedia on the history of notable Indian women which included the life sketches of

पर स्त्री गमन का फल । पृ॰ ५०८ (सर्वांधिकार सुरक्षित)

Figure 9.2 Consequences of adultery (illustration from Yashoda Devi, *Vivah Vigyan Kamshastra: Anand Mandir*, Allahabad, n.d., p. 508)

as many as 500 women. It was a unique exercise keeping in mind the contemporary social structure. Although written in eulogizing terms with an occasional tinge of mythology, this encyclopaedia drew attention to many unsung women of Indian history. It is believed that more than 20,000 copies of the first volume of the encyclopaedia were sold immediately after its publication.[71]

Conclusion

Women's entry into the male-dominated universe of medicine is a story of great struggle even in countries of the so-called West.[72] However, in India, the specific needs of female patients along with the prevalent *purdah* (veil) system and prejudices associated with the mobility of women compelled even the Christian missionaries to induct women into the field of medicine.[73] In order to attain professional status equal to that of male practitioners, women had to overcome many hurdles – social as well as political. Certainly, the official terrain of 'indigenous' medicine was no exception. This brings us back to the question of acceptability of a female practitioner in the field of medicine. The ripples which Yashoda Devi managed to create in the professional circles of Ayurvedic practitioners of North India can be gauged from the fact that she was targeted unceremoniously by some of the male *vaids* despite the commercial success of her remedies and publications. Many of her heterodox arguments largely pertaining to taboo subjects as mentioned above antagonized Ayurveda's normative authority. In fact, cases were filed against Devi, calling her remedies 'inferior' and 'inauthentic'.[74]

The scientific pursuits of women in Western contexts have consistently enjoyed more visibility than those in regions like Africa, Asia or Eastern Europe.[75] Hence, it becomes all the more important to document the life and works of female practitioners like Yashoda Devi. Her prolific writing constitutes a rich source for tracing women's voice in the making of 'modern' Ayurveda. It was practitioners like Devi who brought the benefits of this traditional Indian healing system to those behind the veil. In this context, the popularity which Devi wielded among her clientele had its own significance. After all, unlike nationalism, Ayurveda was not an abstract ideology; it was a healing system producing consumable goods. Hence, it had to create its own market and 'consumers' out of the 'users', besides the ideological and emotional mobilization of the masses. Here Yashoda Devi was successful in creating her own 'consumers' who eventually sustained her presence in a field dominated exclusively by males. This in no way was a mean achievement, keeping in mind the time and space in which Yashoda Devi was trying to carve a niche for herself.

Notes

1 Founded in 1840, the *British Medical Journal* (*BMJ*) is the renowned weekly journal of the British Medical Association. The first editors were P. Hennis Green, lecturer on the diseases of children at the Hunterian School of Medicine, and Robert Streeten of Worcester, a member of the Provincial Medical and Surgical Association Council. In their introductory editorial, Green and Streeten formulated their two main objectives as the advancement of the profession and the dissemination of medical knowledge.
2 'Lady surgeons', *British Medical Journal* (2 April 1870), 338–39.
3 'Lady surgeons', p. 339.
4 J. N. Burstyn, 'Education and sex: The medical case against higher education for women in England, 1870–1900', *Proceedings of the American Philosophical Society*, 117: 2 (1973), 79–89; see also T. N. Bonner, *To the Ends of the Earth: Women's Search for Education in Medicine* (Cambridge, MA: Harvard University Press, 1992).

5 P. Arrizabalaga, R. Abellana, O. Viñas, A. Merino and C. Ascaso, 'Gender inequalities in the medical profession: Are there still barriers to women physicians in the 21st century?' *Gaceta Sanitaria*, 28: 5 (2014), 363–68; R. M. Allen, 'Gender inequality in medicine: Too much evidence to ignore', *Psychiatric Times*, 34: 5 (2017).

6 Yet another exceptional case in North India was that of Prakashvati Devi Jain. However, very little is known about her. She used the title of *Vaidya Visharada*, usually employed by the *vaids* having some sort of institutional training or having passed some kind of qualifying examination conducted by an institute/organization devoted to the cause of promotion of Ayurveda. Also, Prakashvati Devi Jain was the president of the women's department of Amrit Karyalaya located in Agra.

7 C. Gupta, 'Vernacular sexology from the margins: A woman and a Shudra', *South Asia: Journal of South Asian Studies*, 43: 6 (2020), 1113.

8 By Western medicine, I mean biomedicine. This is not to deny the diversity that was integral to Western medicine at least in the nineteenth and early twentieth centuries.

9 By contrast to the metropole, doctors in the colonies were not prohibited from conducting extensive field trials and tests on human bodies. That is why many landmark medical discoveries happened in the colonies and were then transferred to the metropole, e.g., the prophylactic vaccination against cholera and bubonic plague developed by Waldemar Mordecai Haffkine and the discovery of the malarial parasite by Ronald Ross.

10 Anxieties regarding the application of 'unscientific' healing systems were not exclusive to the colonies and Western medicine was not a monolithic entity either. In nineteenth- and twentieth-century Europe, homeopathy, along with phrenology, magnetism, mesmerism, herbalism, hydropathy, naturopathy and chiropractic treatments, came to acquire the status of 'Western medical heterodoxy' vis-à-vis mainstream Western medicine or biomedicine. Furthermore, there were many debates about quackery in Victorian Britain as well. In fact, there were many more quacks than medical doctors as the spectre of unlicensed practice loomed large. See M. Weatherall, 'Making medicine scientific: Empiricism, rationality, and quackery in Mid-Victorian Britain', *Social History of Medicine*, 9: 2 (1996), 175–94.

11 In this regard, Partha Chatterjee persuasively argues that the newly emerging reformist elite of the period tried 'to fashion a "modern" national culture that is nevertheless not western'. While colonized societies had to incorporate material techniques of modern Western civilization in their projects of reorganization, it was imperative for them to maintain a distinctive spiritual essence in order to prevent the erosion of their national identity. See P. Chatterjee, *The Nation and Its Fragment: Colonial and Postcolonial Histories* (Princeton, NJ: Princeton University Press, 1993), p. 6.

12 On medical revivalism in India, see C. Leslie (ed.), *Asian Medical Systems: A Comparative Study* (Berkeley: University of California Press, 1976); K. Sivaramakrishnan, *Old Potions, New Bottles: Recasting Indigenous Medicine*

in Colonial Punjab, 1850–1945 (New Delhi: Orient Longman, 2006); G. N. A. Attewell, *Refiguring Unani Tibb: Plural Healing in Late Colonial India* (Hyderabad: Orient Longman, 2007).

13 Recent research has convincingly shown that the Ayurveda revived during this time was significantly different from its classical version. For details, see R. Berger, *Ayurveda Made Modern: Political Histories of Indigenous Medicine in North India, 1900–1955* (Hampshire: Palgrave MacMillan, 2013); M. Banerjee, *Power, Knowledge, Medicine: Ayurvedic Pharmaceuticals at Home and in the World* (Hyderabad: Orient Blackswan, 2009).

14 I have discussed this search for an appropriate 'indigenous' alternative to Western medicine elsewhere. See S. K. Rai, 'In search of indigenous medicine: Medical pluralism and the Ayurvedic movement in colonial India', Occasional Paper, *History and Society (New Series)*, no. 104 (New Delhi: Nehru Memorial Museum and Library, 2020).

15 Established in 1907, All India Vaidya Sammelan was the largest organization of the Ayurvedic practitioners devoted to the cause of Ayurvedic revival. Its first session was held in Nasik with the twin objectives of streamlining and standardizing the Ayurvedic discourse/movement and bringing all *vaids* under one roof, thereby giving them a collective corporate identity.

16 For some of these biases as exhibited by late-colonial Ayurvedic discourse, see S. K. Rai, 'Gendering late colonial Ayurvedic discourse: United Provinces, c.1890–1937', *History and Sociology of South Asia*, 10: 1 (2016), 21–34; S. K. Rai, 'Invoking "Hindu" Ayurveda: Communalisation of the late colonial Ayurvedic discourse', *Indian Economic and Social History Review*, 56: 3 (2019), 411–26; S. K. Rai, 'Brahmanizing Ayurveda: Caste and class dimensions of late Colonial Ayurvedic movement in Upper India', *Summerhill: IIAS Review*, 25: 2 (2019), 4–9.

17 R. P. Chand, *Plague Darpan* (Patna City: Satya Sudhakar Press, 1916), p. 138 (translation mine).

18 Numerous didactic tracts, journals and cartoons were produced during this period which exhibited the efforts of the middle class to command the sexuality and mobility of 'their' women or the so-called respectable women by invoking the threat of the lustful and lecherous male gaze. For details, see S. K. Rai, 'Gazing at the woman's body: Historicising lust and lechery in a patriarchal society', *Social Scientist*, 47: 1–2 (2019), 49–62.

19 E.g., M. Pandey, *Hamare Bachche* [Our Children] (Prayag, 1931) and J. Sharma, *Santan Palan* [The Rearing of Progeny] (Benares, 1933).

20 See S. Shah, 'Representation of female sexuality in the Ayurvedic discourse of the early medieval period', *Studies in History*, 22: 1 (2006), 45–58.

21 See L. S. Vishwanath, 'Female infanticide: The colonial experience', *Economic and Political Weekly*, 39: 22 (2004), 2313–18. According to Vishwanath, Hindu castes, usually those higher in the hierarchy, which practised female infanticide justified it on grounds that they could not afford the huge dowries or the incalculable marriage expenses having a daughter entailed.

22 According to Hindu Shastric traditions, only the male child can perform the parent's last rites.
23 J. S. Shrivastava, *Vaidya Priya* (Lucknow: Naval Kishore Press, 1924), pp. 241–42. The text suggested applying a paste made by mixing pigeon's excrement and borax (*suhaga*) on the penis before copulation to beget a male child. Similarly, it advised pregnant women to eat cannabis seeds to fulfil their desire for a male child.
24 Published by the Ayurvedic pharmacy of the same name located in Aligarh (United Provinces), *Dhanvantari* was one of the widely circulated Ayurvedic journals of the early twentieth century. It began publication at the beginning of 1923 and was the first Ayurvedic journal to publish a special number in the same year. It was registered under No. A 1215 and was the only medical journal approved for the school libraries by the Text-Books Committee of the Education Department of the Central Provinces and Berar.
25 V. G. P. Varshaneya, 'Ichchanusar putra ya kanya hona', *Dhanvantari*, 10: 4 (September 1934), 388–90.
26 *Sutika griha* or *antur ghar* was a small, dark, unventilated room or hut usually built outside the main quarters, where the woman used to spend the last few days of her pregnancy just before the beginning of labour and ten to twelve days after birth for ritual cleansing. All the routine activities of the pregnant woman during this period of confinement, such as cleaning her clothes, feeding her and all the menial jobs, were performed by a *dai* as only the birthing mother and the *dai* were allowed inside the *sutika griha*.
27 S. Guha, 'From dais to doctors: The medicalisation of childbirth in colonial India', in L. Lingam (ed.), *Understanding Women's Health Issues: A Reader* (New Delhi: Kali for Women, 1998), p. 146.
28 For details, see A. Malhotra, *Gender, Caste, and Religious Identities: Restructuring Caste in Colonial Punjab* (New Delhi: Oxford University Press, 2002), especially chapter 5; C. Gupta, *Sexuality, Obscenity, Community: Women, Muslims, and the Hindu Public in Colonial India* (Delhi: Permanent Black, 2001), especially chapter 7.
29 In the late 1980s and 1990s, many historians, particularly feminist scholars, extensively analysed the discourses associated with the 'female body' and 'sexuality' and their linkages with the social reform projects and nationalist politics. For instance, see P. Chatterjee, 'The nationalist resolution of the women's question', in K. Sangari and S. Vaid (eds), *Recasting Women: Essays in Indian Colonial History* (New Brunswick, NJ: Rutgers University Press, 1990), pp. 233–53 and L. Mani, 'Contentious traditions: The debate on sati in colonial India', *Cultural Critique* 7 (1987), 119–56. Although similar discussions over the 'male body' started around the same period, this theme failed to attract ample historiographical attention. Nevertheless, it is a historical fact that 'male sexuality' also came under scrutiny and restrictions were imposed on it during the late nineteenth and early twentieth centuries.
30 In fact, discourse over *Brahmacharya* or sexual self-control became a key feature of nationalist as well as communal discourse during this period.

Mahatma Gandhi brought the issue of *Brahmacharya* to the forefront of nationalist politics. According to Gandhi, those who wanted to perform national service, or to have a gleam of the real religious life, must lead a celibate life, whether married or unmarried. This constituted one of the vows drawn up by Gandhi for those seeking the membership of his Ashram at Sabarmati. See C. F. Andrews, *Mahatma Gandhi's Ideas* (London: Allen and Unwin, 1929), pp. 101–11. Similarly, giving *Brahmacharya* a more institutionalized shape, the Rashtriya Svayamsevak Sangh (RSS) and its various affiliated organizations made sexual self-control the 'supreme characteristic' of a man.
31 J. Sharma, *Arogya Darpan* [Health Mirror], vol. III (Prayag, 1898), p. 9.
32 Sharma, *Arogya Darpan*, vol. III, p. 29. Thus, the 'beauty' of a girl belonging to the lower caste often inflated caste insecurities at the upper echelons, leading to the character assassination of the girl.
33 C. Shastri, *Brahmacharya Sadhan* [The Practice of Celibacy] (Lucknow, 1928), p. 75.
34 *Vaid* or *vaidya* is a generic term used for Ayurvedic practitioners.
35 Later, Yashoda Devi republished many of the books written by Satyabhama Bai, such as *Jeevan Raksha, Dhatri Vidya, Bal Vaidyak, Ratna Sangrah* (various volumes) and so on under the banner of Stri Shiksha Pustakalaya.
36 Literally, 'A guide to the proper code of conduct of women'.
37 G. Prakash, *Another Reason: Science and Imagination of Modern India* (Princeton, NJ: Princeton University Press, 1999), pp. 148–49. The preservation of health and physical well-being of the family members constituted one of the fundamental topics to be covered under the rubric of home science, a discipline immensely popular with twentieth-century Indian women. For some interesting insights into the evolution of home science in India and its complicity with the idea of producing 'good wives' and mothers, see Anne Hardgrove's chapter in this volume.
38 According to Hindu scriptures, an individual's life passes through four different but sequential stages or *ashramas*, namely *Brahmacharya* (student), *Grihastha* (householder), *Vanaprastha* (forest walker/forest dweller) and *Sanyasa* (renunciator).
39 Rachel Berger points out that by referring to her cookbooks as *Pakshastras* (literally, the written scriptures of nutrition), Yashoda Devi gave historic gravitas to her writings as textually authoritative and backed by the sense of 'timeless wisdom' as delineated in classical texts of the Hindu tradition. See R. Berger, 'Between digestion and desire: Genealogies of food in nationalist North India', *Modern Asian Studies*, 47: 5 (2013), 1622–43.
40 Prakash, *Another Reason*, p. 149.
41 Y. Devi, *Santan Palan* [The Rearing of Progeny] (Allahabad, 1913).
42 Y. Devi, *Shishu Raksha Vidhan arthat Balrog Chikitsa* [Manual for the Protection of Children] (Allahabad, 1912).
43 Y. Devi, *Vivah Vigyan Kamshastra: Anand Mandir* [Science of Marriage and Sex] (Allahabad, n.d.), p. 146.

44 Y. Devi, *Vivah Vigyan Kamshastra: Anand Mandir* (Allahabad, n.d.), p. 146 (translation mine).
45 Y. Devi, *Dampati Arogyata Jeevanshashtra* [The Science of a Healthy Conjugal Life] (Allahabad, 1927), p. 49.
46 Devi, *Dampati Arogyata Jeevanshastra*, p. 20.
47 Devi, *Dampati Arogyata Jeevanshastra*, p. 66.
48 See A. H. Miller and J. E. Adams (eds), *Sexualities in Victorian Britain* (Bloomington: Indiana University Press, 1996), p. 63.
49 On the Ayurvedic concept of *Ashta-maithun*, see Rai, 'Gendering the late colonial Ayurvedic discourse', pp. 30–31.
50 A holy fair held at *Sangam* (confluence of the rivers Ganga, Yamuna and a mythical Saraswati) in Allahabad around January–February of every year.
51 Devi, *Dampati Arogyata Jeevanshastra*, pp. 6–7.
52 It is in this context that Charu Gupta, drawing on Roger Chartier's work on print culture, argues that 'people could limit and frame syllabi, they could order prescriptive texts, but once women were educated it was difficult to control what they read and the uses to which they put their knowledge'. Gupta, *Sexuality, Obscenity, Community*, p. 174. In other words, education for women was a kind of double-edged sword that swung both upwards as well as downwards in the existing structure and norms of the Indian society.
53 Devi, *Dampati Arogyata Jeevanshastra*, p. 5.
54 This appears to be particularly significant in the Indian context where *purdah* was a norm, especially among women belonging to upper-caste, middle-class families. In a different context, Kathryn Edgerton-Tarpley (this volume) notes that women patients preferred female doctors. According to Tarpley, there were occasions when women flatly refused to be treated by male practitioners, especially in the case of 'women's illnesses'.
55 In fact, one of the self-healing manuals written by Devi, *Ghar ka Vaidya* [Home Doctor] (Allahabad, 1912), became quite popular among householders and women.
56 According to a source, the 108 books and booklets written by Yashoda Devi registered the sale of as many as 50,000 copies, which was impressive given contemporary readership.
57 C. Gupta, 'Procreation and pleasure: Writings of a woman Ayurvedic practitioner in colonial North India', *Studies in History*, 21: 1 (2005), 26. Unfortunately, according to Gupta, most of these letters written to Yashoda Devi by ordinary women are no longer available. Had these letters been available, they would have been a wonderful source to shed light on the health and domestic concerns of women of the period under discussion.
58 Gupta, 'Procreation and pleasure', p. 38.
59 Y. Devi, *Nari Sharir Vigyan Stri Chikitsa Sagar: Sambhog Vigyan* [Women's Physiology and Medical Treatment: Science of Intercourse] (Allahabad, 1938), p. 696.
60 A genre of Sanskrit texts on law and day-to-day conduct.
61 Refers to the act of bathing after menstruation.

62 Devi, *Vivah Vigyan Kamshastra*, p. 176.
63 Hindi term that stands for sexual intercourse.
64 Devi, *Vivah Vigyan Kamshastra*, p. 150.
65 Devi, *Nari Sharir Vigyan*, p. 147.
66 Y. Devi, *Dampatya Prem aur Ratikriya ka Gupt Rahasya* [Conjugal Love and Secrets of Sexual Intercourse] (Allahabad, 1933), p. 218.
67 Devi, *Dampatya Prem*, pp. 359–60.
68 Devi, *Vivah Vigyan Kamshastra*, p. 476.
69 Devi, *Vivah Vigyan Kamshastra*, p. 443. Here it should be noted that the Child Marriage Restraint Act of 1929, popularly known as the Sarda Act, fixed the age of marriage in India for girls at fourteen years and boys at eighteen years. In 1949, after India's independence, it was amended to fifteen for girls, and in 1978, to eighteen for girls and twenty-one for boys. If we compare these official fixtures with the opinion expressed by Yashoda Devi regarding the proper age of marriage, one can easily see how ahead of her time she was.
70 Devi, *Vivah Vigyan Kamshastra*, pp. 444–45. For Yashoda Devi the difference between the age of husband and wife should not be more than one and a quarter (*sawai*).
71 Y. Devi, *Samsar ka Nari Itihas (Bharat ka Nari Itihas)* [Women's History of the World (Women's History of India)], Vol. 2 (Allahabad, 1922), p. 142.
72 M. P. Behringer, 'Women's role and status in the sciences: An historical perspective', in J. B. Kahle (ed.), *Women in Science: A Report from the Field* (Philadelphia, PA: The Falmer Press, 1985), pp. 4–26.
73 As Samiksha Sehrawat argues, 'the use of the figure of the *zenana* patient – the woman confined to seclusion and prevented from seeing male physicians – as a signifier for all Indian women created considerable dissonance in discussions about how to provide Indian women with medical care'. Inducting women into the field of medicine was an important way out in this connection. See S. Sehrawat, *Colonial Medical Care in North India: Gender, State and Society, c. 1840–1920* (New Delhi: Oxford University Press, 2013), p. 109.
74 Gupta, 'Vernacular sexology from the margins', p. 1113.
75 A. H. Koblitz, *A Convergence of Lives: Sofia Kovalevskaia, Scientist, Writer, Revolutionary* (New Brunswick, NJ: Rutgers University Press, 1993), pp. xiv–xv, xxv–xxvi.

10

Lady Irwin College: Domestic science post-secondary education for three women graduates in India

Anne Hardgrove

Introduction

This chapter considers the impact of domestic science education at the post-secondary level for three women graduates in twentieth-century India. Before considering the intimate life-histories of the graduates, I first discuss how the development of higher-education opportunities for women in home science came about. The analysis begins with the reasons behind the professionalization of housekeeping under the conditions of colonial rule. As part of the women's movement in anticolonial nationalism, Lady Irwin College in New Delhi was established as the first and premier institution in India to offer college-level instruction in home science. What impact did domestic science college education have on women graduates of Lady Irwin, in their lives and in their careers? What was the public perception of home science and how did it open opportunities for middle-class Indian women? More intimately, how do women graduates themselves reflect back upon their higher education in home science and its role in changing their life circumstances? To answer these questions, I use a close-up scale of analysis and methods of narrative and oral history to examine the lives of three Lady Irwin graduates. One woman's life is in the public record. For the others, I turn to oral histories I conducted with two women whose education at Lady Irwin was caught up in the historical events of independence and partition. What these women have to say about their own education is a microcosm of the impact of domestic science college for women in India, from the first-hand perspective of women themselves, as opposed to considering only top-down colonial and national discourses about the expected role of domestic education in the lives of women. These oral history narratives are also caught up in nationalist ideologies and personal perspectives yet are rich sources of everyday meaning from the perspective of home science graduates themselves. This approach to history-writing helps elucidate divergent understandings of the past as memories of lived experience.

Brief historical background of home science

Women's education in domestic science in India began in the nineteenth century as result of two separate but related colonial criticisms of Indian society: first, that the Indian home was inadequate to prepare Indians for modern social life, in contrast to traditional household management;[1] and second, there was a perceived absence of female education.[2] Under the missionary and colonial critique, the Indian home – seen as the repository of indigenous tradition – became one of the primary sites for social reform and modernization.[3] Yet this critique of Indian home life led to the formation of self-disciplinary domestic science texts. Missionary women and social reform associations began to provide domestic education for middle-class Indian women.[4] Indian society in the nineteenth century gave considerable resistance to formal female education. Even domestic education could be seen as at odds with the notion of the household being the cradle of tradition, such as in evidence gathered from local newspapers. In a statement about the domestic science curriculum in the Bengali newspaper *Naba Bibhakar Sadharani* (8 April 1889), a writer commented that it was not desirable that the Hindu girl should be given the denationalizing English education which is given to the Hindu boy.[5]

Nationalist concerns from the 1920s onwards championed home science as a mode of preparing Indian wives and mothers to establish homes suitable for future Indian leadership. In 1932, the All India Women's Conference recommended home science as a new academic discipline which catered not to the educational needs of boys, but addressed the special curriculum needs of Indian women, destined to become wives and mothers. Home science was institutionalized into higher education first in Lady Irwin College, then in other universities and home science institutes in India. Home science in India developed in tandem with American colleges offering the subject of domestic education, which in the US became known as home economics.[6] During the 1950s–1960s, collaborative arrangements were made with American universities, including the University of Tennessee-Knoxville.

The development of appropriate roles for the modern Indian woman and the reform of the household played a major part in the anticolonial and social reform movement. Radha Kumar writes that, 'instead of being a sanctuary [from the world] ... the home began to represent the dead weight of traditions which were scorned as bigoted or barbaric'.[7] As Partha Chatterjee argues, India was faced with the challenge of overcoming British colonial domination by cultivating 'the material techniques of modern Western civilization while retaining and strengthening the distinctive spiritual essence of the national culture'.[8] The Indian home, seen as embodying the source and safe harbour of indigenous tradition, became one of the primary sites for social reform and modernization. By the 1920s, the invention of the scientifically trained 'housewife' became an answer to how to professionalize women's work in modern life. Besides giving women a place in formal education, women's education was seen as a positive way to instil in women specific values and virtues of the respectable middle classes. Chatterjee describes these values as the:

> typically bourgeois virtues characteristic of the new social forms of 'disciplining' – of orderliness, thrift, cleanliness, and a personal sense of responsibility, the practical skills of literacy, accounting and hygiene and the ability to run the household according to the new physical and economic conditions set by the outside world.[9]

Home science became embroiled in nationalist discourse, with the implicit understanding that modern homes would cultivate modern citizens.[10]

The professionalization and institutionalization of home science as an academic discipline was not at all unique to India.[11] Almost all urbanized societies faced the question of how women should find their place in a modernity and public life typically gendered as masculine. Scientific education and careers in science were seen as restricted to men. Science had come to prominence as a man's field and almost universally restricted women's access. Natural science was both defined by and defined as an exclusively male realm. As Neelum Kumar notes, science was deeply identified with masculinity, especially in the medical field, but across all its branches.[12] As such, Kumar rightly contends that women scientists faced a 'triple threat' of an outside job taking a woman away from her marriage and duties at home, while the field of science itself and scientific careers were exclusionary towards women.[13]

Finding a way to professionalize women's roles and finding education and careers deemed appropriate for women needed to take into consideration women's supposedly unique, natural qualities of caregiving and household management. Cultural differences and institutional differences helped dictate the national form that domestic studies would assume. The US, for instance, developed home economics, while Japan featured home education. In the US, domestic education courses were renamed home economics in the 1920s, though the curriculum was still seen as having a scientific basis. The founders of American home economics intentionally chose a Dewey Decimal library system classification number that would group the subject with sociology rather than needlework. Americans behind the home economics movement specifically chose to recast the classification of 'pauperism' in the lofty hope that home economics would alleviate poverty. As historian Shizuko Koyama notes for Japan, domestic education lay claims to 'presumably scientific fields' which would be 'validated in the public eye by the authority of science'.[14]

For these reasons, and as referenced in the oral history sources to come, I argue that domestic science became the perfect 'back door' to women seeking careers and an innovative way of bringing women into both higher education and scientific fields. Domestic science gave women access to a scientific education that stood up against the triple threat identified by Kumar: a scientific curriculum specially designed and carved out for women that did not interrupt the male prerogative in science, and seen as in support of, and not at counterpoint to, women's roles as good wives and mothers. Of course, and perhaps by necessity, domestic education trafficked in more traditional and conservative stereotypes of appropriate activities and skills for women. But at the same time, the subject also provided an opening for women to obtain an education that was socially acceptable.

The curriculum of home science in India has, unsurprisingly, changed over time just as it has in the United States. Generally speaking, what started out in both countries as an education in cooking, sewing and child-rearing evolved to be more community- and policy-based. In India, domestic science has contributed significantly to knowledge and practice in rural development and public health. In the United States, the roots of home economics from educational curriculum improvements at historically Black colleges and universities (HBCUs) is understudied.[15] The idea behind home economics, for white women, was to systematize and reduce the amount of time and energy women spent on household chores. For Black leaders of the home economics movement like Mary Murray Washington, wife of Booker T., home economics promised to help end racial inequalities of the domestic sphere.[16]

The urban and rural context of domestic education is important. In the US, home economics helped chart a path towards establishing the federal poverty level, consumer protections and the school lunch programme. The author of this chapter was herself educated in this tradition. In small-town Midwestern United States in the 1970s and 1980s, both boys and girls were required to take one semester of home economics in the seventh grade, with 'shop' being the counterpart in the other semester. In eighth grade, students chose, with almost all girls opting for home economics (cooking and sewing), while almost all boys turned to shop and auto-mechanics. This curriculum was geared both towards rural development and practical skills for homemaking.

The question remains of the extent to which domestic science is recognized in hindsight by historians of science. Historians of science in the past made the same narrow judgements as their predecessors did and ignored the role of home education by excluding it from 'real science'. Studies of women and science, or just of science in general, in the past have not included examples of domestic science. For India, Deepak Kumar's *Science and the Raj* (1995) is one example. Kumar's chosen chronological period of 1857–1905 is admittedly too early for the formal institutionalization of domestic science, yet he does not look at the informal domestic education efforts for women that might have otherwise been considered as scientific precursors. This edited volume builds upon the contributions of historians of gender who argue that domestic science is important to the history of science in general.[17]

Establishment of Lady Irwin College

The second All India Women's Conference (hereafter, AIWC) helped establish the idea for Lady Irwin College. It was named for its official patron Lady Dorothy Irwin, wife of India's viceroy, who was an active member of the AIWC. The AIWC began in 1927 as a major reform association founded by progressive British and Indian women and exists to this day. The goal of the AIWC for women's education for the first twenty-some years, up until independence, was to train women to be good wives and mothers. After 1947, the AIWC worked to promote women's education for a variety of career paths. Indian educator Hannah Sen was chosen to be the first director of Lady Irwin College from 1932 to 1947. The daughter of a Jewish mother and Hindu father who converted to Judaism, Sen was an experienced teacher with undergraduate degrees from Calcutta and a graduate degree obtained after her marriage from the University of London. She was invited to return to India to take up this role.[18]

One important thing to understand about women's education in modern India is its fraught history. In 1913, literacy rates for women in India were estimated to be less than 1 per cent.[19] Until the 1920s and 1930s, there was a strongly held Hindu belief in North India that an educated girl would become a widow. Girls were not meant to study, especially in coeducational environments. Middle- and upper-class families had preferred that their daughters have informal education in the security of their own homes, to assure the daughters' modesty and reputation until they could be married. One very real economic factor for female education has been dowry. The more educated the girl, the more dowry would be demanded by the groom and his family. The basic logic for this assumption has been that educated girls would be less compliant in terms of melding into the expectations of the husband's patrilocal family. There were two tensions in particular: first, that the girl should not be better educated than the boy; and second, that the girl should not be too emboldened to show disrespect to her less-educated mother-in-law. Home science became a way for young adult women to pursue a degree that was appropriate for women, without threatening their opportunities in the dowry-driven market of marriage.

Lady Irwin College: Domestic science in India 241

One of the issues I explore in this chapter is the rather odd decontextualization of home science from the local Indian environment. The adoption of 'science' as the answer to curing the woes of the home stemmed from several places. Yellow journalism in the 1920s, such as that practised by American Katherine Mayo, portrayed the much-exaggerated ills of Indian society as stemming from unregulated, undisciplined and unbridled Hindu male sexuality.[20] Utopian socialism, very popular in British intellectual circles in the 1920s, stressed the value of science for modelling human behaviour. The Indian home was seen as the site of tradition but needed purported modernization to meet the needs of the modern nation-state. Like science in general, domestic science too was at once nationalized and decontextualized. As such, domestic science fits squarely into the Nehruvian scientific vision for modern India, even though, as I explore through my case studies, it remains at odds with some cultural realities. As Deepak Kumar writes, we must consider how a supposedly 'benevolent, apolitical, and value-neutral' subject is adapted into a society under colonialism. I am interested in how the near-universal yet illiberal presence of domestic servants meshed with bourgeois ideas of the housewife as the professional housekeeper.

During a year I spent in New Delhi affiliated to Lady Irwin College in the 1990s, I found that many of the girls enrolled at Lady Irwin College came from business families, particularly the very financially successful Marwari migrant traders and industrialists. Many Marwari teenagers and young women from conservative families are steered into home science, as a way of gaining higher education while at the same time mediating social expectations for the marriage market, including dowry. Marwari friends told me funny stories of hiding their higher education from their fathers, who often worked very long hours and were hardly at home to observe the whereabouts of their daughters during the day. One friend finished her BA in Hindi literature with her father's blessing but did not tell him that she had actually started work on her master's. When he inadvertently found out, he shouted at her, 'Are you trying to kill me? Do you know how much more dowry I will have to give?' Eventually, they negotiated a deal where she would agree to a marriage match before completing her PhD.

Female education in India has been associated, usually negatively, with marriageability and marriage prospects. This was true for all communities, despite limited early nineteenth-century efforts by reform movements such as Brahmo Samaj which promoted literacy and education among girls and young women. Indeed, domestic science still targets middle-class young women, who have access to higher education. Women's education has presented several thorny issues for women and their families in India. Some of the issues stem from the question of a woman's marriageability, in managing the expectations of dowry and compatibility with in-laws. For other women, issues of marriage may not be imminent, but considerations of college choice, safety and reputation were paramount.

Looking at three women who graduated from Lady Irwin helps us to establish how domestic science education fit into a twentieth-century life course. How did domestic science fit into the family expectations and future lives of Indian women? The women I describe came to Lady Irwin from South India (1960s) and, in the 1940s, from Sindh in the west (post-Partition Pakistan) and from Kolkata (then Calcutta) in the east. Lady Irwin College held considerable prestige of being located in the nation's capital, New Delhi, and was often featured as a showcase of national achievement and pride during the nationalist movement. For each of these women, there was both ambivalence and acceptance about studying home science at Lady Irwin. Each was grateful for the opportunities that her education gave her, even if she knew she was not receiving an equivalent education to male family members. Still, Lady Irwin provided a platform for their future goals and plans.

Case study 1: Shyamala Gopalan

I would like to briefly tell the story of a 1960s graduate of Lady Irwin College, who has achieved renown posthumously, as the mother of a prominent American politician. Shyamala Gopalan was born in 1938 in Chennai (formerly Madras), a major city in South India, a city and region acclaimed for its strong traditions in education. She died in Oakland, California, in 2009. Shyamala is perhaps remembered best by her married name in the US, Harris, whose eldest daughter Kamala was elected vice-president of the United States in 2020. Shyamala wanted to study biochemistry and eventually did earn a PhD in the subject at the University of California, Berkeley. While not every graduate of Lady Irwin College will go down in the history books as Shyamala will, part of her story is very telling about the College, the discipline of home science and its place in her remarkable career as a breast-cancer researcher. After completing her secondary education, Shyamala enrolled in post-secondary home science at Lady Irwin College. I contend that this became one path that women used to access a scientific education, despite teasing and uncertainty by family members about the actual curriculum.

Newspaper accounts give us recollections of what Shyamala's family members thought about her enrolment in home science. 'I had to go there.' As reported by Ellen Barry in the *New York Times*,

> Ms. Gopalan, the oldest child in a high achieving Tamil Brahmin family, wanted to be a biochemist. But at Lady Irwin College, founded by the British to provide an education in science to Indian women, she had been forced to settle for a degree in home science. Her brother and her

father thought it was hilarious. 'My father and I used to tease her like nobody's business,' said her brother, Gopalan Balachandran, who would go on to earn a Ph.D. in computer science and economics. 'We would say, "What do you study in home science? Do they teach you to set up plates for dinner?" She used to get angry and laugh. She would say, "You don't know what I'm studying."' His sister died in 2009. But in retrospect, he realizes she must have been seething. 'She would have been frustrated like hell,' he said.[21]

Her father also ribbed her about studying at Lady Irwin. In an article in the *Los Angeles Times*, the same brother recalled their father saying, 'What is home science? ... Are you learning how to invite guests?'[22] Perhaps Shyamala had the last laugh when she applied to Berkeley for biochemistry without informing her father, only showing him the letter of acceptance when it arrived. He then agreed to pay for her first year, leading to her career as an accomplished scientific researcher in breast cancer. But as the mother of a future US senator and vice-president, Shyamala became a public figure through her daughter Kamala's achievements.

I want to pause with one aspect of Shyamala's story, which leads me to my argument for this paper. Like Kamala Harris' own uncle and grandfather, we might be tempted to look at the implementation of home science in India as retrograde. We might also chuckle and consider it to be a laughable, non-academic subject whose only role was to reinforce the colonial cult of domesticity and keep women in their places. Indeed, that is the popular reputation of home science in India. Certainly, there were much earlier women trailblazers, who become doctors and politicians and lawyers, among other professions, regardless of whether that was not the mainstream objective for middle-class families who sent their daughters to Lady Irwin.

I now turn to the other two graduates of Lady Irwin College whom I met personally and had the opportunity to interview. My own oral history interviews with women graduates of Lady Irwin College feature in Case Studies 2 and 3. These conversations allowed me to connect with women's impressions of being at the college, especially regarding their memories of the curriculum. These interviews were conducted about twenty-five years apart, in 1996 and in 2022. Ritu and Flower, the two narrators or interviewees, were both at Lady Irwin during the late 1940s. This was a heady time for Indian nationalism and the Partition of India into India and Pakistan. Each woman came to the College for different reasons. Both women became active in nationalist politics as students at Lady Irwin College, leaving behind the more provincial backgrounds in which they were raised. Unlike Shyamala, both remained in India for much of their adult lives, using their degrees to work as teachers in a variety of contexts.

Case study 2: 'Ritu', a pseudonym

Ritu, a Sindhi woman, did two years of 'intermediate' university at Bombay University at Karachi (still British India), studying English, political science and French, before joining Lady Irwin College.[23] In Sindh, education for women was not very developed. Her mother never learned English and died of tuberculosis when Ritu was three. Her father remarried a seventeen-year-old who struggled to take care of the stepchildren, who were mostly boys. Interested in going into public administration, Ritu had heard about Lady Irwin College because dignitaries such as Gandhi and Nehru had presided over convocations there. Ritu was excited about political activities in New Delhi, relishing stories she heard about student picketing. At her St Joseph convent school, girls were not allowed to engage in political activities or picketing. Her father agreed that she could attend Lady Irwin College, because she could go with a female cousin. Four students travelled together from Karachi to New Delhi. She came to Lady Irwin in 1947 and did a diploma.

Ritu spoke at length about the capstone experience at Lady Irwin College. One of the centrepieces of Lady Irwin's curriculum was the cottage system. In groups of seven, students were sent to live in cottages for six weeks and oversee all domestic work – cleaning, cooking and washing clothes and dishes. Most importantly, they had to learn the art of conversation at the dinner table. Prominent guests would come, including freedom fighters, to sit and eat with them at the table. Ritu mentioned that students from other colleges joked that Lady Irwin students were only there to be trained to marry rich husbands. Yet I will note that while the 'Mrs Degree' has been a long-held stereotypical part of women's forays into higher education, in the context of India, this can be seen as progressive. Meeting non-related male visitors without direct parental supervision provided a sanctioned way for young women at Lady Irwin to learn to socialize outside of the extended family, a privilege that they would probably not have enjoyed outside of a residential educational experience.

Ritu recounted a couple of memories which included Mahatma Gandhi, who made regular visits to the College for political meetings. The College put on a performance of the *Ramayana*, with Gandhi being the chief guest. Ritu was one of three violinists providing music for the evening, and at the end of the evening she took a turn in honouring Gandhi by ceremonial touching of his feet. Ritu's father came from Karachi to see the performance, like many parents who wanted to meet Gandhi as the chief guest. Gandhi's last visit to the College came in the form of his funeral procession, which passed by the College. The College provided ladders for the students to see the procession over the tall walls of the institution. The College had received the news of

Gandhi's assassination at the convocation. Classes were cancelled and no food was served that day. In the days and weeks that followed, the College hosted sequential prayer services in the different religions of India.

Ritu's family moved to Kolkata during the Partition, while she was in New Delhi. She finished her diploma and went to join her family in Alipore, a posh neighbourhood of Kolkata. She had hoped to go back to Lady Irwin for an additional year to earn her teaching credentials, but her family had other plans. Her marriage had been arranged to a man from another Sindhi family from Karachi. Her fiancé had been tasked with starting a bank by G. D. Birla, one of India's leading businessmen and industrialists. He was fond of Western music and had had several girlfriends. They met for a movie, with her family on one side and his family on the other. Ritu's stepmother insisted that they have a traditional 'showing' party in the garden, where Ritu would dress up, wear make-up and make desserts. The couple, supervised, would make polite conversation. Ritu mentioned that she felt silenced at these events and would have preferred to meet separately at a restaurant for coffee. But that did not happen. The man's family was very impressed with her education at Lady Irwin College and wanted to seal the deal. After the movie and the garden party, they were engaged.

Ritu was not happy living with her in-laws after the wedding. After six years she left and returned to her father's household, which was then in Patna. She had hoped to have a degree of independence, and perhaps return to Lady Irwin for the year of teacher training, but her stepmother was quick to pull her into domestic tasks. Ritu returned to Lady Irwin to ask for a job. Tarabai was now the head of the school and praised her work as a student. If only she had stayed on to do her teacher training after her diploma. She helped Ritu secure a job at Ranchi Women's College, where she taught between 1955 and 1959. She briefly reconciled with her husband. When that relationship soured again, she completed her BA in sociology and political science in New Delhi. Ultimately, she resigned her position in Ranchi to go back to her husband, teaching at Calcutta's Modern High School for girls and helping to raise a niece. An industrialist's wife contacted her about starting a sort of finishing school for Marwari girls. The community had become wealthy and began to see the value of becoming more progressive in social norms. Ritu founded Saraswati Niketan in 1967. The curriculum included some Westernized, finishing-school-type subjects like poise, dinner party seating, serving food (because bearers had become too expensive) and etiquette, following American First Lady Eleanor Roosevelt's advice. Ritu especially wanted to teach the girls to become more attentive to the world around them and included poetry and current events. She recounted that a student might be able to read a news article but had no idea what it meant. The students, who typically ranged in age from sixteen to twenty-four, included both unmarried and married women, who served as mentors.

Ritu's life gives us an interesting example of the accomplishments of a Lady Irwin student. Her opportunity to study there completely changed her life. She escaped the cruelty of a stepmother and was able to develop her personality in the independent and safe atmosphere of a women's residential college. Having this education gave her options in life. She was able to opt for a career when she decided to leave her husband's house, and her career provided her with fulfilment during her twenty-year period of running the finishing school. Ritu had developed the inner strength to venture out on her own, to take jobs, to persevere even when her life seemed full of troubles. She was able to share this knowledge with hundreds of girls at the finishing school that she had started.

Case study 3: Flower Silliman

My year looking into the creation of home science influenced my fascinations with Indian history. During my first year teaching history full-time at the University of Iowa, I became good friends with Jael Silliman, whose mother Flower graduated from Lady Irwin College in the 1940s. Flower does not see herself as a 'typical' graduate of Lady Irwin College. This made our interview one of the most joyous parts of researching the history of home science. Sometimes, a viewpoint far away from the centre can be more revealing than a mainstream perspective. I sent Flower a list of questions that I would be asking her in advance. I intentionally did not plan to conduct the interview in any kind of linear, chronological fashion. Oral history methodology is at its best, after all, when trying to get at attitudes, beliefs and the meaning of experiences to people, without focusing needlessly on nailing down elements of 'fact'.[24] Here I join digital humanities scholars in trying to decolonize oral history methodology, by pushing against the limits of spatial and temporal boundaries. My approach follows that of South Indian Dalit (ex-untouchable) writer Bama in her autobiography *Karukku*, in which she selected a series of topics to discuss that were not bound by chronology but allowed her to revisit episodes of her life through a variety of different thematic lenses.[25]

Born into a Baghdadi Jewish family in Kolkata, Flower grew up in a tiny community within the cultural complexity that is Bengal and Kolkata. Pre-1947 Independence cities in Bengal, including Dhaka and Kolkata (formerly Calcutta), were characterized by their cosmopolitanism, in the realms of religion, ethnic and linguistic affiliations, along with caste and class. People of Hindu, Muslim, Christian, Jewish, Chinese, Anglo-Indian and Armenian backgrounds, among many others, mingled at schools, marketplaces, coffee houses and cinema halls.[26] Then, as today, it was not unusual for non-Christian middle and upper-class children to be educated at Catholic schools, as these were regarded to be among the premier institutions of learning. People gave genuinely friendly greetings to neighbours from other faiths and ethnicities, knowing their names and the basic facts about their families, even if they may not have socialized together much inside their homes. Flower recalled listening to record albums with other children and teenagers, singing and dancing to the latest hits of Frank Sinatra.[27] Though we might characterize Flower's childhood as appearing Westernized, at least in contrast to other communities in Calcutta, the family was never wealthy. Her parents both worked and the family moved around from rented apartment to rented apartment, never able to purchase a home.

Her family's decision for Flower to go to Lady Irwin College was, in a sense, community-driven. Flower noted that her neighbourhood was Jewish, her teachers were Jewish, her relatives and friends were Jewish, and they tried not to mix with others. She commented that in hindsight she felt like a butterfly trying to get out of a cocoon of Jewishness. Flower attended a Jewish girls' school, with six girls in her graduating class. The school followed the Junior and Senior Cambridge Curriculum (like O and A Levels today), where students finishing their O Levels went to work and A Level students entered university. Her older brothers, on the other hand, went to the non-Jewish Calcutta Boys School. They chose matriculation with World War II breaking out. One worked for the US Army and the other became a silkscreen artist. The senior Cambridge included 'housecraft' but because of complicated dietary laws (kosher), they did not teach cooking. Flower reported that she did not like this course very much and did not do well in it. Instead, they did needlework and laundry and housecraft. Lady Irwin College was the only college her mother would consider for her. Flower had hoped to do a BA degree in Lucknow. A couple of Jewish girls had gone there and came back 'unscathed'. Flower's Jewish housecraft teachers had themselves gone there. The founder of Lady Irwin College, Hannah Sen, was a Jewish woman, and therefore her institution was deemed trustworthy for Flower. The family had evacuated to New Delhi during World War II, when Kolkata was bombed. By this time, 1946, the war was over.

There was one other Jewish girl from Bombay at Lady Irwin, but she reported that this was her first encounter with Hinduism, Sikhism, Islam and other religions. She reports taking on an interfaith attitude from that time, being thrust into a truly All-Indian atmosphere, influenced greatly by Mahatma Gandhi. She described the atmosphere of Lady Irwin as very tolerant of all faiths, as she celebrated Guru Nanak's birthday and the Hindu Holi. Flower packed her Kolkata wardrobe full of pants and skirts and tops, but once at college she realized that everyone else wore *salwar-kameez* (tunic over loose pants). When she brought her *salwar-kameez* home to Kolkata, her mother refused to let her wear them. Flower sometimes borrowed saris to wear at college, but she said her mother did not know.

Flower reported to me that she was never that interested in the Lady Irwin College curriculum but was fascinated by everything going on in the outside world. The College provided a forum for leading political figures to attend meetings, including Gandhi, Mountbatten, Sarojini Naidu, exposing the students to much of the Indian politics shaping the future of India. Flower received her first diploma from Mountbatten and her second from Nehru. As a student, she described herself as very ordinary. She felt that subjects like sewing and laundry were 'stupid'. She recounted that the students were assigned as a group a patch to do gardening, but she never went, feeling that the work was beneath her. Another of the girls in her group enjoyed gardening, doing all the work, with their group actually winning a prize for the best garden. Flower acknowledged that the other girls took home science more seriously, their marriages were being arranged and they wanted to be educated in the craft of home science. Flower noted that people in India never cultivated their own garden; they would get some servant to do the work. In those days there were no part-time servants, and her own family had three full-time employees: a cook, a valet and a *jamindar* (guard).

Flower was a student at Lady Irwin during the Partition, and saw refugees coming in on trains, blood dripping from the train cars. She described Christian friends travelling to Lahore for Christmas coming back as refugees, because of the formation of Pakistan. The College and many urban households organized 'Operation Chappati' where they sat up all night cooking for the refugees pouring into India without access to food. Sacks of chappatis were airlifted to people coming across the border so that they had access to basic food during their journey. Flower remembered a photograph of herself making chappatis on a coal stove. She explained that due to the hundreds of refugees moving from Lahore to Delhi, the extreme housing shortage caused many girls to be put into the Lady Irwin College dormitories, leading to overcrowding. Lady Irwin was in an upper-middle-class Hindu area. Muslim girls in the dorms – including wards of the sultan of Hyderabad – lived in fear of the chanting rioters in the streets outside. The girls learned *lathi* charging, which Flower described as a kind of self-defence with a stick.

All the students were from high-status families and Flower described how there were few elite options for the daughters of these families. The few lady colleges, Flower said, meant that you were dumped into home science. Jewish girls were considered white and there was uncertainty about their status once India changed from being British India to the Indian Republic. About a third of the Jewish families left the country. One of Flower's brothers went to London and another brother went to Australia, with their parents emigrating to Australia as well. Like Ritu, Flower commented on the unit where the girls lived together in the house as one of the highlights of her time at the home science college. 'We did everything', she said. 'We cooked, we cleaned, we washed clothes and we sewed and we ironed'.

How did Lady Irwin College benefit Flower? What opportunities did it provide her? What thoughts or regrets did she have about her educational trajectory? Early on in the interview, Flower noted that she did not think highly of home science in terms of what it could teach her as an academic discipline, noting that home science taught a number of household tasks which were actually things that servants typically did. Although she did not grow up in wealth, her status as a middle-class Indian gave her the confidence that she would probably never be asked to do the tasks that she learned in her home science training. Flower's remark reminds me of the inevitable collision between 'value-free' science of the 1920s with the realities of domestic science emerging in societies defined by social, class, caste and racial inequalities. At certain points in the interview, Flower expressed dismay that she had gone to Lady Irwin, saying that she could have done a degree in another subject, like other girls of her generation. But in the circular, non-linear fashion that we did the interview, she changed her mind, and concluded that she was glad that she had graduated from Lady Irwin. Outside of home science, degrees in teaching, literature and philosophy were considered most appropriate for young women, compatible with social expectations of womanhood. By using a flexible methodological approach to the interview, Flower was able to express contradictory thoughts about her education, reflecting back across several decades in making meaning about her life.

Understanding the implementation of a self-professed apolitical, supposedly neutral subject like science into societies demarcated by class, let alone caste and race, raises interesting questions. One thing that has always puzzled me about home science as a discipline and as a field of study is the very thorny, very illiberal question of household servants, and how home science does or does not anticipate or accommodate these ubiquitous employees, without whom few middle- and upper-class households would function, yet who were almost always rendered invisible in representations of the modern family. Family studies as part of a nationalist home science never had much to say about the place of servants in the culture. The culture of servants in colonial India was more thorough, with Hindi words like 'shampoo' (originally, the command 'to massage') entering the global mainstream of multilingual English.[28] Historical studies on servants in Indian households are sparse, except for the pioneering work of Swapna Banerjea, Nitin Sinha and Nitin Varma.[29] Historian Swapna Banerjee's beautifully written *Men,*

Women and Domestics: Articulating Middle-Class Identity in Colonial Bengal looks at how employer–servant relationships helped define both nationalism and middle-class status.[30] In the United States, servants were more ubiquitous until World War II and even longer. The idea behind home economics, for white women, was to systematize and reduce the amount of time and energy women spent on household chores. The question of how domestic education meshed with the presence of household servants deserves fuller consideration.

Post-independence development period

The British left India in 1947 and India became a republic in 1950. Though technically unaligned in terms of the Cold War between the Soviets and the Americans, India drew on both Soviet and American models in developing itself into a modern nation. The US-based Ford Foundation established a collaboration between the University of Tennessee and Indian educators to develop undergraduate and graduate programmes of Home Science. The colleges teaching home science in India included institutions in Bangalore, Baroda, Madras (now Chennai), Lady Irwin in New Delhi and SNDT in Bombay, now Mumbai. Lady Irwin continued to be the model institution, with Queen Elizabeth visiting in 1961. Experts in home science from the University of Tennessee travelled to the institutions to help establish programmes. They began to meet each other for three-day conferences each quarter, to compare notes and meet with Indian home science experts. Of course, the American home differed substantially from the home in India. American experts had a lot to learn about Indian cultures and society. Major differences included the extended family kinship system, but the urban–rural gap was also significant. The way that women cooked meals and practised child-rearing differed. One specialist was quoted as saying, 'We moved into our own house – an excellent first-hand experience of what it is to establish a home in India. Our experiences with purchasing supplies and equipment, with hiring servants and with organizing the operation of a household gives us a better understanding of some of the Indian homemaker's problems.'[31]

The tension that the home science specialist pointed out – between Western and Indian ways of keeping house – is a theme that I see running throughout the curriculum of home science, both explicitly and implicitly. Let us consider a few examples. The Home Science Association of India published an overview, 'Home Sciences in Colleges and Universities in India', in 1958. A photograph appears near the beginning of the publication, between the table of contents and the introduction. The photo caption reads, 'The focus of home science is on homes and families.' The photograph displayed pictures of a posed Indian family seated near the corner of a living room, presumably theirs. The mother/wife sits in the middle, with her family circled around her. Her husband is at twelve o'clock, a daughter at three, the youngest child and daughter at five, closest to the mother, and a son at nine. The children sit on wicker ottomans, while the wife appears to sit on a larger chair, and the husband sits behind her. A Rajasthani mirrorwork cloth is draped on the left, in the centre is a reading lamp and to the right are five shelves, holding carefully arranged figurines and books. Perhaps most remarkable about the photo is the clothing shown on each family member. The girls wear Western-style dresses, the older daughter sporting Mary Janes (shoes). The brother wears a button-up shirt and flared trousers, along with shoes and socks. The father wears a business suit and a tie, and even a kerchief in his upper-left jacket pocket. The mother is the only one wearing Indian-style clothing, a sari draped over a dark-coloured, modest-cut, elbow-length sari blouse. On her forehead is a bindi, the so-called 'dot' that Hindu women wear to show that their husband is alive (and they are not widowed). The photograph reflects that fashion fact that professional dress for women is a sari, whereas there are multiple choices for other family members. The photo speaks volumes about the woman being the anchor of Indian tradition for her family. Though well-versed in Western conventions of clothing, furniture and a reading lamp, the heartbeat of the home is unmistakably Indian.

The nationalist dimension of home science continued well into post-independence. My second example comes from a 1979 book on the teaching of home science. The book is instructive on how the subject is situated within a nationalist framework. In the introduction, authors R. R. Das and Binita Ray write, 'A congenial and happy home has bred many a successful personality in varied fields while too often broken unhappy homes have turned out misfits, failures and even criminals.'[32] The co-authors of *Teaching of Home Science*, Das and Ray, both have Bengali surnames, and are based in institutions in Orissa, the considerably poorer, less-developed and less-well-educated state to the south of Bengal. Das and Ray stress that the extent of home science is not limited to the four walls of the house but has national aims. They write that serving the nation and 'promot[ing] international goodwill and understanding is the most important contribution that Home Science has made'.[33] Like much national discourse of the post-independence state, the authors' writing about home science is at once ancient and modern. Homemaking, they claim, has roots in ancient history, but has been adapted for individualistic thinking. Problems plaguing the modern household include smaller families, the weakening of the extended family structure, declining authority for the family head and the increasing independence of women.[34] The focus on the individual that is reflected in

Das and Ray's textbook is at once predicable and strange. The Western liberalism within the development discourse is reflected in the authors' stress on decision-making, independent thinking and self-judgement. Yet Indian society, like most societies, is based on relational thinking, the family and the collective.

A future agenda for research is to think about some of the gaps that exist between Indian and Western systems of housekeeping. I have already mentioned the question of the individual over the family. How do modern machines, such as the washing machine, fit into actual middle-class homes, most of which have either 'part-time' or live-in servants. Over the years that I have done research in India and come to know several Indian families, servants have been an inevitable part of almost every household. In some households, generations of servants work for generations of householders. In the 1980s, before liberalization and inflation, part-time servants might earn one or two dollars a month. Now payments are much higher. The cruellest part of the servant tradition is that while servants make possible a middle-class domesticity, convenient and comfortable for middle-class families, they are often separated by their own biological families and kin for months or years on end, sending remittances. This is not just in India, of course. Thousands of nannies worldwide work for one family to achieve a middle-class existence for their own children, though separated. This gap between home science's idealized households and Indian and Western realities is complex. The class dimension of domestic science is assumed but implicit. Women who study domestic science have historically come from the upper-middle classes, who could afford the tuition to send their daughters away for a college education.

It is instructive to look at representations of home science within Indian society generally. I would like to move towards a conclusion with a brief scene from one of India's most acclaimed filmmakers, Satyajit Ray, which comments on the place of home science. Film sources give a sense of how domestic science has been viewed in various regions of India. In Satyajit Ray's film *Mahanagar (The Big City)*, the younger sister in the middle-class Bengali family, Bani, studies domestic science. The setting of the film is the 1950s, and Bengal is flooded with refugees from East Pakistan after the Partition. There is much more economic pressure on families to meet basic expenses and the major tension of the film is in the daughter-in-law, Aroti, going out to get a job to help earn money for the family. Husband Subrato reluctantly agrees, even though he feels strongly that 'A woman's place is in the home', a saying that he repeats to his wife in English, suggesting that he is maintaining Western standards.

Within the Mazumdar family, we see progressive levels of education in both generation and in age group in the film. The grandmother's educational attainment is not mentioned in the film, but I would guess that she has either finished grade ten or perhaps high school. The daughter-in-law and star of the film, Arati, finished one year of college before dropping out to get married to Subrato. Subrato's unmarried teenage sister, Bani, is in college. The question about Bani's studies concerns her fees in this cash-strapped family, more so than the choice of subject, as the following dialogue suggests:

Subrato: What do you study?
Bani: Home science.
Subrato: Laughing. [But] you will just end up in the kitchen like Arati!

Later in the film, teenage sister Bani offers to leave college for the extended family to save on her college fees. But the family agrees she should stay in college and that decision is important for her future. The conservative family sees home science as the proper subject for Bani and she should not leave it.

Conclusion

My paper raises the question of how home science fits into the Indian educational system for women. It was never, of course, the only subject that Indian women studied, but it is one that many women did study and continue to study today. Women from very progressive families did enter the traditional scientific professions at the same time that more conservative families had their daughters enter domestic science. This is not to say that home science has not produced useful knowledge. Published research papers come out of Lady Irwin College on rural development, vaccine awareness and environment and conservation. There is no question that home science has continued to change its focus, evolve and incorporate many important areas of knowledge. The subject of domestic science allowed women access to higher education, and a scientific education at that, at a time when this was far from assured during a woman's life. This is particularly true of Gopalan's experience.

In the 1920s, Lady Irwin College in New Delhi, India was established as an institution dedicated to higher education for well-to-do Indian women. The college focus began as home science, in which students would learn and use a scientific approach to home management and domesticity. Home science in India developed in tandem with home economics in the United States, with an exchange of experts from the University of Tennessee at Knoxville. Starting in about the 1960s, the College began to attract women from business communities, who sent their daughters for higher education in subjects that would match the expectations of their husbands' families. Rather than see women's participation in home science as a retrograde choice, I argue that home science was an innovative way to bring women into higher education, which was otherwise seen as a negative in India's all-important marriage markets. The graduates I interviewed contributed in various ways to the education of other girls. For Ritu and for Flower, being at Lady Irwin College involved them in nationalist politics, giving them a stronger sense of India's unity in diversity. Much closer to my home in the US is the mother of Kamala Harris. Shyamala's higher education at Lady Irwin helped launch her career as a biochemist and mother of the United States' first female – and female of colour – vice-president. Women studying home science in the three case studies I present were not unaware that they had fewer opportunities than their brothers or even other young women in India. At the same time, Lady Irwin College gave them each a platform to become interested in politics and forge productive lives and careers. Each of them experienced periods of troubled marital life, but their education gave them a basis to pursue jobs and relative independence.

Notes

1 A. King, *The Bungalow* (London: Routledge & Kegan Paul, 1984).
2 D. Fleming, *Schools with a Message in India* (London: H. Milford, Oxford University Press, 1921); M. A. Laird, *Missionaries and Education in Bengal* (Oxford: Clarendon Press, 1972).
3 M. Borthwick, *The Changing Role of Women in Bengal* (Princeton, NJ: Princeton University Press, 1984); D. Chakrabarty, 'The difference-deferral of (a) colonial modernity: Public debates on domesticity in British Bengal', *History Workshop*, 36 (1993), 1–34; P. Chatterjee, 'Colonialism, nationalism, and colonized women: the contest in India', *American Ethnologist*, 16: 4 (1989), 609–21; N. B. Dirks, 'The home and the world: The invention of modernity in colonial India', *Visual Anthropology Review*, 9: 2 (1993), 19–31; King, *The Bungalow*.
4 M. Kishwar, 'Arya Samaj and women's education', *Economic and Political Weekly*, 21: 17 (1986), WS9–WS24; N. Kumar, 'Widows, education and social change in twentieth-century Banaras', *Economic and Political Weekly*, 26: 17 (1991), WS19–25; K. I. Leonard and J. G. Leonard, 'Social reform and women's participation in political culture: Andhra and Madras', in G. Minault (ed.), *The Extended Family: Women and Political Participation in India and Pakistan* (Delhi: Chanakya Publications, 1981), pp. 19–45.
5 Translated and quoted by Borthwick, *The Changing Role of Women in Bengal*, p. 92.
6 S. J. Stage and V. B. Vincenti (eds), *Rethinking Home Economics: Women and the History of a Profession* (Ithaca, NY: Cornell University Press, 1997); S. Y.

Nichols and G. Kay (eds), *Remaking Home Economics: Resourcefulness and Innovation in Changing Times* (Athens: University of Georgia Press, 2015).

7 R. Kumar, *The History of Doing: An Illustrated Account of Movements for Women's Rights and Feminism in India, 1800–1990* (New Delhi: Kali for Women, 1993).

8 P. Chatterjee, *The Nation and Its Fragments* (Princeton, NJ: Princeton University Press, 1993).

9 Chatterjee, 'Colonialism, nationalism, and colonized women'.

10 M. Hancock, 'Home science and the nationalization of domesticity in colonial India', *Modern Asian Studies* 35: 4 (2001), 871–903.

11 On Britain, see N. L. Blakestad, 'King's College of Household & Social Science and the household science movement in English higher education c. 1908–1939' (PhD dissertation, University of Oxford, 1994).

12 N. Kumar, *Women and Science in India: A Reader* (New Delhi: Oxford University Press, 2009).

13 Kumar, *Women and Science in India*, p. xxiii.

14 S. Koyama, *Ryosai Kenbo: Educational Ideal of 'Good Wife, Wise Mother' in Modern Japan* (Boston, MA: Brill, 2012), p. 59.

15 D. Dreilinger, *The Secret History of Home Economics* (New York: Norton, 2021), p. x. Interestingly, my inter-library loan arrived from St Philip's College, a historically Black community college in San Antonio that started as a school of domestic science. This coincidence mirrors Dreilinger's argument that home economics has roots in African American educational reform.

16 Dreilinger, *The Secret History of Home Economics*, p. ix.

17 See, for example, M. W. Rossiter's classic text, *Women Scientists in America* (Baltimore, MD: The Johns Hopkins University Press, 1995); D. L. Opitz, 'Domesticities and the sciences', *Histories*, 2: 3 (2022), 259–69, https://doi.org/10.3390/histories2030020, esp. note 5. For a recent example of such a study, see D. P. Munns, ' "Not by a decree of fate": Ellen Richards, euthenics, and the environment in the Progressive Era', *Journal of the History of Biology*, 56: 3 (2023), 525–57, https://doi.org/10.1007/s10739-023-09733-9.

18 J. G. Roland, T. M. Gubbay and J. Silliman, 'Baghdadi Jewish Women in India', in *The Shalvi/Hyman Encyclopedia of Jewish Women*, last updated 2022, https://jwa.org/encyclopedia/article/baghdadi-jewish-women-in-india.

19 H. H. Risley and E. A. Gait, *Census of India*, vol. 1: India, part I: Report (Calcutta: Office of the Superintendent of Government Printing, India, 1908), p. 158, http://piketty.pse.ens.fr/files/ideologie/data/CensusIndia/CensusIndia1901/CensusIndia1901IndiaReport.pdf (accessed 13 August 2024).

20 M. Sinha, 'Introduction', in K. Mayo, *Mother India: Selections from the Controversial 1927 Text, Edited and with an Introduction by Mrinalini Sinha* (Ann Arbor: University of Michigan Press, 2000).

21 E. Barry, 'How Kamala Harris's immigrant parents found a home, and each other, in a Black study group', *The New York Times* (13 September 2020).

22 S. Bengali and M. Mason, 'The progressive Indian grandfather who inspired Kamala Harris', *Los Angeles Times* (25 October 2019).

23 This section draws on an interview with 'Ritu', a pseudonym I use to protect her identity. The interview was conducted in person in Kolkata in spring 1996.
24 Zoom interview with Flower Silliman, 5 April 2022.
25 Bama, *Karukku* (New Delhi: Oxford University Press, 2014).
26 S. N. Hashim, 'The cultural mix of Old Dhaka', in M. Guhathakurta and W. van Schendel (eds), *The Bangladesh Reader: History, Culture, Politics* (Durham, NC: Duke University Press, 2013), p. 147.
27 I note that Sinatra was predominantly a 1950s singer, illustrating that memory is not always cosympathetic with history. In any case, the point is that Flower listened to Western music as a form of entertainment.
28 See the definition of 'shampoo' in Henry Yule's *Hobson-Jobson* dictionary of Anglo-Indian words: 'Shampoo, v. To knead and press the muscles with the view of relieving fatigue, &c. The word has now long been familiarly used in England. The Hind. verb is chāmpnā, from the imperative of which, chāmpō, this is most probably a corruption, as in the case of Bunow, Puckerow, &c. The process is described, though not named, by Terry, in 1616: "Taking thus their ease, they often call their Barbers, who tenderly gripe and smite their Armes and other parts of their bodies instead of exercise, to stirre the bloud. It is a pleasing wantonnesse, and much valued in these hot climes." (In Purchas, ii. 1475).' https://dsal.uchicago.edu/cgi-bin/app/hobsonjobson_query.py?qs=SHAMPOO&searchhws=yes&matchtype=exact (accessed 13 August 2024).
29 S. M. Banerjee, *Men, Women and Domestics: Articulating Middle-Class Identity in Colonial Bengal* (New Delhi: Oxford University Press, 2004); N. Sinha and P. Kumar (trans.), *Lesser Lives: Stories of Domestic Servants in India* (New Delhi: Pan McMillian, 2021); N. Varma, N. Sinha and P. Jha (eds), *Servants' Pasts: Sixteenth to Eighteenth-Century South Asia* (Hyderabad: Orient Blackswan, 2019).
30 Banerjee, *Men, Women and Domestics*.
31 United States, International Cooperation Administration, Office of Public Reports, *Technical Cooperation through American Universities* (Washington, DC, 1957), pp. 16–17, https://babel.hathitrust.org/cgi/pt?id=mdp.39015030467800&seq=7 (accessed 13 August 2024).
32 R. R. Das and B. Ray, 'Introduction' to *Teaching of Home Science* (New Delhi: Sterling Publishers Private Limited, 1983).
33 Das and Ray, *Teaching of Home Science*, p. 16.
34 Das and Ray, *Teaching of Home Science*, p. 2.

11

Clara Park: A mother's intimate knowledge and child science

Marga Vicedo

Women's contributions to science have been underestimated by the predominant focus on research pursued in academic institutions and professional settings such as laboratories. Studies of collaborative couples in science and of scientists' family lives have revealed that wives, sisters and daughters often carried out tasks important for the scientific pursuits of male scientists in their households. Thus, several historians have called for recognizing the significance of domestic spaces as sites of knowledge production. Looking at those spaces has helped illuminate women's hidden contributions to science.[1] Few accounts, however, have focused on mothers.

Mothers are not completely absent from history, but they rarely appear as contributors to scientific knowledge. Historians have examined changing conceptions of motherhood, maternal care and love.[2] Scholars of technology have explored the impact of technological developments on mothers, both as users and as contributors to innovations in different areas.[3] Mothers have also been scientific 'objects' of study and historians of science and of medicine have shown how they were blamed for whatever ailments their children suffered.[4] Furthermore, modern science has played a key role in shaping views of good mothering and several studies illuminate the rise of 'scientific motherhood' in the twentieth century.[5] Yet, many histories of women in science do not examine their roles as mothers or the impact of motherhood on their scientific activities.

It is difficult to recover mothers' contributions because their voices were excluded from modern science even in the one area in which they were heralded as natural experts: children. Women's maternal instincts allegedly prepared them to care for children but also prevented them from observing them objectively. Nevertheless, a few scholars have recovered the role of women as agents, not subjects, of scientific research in child studies. One must mention the work of Sally Shuttleworth and Christine von Oertzen, who have examined mothers' contributions to early child studies and debates about whether mothers were qualified to offer authoritative knowledge about children.[6]

In my view, the exclusion of mothers from the sciences of childhood was connected to the long-standing belief that emotions are antagonistic to scientific pursuit. Take the case of mothers who observed their children's development. First-hand experience has been considered crucial for obtaining knowledge in many realms. Yet, personal experience has also been seen as suspicious since the knower's subjectivity threatened what many sciences saw as necessary for objective knowledge: detachment from one's object or subject of study. Scholars proposing a feminist epistemology have argued that this detachment is neither possible nor desirable.[7] But the modern conception of objectivity as the absence of subjective emotional attachment has been predominant in science and prevented mothers for participating in knowledge production. Mothers who observed and studied their children were not detached from their subjects of study and their maternal affects could bias their perceptions.

This paper examines how Massachusetts writer and homemaker Clara Park challenged the view that emotional involvement is incompatible with objectivity in her writings about raising her autistic daughter Jessica. In her 1967 book *The Siege* and other publications, Park called upon scientists to recognize that intelligence and love are not enemies. Therefore, a mother could provide reliable information about her child's development. Furthermore, she claimed that daily contact with their children allowed parents to acquire 'deep knowledge of the child in context'. Arguing that lived experience could provide valuable insights to complement clinical and research work, Park fought to have a mother's voice recognized as a legitimate source of expertise. In doing so, she challenged us to rethink the role of subjectivity in the construction of scientific knowledge, which opens new possibilities for valuing women's insights in the domestic space. An examination of Clara Park's writings and her extensive private archive containing her notes, correspondence with other mothers as well as with scientists working on autism reveals the crucial ways in which mothers in the United States contributed to shaping the discussion about autism.[8]

Clara Claiborne Park: Mothering an autistic child

Born in 1923, Clara Claiborne grew up between Virginia and New York, between a racist and upper-class world she despised and the progressive intellectual milieu of her New York City school friends. In New York, Claiborne attended the Dalton School, before entering Radcliffe College in 1940 with a scholarship.

Though connected to Harvard University, at the time Radcliffe was a female college that many Harvard professors did not see as being on a par with its prestigious all-male counterpart. Believing that women were unable to meet the rigors of intellectual pursuits and demanding careers, some Harvard professors did not want to teach women. In this sexist environment, Claiborne pursued studies in English literature, which would remain her lifelong passion. In 1944, she graduated with a degree in English and aspired to write essays and poetry as well as form a family with David Park, a Harvard physics graduate.

As many of her classmates did, Claiborne followed contemporary social expectations in choosing motherhood over a career. Initially, while David Park served in England during World War II, she worked as an editorial assistant in New York City, but only for a year. When the war ended, they married in New York City. Clara Claiborne Park then followed her husband to the University of Michigan at Ann Arbor, where he obtained a PhD in physics and she pursued a master's degree in literature. Next, the Parks moved to the Institute for Advanced Studies in Princeton for David's postdoctoral studies. In 1951, they settled in Williamstown, Massachusetts, because David had accepted a position at the Physics Department at Williams College. Here, David focused on his research and writing while Clara devoted herself to raising their three children, Katherine, Rachel and Paul. Many educated women who were married to faculty members led similar lives because of social expectations and anti-nepotism laws that barred them from pursuing academic careers at the same institution where their husbands worked. Though introduced to avoid situations where personal preferences could unduly influence university hires, in practice these rules prevented many women with advanced degrees from combining motherhood, or even marriage, and an academic career.

In the late 1950s, Park felt immensely proud of her children and her mothering work, but she longed to devote some energy to her own intellectual development by writing and teaching English literature. She had to keep that hope on hold when she unexpectedly became pregnant again. Her fourth and last child, Jessica Park, was born in July 1958 (see Figure 11.1). Jessica cried incessantly for several months as an infant, but Park did not trouble herself about that because her other children had been through colicky periods as well. But, as an experienced mother, Park noticed some features in her daughter's behaviour that concerned her. Jessica seemed quite detached from her surroundings. She did not reach for any objects. She also showed little interest in people. She was content to sit by herself for hours. Following the advice of her paediatrician, the Parks consulted a specialist, who diagnosed Jessica as autistic in 1961, at three years of age.[9] Though at the time parents of disabled children were advised to leave them in an institution, the Parks did not want to separate Jessica from the rest of the family and decided to raise her at home.

Eager to understand her daughter's condition, Clara Park kept a notebook about Jessica's development. As a good scientist would do, she observed her systematically, took copious notes and separated her observations in different categories, such as language, social interactions and so on. She also tried different activities with toys and games to entice Jessica to interact with her surroundings. She approached these as 'trials' or 'experiments' to find out what activities Jessica enjoyed so they could pursue them together. Park also kept records of Jessica's progress and her struggles. When Park brought Jessica to the experts in children's minds – psychiatrists, psychologists and psychoanalysts – she thought her notes would be useful for her treatment. She also thought her insights could contribute to a better scientific understanding of autism. However, her notes were used for a completely different purpose: to diagnose her as a bad mother and the culprit of her daughter's affliction.

By presenting data about her daughter and by trying to analyse her condition systematically, Park only proved to these experts that she was an 'intellectual' mother. For them, this meant a mother who was unable to love her child 'naturally', a mother who followed her intellect rather than her maternal instincts. She was the kind of mother whom researchers at the time accused of driving their infants into autism.

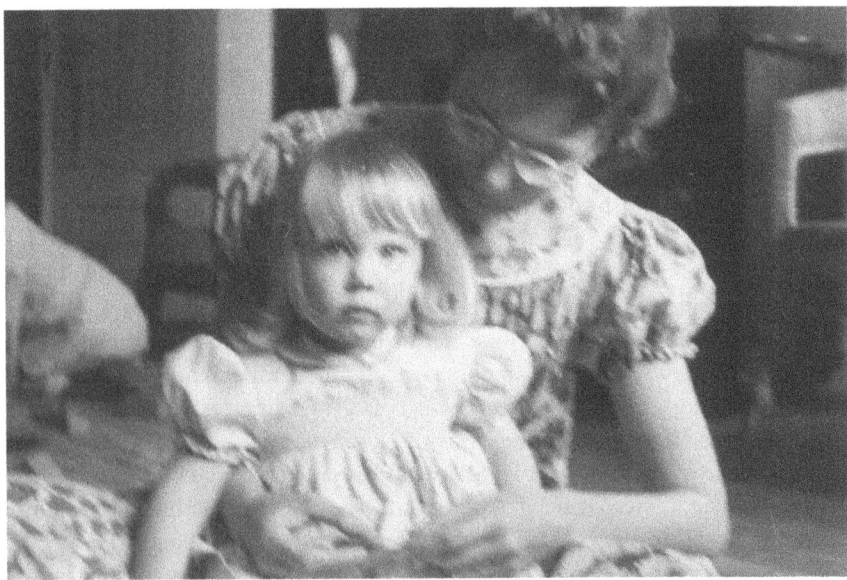

Figure 11.1 Clara Park and her daughter Jessica, c.1960 (courtesy of Clara Park)

This emphasis on the mother as the cause of autism arose from a remarkable turn of events in medical understanding. Psychiatrists had used the term 'autism' as a symptom of different conditions before Hans Asperger in Austria and Leo Kanner in the United States employed it to identify a unique condition in the early 1940s. At Johns Hopkins University, Kanner identified 'infantile autism' as a unique syndrome in two influential 1943 and 1944 papers. In them, he described children who did not fit standard psychiatric diagnoses. These children tended to self-isolate, enjoyed repetition and sameness, and in many cases displayed a peculiar use of language. Kanner, relying heavily on notes provided by the parents, pointed out that the children he categorized as autistic had been so from birth.[10] He concluded that autism was an inborn condition.

But Kanner changed his mind and, together with other psychiatrists and psychoanalysts, argued that mothers played a key role in causing autism. In the 1950s, the work of psychoanalysts René Spitz and John Bowlby on the detrimental effects of maternal deprivation or separation became widely known. They argued that the absence of sensitive maternal love caused a range of psychopathologies in children.[11] Influenced by these views, Kanner said that parents of autistic children left them in refrigerators that did not defrost.[12] This statement led to the concept of the 'refrigerator mother', the idea that cold, intellectual mothers led their children to retreat into autism. In his bestselling 1967 book *The Empty Fortress: Infantile Autism and the Birth of the Self*, psychoanalyst Bruno Bettelheim also blamed rejecting mothers for their children's autism.[13] Bettelheim is often presented as the originator of the refrigerator-mother metaphor, though he never used the term itself. Still, his views exerted a strong sway in the popular culture and encouraged the wave of mother-blame at the time.

In this context Clara Park was blamed for her child's condition. She first felt she was 'on trial' when she went to the Putnam Clinic in Boston, a major psychoanalytic centre for children in the United States. When the Parks took Jessica there in 1963, Clara brought her notebooks with her, hoping they could be of help to the staff at the Centre. But only a social worker interviewed Clara, alone. A psychiatrist interviewed David. The same psychiatrist saw Jessica separately. Jessica was diagnosed as suffering from 'atypical development', the Centre's term for autistic children, a condition that allegedly resulted from disturbances in maternal affects. They recommended therapy for Jessica and for Clara. Later that year, when the Parks moved to Cambridge, England because David had research leave, they took Jessica to the Hampstead Clinic, founded and directed by Anna Freud in London. Here, Clara's efforts to document Jessica's development were seen as proof that she approached her child 'as a project'. The experts there also recommended therapy for Clara.[14] Her encounter with professionals who blamed her for Jessica's disabilities was devastating for Clara, who struggled with depression during her year in England.

Fortunately, Park found encouragement after reading Bernard Rimland's 1964 book *Infantile Autism*.[15] A US Navy psychologist who began to investigate autism after his own son was diagnosed with the condition, Rimland criticized the psychoanalytic accounts that blamed mothers while defending the position that autism had an organic origin and mainly concerned cognitive abilities. The first letter he received after his book's publication was from Clara Park. In a five-page letter, she praised his contribution. Yet, Park told him she disagreed with his belief that autism was mainly a cognitive condition.[16] For her, the affective aspects of autism were primary. She came to that conclusion by reflecting on Jessica's use of language. Park had noticed that Jessica could easily learn difficult words such as dodecagon and trapezoid, even after hearing them only once. However, she had difficulties in learning concepts related to emotions, moods and feelings. And she showed no inclination for social contact. Though Park had no scientific credentials, Rimland appreciated her insights and encouraged her efforts to challenge mother-blame in autism research. Park decided to write about her experiences in raising Jessica.

Park's book, *The Siege*, the richest record of raising an autistic child to date, appeared in 1967.[17] The book was unique. It provided a detailed record of Jessica's development and an examination of the available evidence for different views about autism. It was a hybrid, both memoir and analytic essay. At its core, Park presented a defence of the value of intelligent love for motherhood and for science. 'Intelligence and love are not natural enemies', Clara Park claimed.[18] Rejecting the separation of thinking and feeling prevalent in contemporary conceptions of good mothering and good science, Park aimed to show that intelligence was not incompatible with love and that both were valuable in mothering. Further, intelligent love could lead to reliable knowledge. In arguing that intelligence and love were not natural enemies, Park contested dominant notions of maternal love and scientific epistemology.

Park challenged prevalent notions of maternal love because at the time most child psychoanalysts and psychologists, including Erik Erikson, Margaret Ribble, Bowlby, Bettelheim and Spitz, claimed that what a child needed to develop into a psychologically healthy individual was 'natural' love, some kind of raw feeling that should not be sullied by a mother's intellectual interests or other preoccupations besides her child.[19] But none of those researchers had explained why this type of love was better or more nurturing for the child than other types of love, including intelligent love. For Park, a mother's intellect could contribute to good mothering. Clara's account of family life in *The Siege* showed how her intellectual mind was not an obstacle to loving her daughter Jessica. Her analytic capacity should be seen as an asset, not a hindrance, to being a good mother. Furthermore, Park argued that love could be helpful in generating reliable observations and reaching authoritative knowledge. In a chapter entitled 'The amateurs', she wrote that parents could contribute to advancing scientific knowledge about autism because their love for their children was not an impediment to understanding their condition.

Yet, the prevailing belief in the separation between subjective feelings and objective thinking could be seen in the response Park received from editors and reviewers who often questioned that she could be a good mother while being objective about her own child. For example, one reviewer said: 'The account is so successful because Mrs. Park has the unusual capacity to stand aside and look at herself and her child and interpersonal relationships objectively and analytically. Tragically, it is perhaps this very quality which helped to produce a child of Elly's type.'[20] (Elly was the name Clara Park used in her book for Jessica.) So, a mother could not be objective and still be a good mother.

That reviewer's response was not surprising because Park was challenging a central tenet in scientific epistemology. Historians and philosophers have shown that objectivity is not a monolithic concept and has changed over time. In modern science, as Lorraine Daston, Peter Galison and other historians have documented, objectivity came to be defined as the absence of subjectivity.[21] And what could be more subjective than a mother's feelings? As such, many researchers thought a mother's love would bias her perspective. Indeed, the question of whether mothers could provide reliable observations about their children had been a matter of debate since the start of child studies in the late nineteenth century.

A mother's siege: The power of intimate knowledge

As Sally Shuttleworth has shown, the early days of child psychology and child psychiatry witnessed fierce debates concerning who was qualified to provide reliable observations about children's minds and behaviour. The establishment of the British journal *Mind: A Quarterly Review of Psychology and Philosophy* in 1876 sparked a wave of interest about child studies among leading British scientists and literary scholars, including G. H. Lewes and James Sully. From its inception, *Mind* published articles arguing that mothers were not competent to observe children, though some mothers challenged those views. Self-proclaimed male experts in the new science of childhood, such as Sully, argued that observations gathered by women were not to be trusted because they were inevitably coloured by their maternal instincts.[22]

Encouraged by the interest in child studies, naturalist Charles Darwin published in *Mind* his 1877 paper 'A biographical sketch of an infant', based on his observations of one of his own children years earlier. Now, this paper is considered one of the founding texts of child psychology. Although Darwin did not include the observations that his wife Emma had carried out, he did collect the observations of mothers for his studies about children and about the expression of emotions. 'Ask her to look out (for no. 5) when one of her children is struggling & just going to burst out crying.' So wrote Darwin in a letter in which he asked his fellow scientist Thomas Henry Huxley to pass on a questionnaire regarding children's facial expressions to his wife Henrietta. Trusting mothers to provide reliable observations, he also asked other women friends to do the same.[23]

Though worries over scientific authority continued, in the early twentieth century some mothers published baby biographies that reached broad audiences. Women could be good mothers while also observing their children and writing about their development. In 'Science in the cradle', Christine von Oertzen has examined the history of these baby diaries and of women's early contributions to understanding children's development.[24]

Over the course of the twentieth century, however, it became difficult for women to contribute to the child sciences for several reasons. First, the professionalization and institutionalization of modern science in the early twentieth century increased the long-standing dismissal of women's contributions to knowledge-making because women had little access to formal education, were unable to obtain jobs in many academic institutions and were often relegated to positions of assistants to male scientists in professional settings or unpaid and invisible assistants to their husbands and brothers in domestic settings. Second, not only were women kept out of science, but science gradually disregarded skills that were associated with women and, especially, with mothers.

Legitimizing child developmental science led to a rejection of maternalist traditions that had been important at the beginning of child studies in the early twentieth century. As the fields of psychology and child development became professionalized, they dismissed patience, empathy, sympathy and other traits associated with maternal care as interfering with scientific discipline. The scientific credentials of the observer grounded in academic studies (not domestic experiences) guaranteed the quality of the observations. In addition, experimental work gained prominence over observational studies.[25] These tendencies hardened with the spread of a 'scientistic' stance in the social science disciplines after World War II, which historian Mark Solovey has shown was also favoured by private and public funding agencies in the United States.[26] As a result of these trends, emotions in general and maternal subjectivity in particular were perceived as detrimental to the pursuit of objectivity in child development studies, child psychology and child psychiatry.

Clara Park, however, not only disagreed with the view that maternal love prevented her from understanding her daughter's condition, but also believed that mothers had much to contribute to the science of child development. But what exactly could parents bring to the table? Park pointed out that parents, especially mothers, could observe their children in a variety of daily situations, contrary to the experts who only interacted with them during clinical or therapeutic visits. The parents observed the children in their natural environments and in various settings. This gave them a large and varied set of data that went beyond the clinical observations during 'the artificial situation of the therapy hour'. Through their close everyday interactions with their children, parents acquired 'deep knowledge of the child in context'. This, Park argued, was an advantage that could come only from 'intimate knowledge'.[27]

Furthermore, Park claimed that it is easier to understand the meaning of specific acts when one knows a child's history. As the parents have extensive familiarity with their child's development from birth, they can ascertain how the child's past might influence their actions in the present. Contrary to the parents, clinicians and researchers only see the child in the present. Yet, Park noted, sound scientific interpretation depends on accessing 'the details of the child's history to make sense of what [one] has found'. If understanding a child was akin to exploration, Park reasoned that only the parents could offer 'a map that shows the major landmarks'.[28] It is only in such a historical or developmental context that one can interpret adequately what specific actions mean for a child. In a word, the intimate experience of the mothers should not be seen as an impediment to knowledge-making, but as an advantage to understanding their children's psychological development.

Thanks to knowing her child intimately and being able to understand her behaviour in context, Park could contribute to a better conceptualization of autism as well as helping autistic children. First, Park challenged a common view of autism at the time that was erected upon the entrenched dichotomy between the cognitive and affective realms. As mentioned earlier, Park initially agreed with many researchers who considered autism a condition of the affects. The perception of autistic children as being unwilling or unable to engage in social relations encouraged this view. However, over time, as she followed Jessica's development, Park realized that focusing on the affects made it difficult to see the significance of language differences, cognitive capacities, social needs and other aspects of her condition. In the revised edition of *The Siege*, published about a decade after the first and containing an update on Jessica's progress, Park argued that in Jessica's development one could appreciate the deep interrelation between the cognitive and the affective. Therefore, Park called for rethinking the reductionistic categories contemporary science used to interpret autism and many other mental conditions. About autism, she asked: 'Cognitive or affective impairment?' And she answered: 'It is not only unnecessary to choose between them; it is impossible.'[29] Thus, knowing Jessica's history enabled Park to reject categories that were problematic for understanding a child's emotional and psychological development in general and autism in particular.

Park's experience with her daughter also proved of crucial importance when it came to therapeutic approaches, an area in which the experts did not provide much help. When Jessica Park was diagnosed in 1961 as autistic, which was about twenty years after Kanner had identified infantile autism as a unique syndrome, psychiatry still had little to offer children with this diagnosis. Bettelheim recommended 'parentectomy' or radical separation from the parents, especially from the mothers whom he blamed for their child's condition. Many psychoanalytic centres offered psychotherapy for the children *and* their mothers. Kanner advised sending autistic children to a farm, a place where the combination of scheduled activities and little pressure created a restorative environment for them – away from the rigid expectations of refrigerator mothers. In some residential centres, autistic children were treated with electro-shocks, amphetamines and LSD or were simply ignored and received no support for their development.[30] Parents like the Parks who decided to raise their children at home were on their own.

Adopting a pragmatic approach, based on her intimate knowledge of her daughter, Clara Park focused on whatever could help Jessica, which meant picking and choosing from different approaches regardless of their theoretical foundations. For example, consider Park's use of behavioural conditioning, which became widely used with autistic children in the mid-1960s. This approach, promoted by UCLA psychologist Ivar Lovaas, relied on Harvard psychologist B. F. Skinner's view that most behaviours can be learned by reinforcing desired behaviours with rewards and discouraging undesirable ones with punishments. Lovaas' operant-conditioning-based behavioural therapy (later called applied behaviour analysis or ABA) first focused on teaching language to autistic children in a laboratory setting using things such as sweets or small sips of juice as rewards. Lovaas was quite controversial because he also used aversive tools – punishment or unpleasant stimuli – to eliminate children's self-inflicted harms or aggression towards others. Aversive stimuli included slaps on the thigh, shouting at the autistic child and, in some cases, low-level electric shocks.[31] However, Clara transformed ABA into a truly collaborative effort with Jessica. Rather than imposing on her specific behaviours and using aversives, mother and daughter worked together to develop an approach that recognized Jessica's own agency and desires. In fact, Jessica chose herself to use behavioural methods to change some behaviours that frustrated her.

In her adolescence, Jessica met an autistic boy who used a golf-counter to keep track of points that he acquired by following certain rules established by his family. Jessica loved the gadget and wanted to use it as well. After obtaining one, she and her mother created a weekly contract and decided which specific actions were to be rewarded, such as 'saying please' or 'answering when spoken to'. Together, they also decided how many points Jessica would get for each desirable behaviour and what the reward would be. At the beginning, if Jessica reached 100 points by the end of the day, she would get an ice-cream. Jessica tracked the points herself. Her mother administered the rewards. The only 'penalty' they introduced was deducting points for actions that they both found unkind towards other people, such as hitting someone or pushing a person while waiting in line. Jessica enjoyed setting goals for herself and loved the contracts. On some occasions, her brother and friends also designed contracts to reach goals such as to exercise more or to stop saying bad words. Thus, in a collaborative manner, Jessica, her family and her friends used behavioural principles while rejecting the problematic aspects of ABA therapies.[32]

In this way, Clara Park and other mothers who knew their children intimately adapted standard therapies to suit their children and circumstances, in the process changing those therapies and contributing new knowledge about their suitability for autistic people. Developed in the domestic space, their knowledge has often been unappreciated. However, it was valuable and often travelled from family to family as well as from families to therapists and researchers. To wit, Clara and Jessica Park had learned about the system of acquiring points and keeping track of them by using a golf counter from another family with an autistic son. In turn, Park wrote to other families about their use of the gadget and the contracts made collaboratively with their autistic children. Later, she wrote about this method for helping these children in several publications.[33] In short, this knowledge gained in domestic spaces moved out of their homes and contributed to a better understanding and implementation of therapeutic practices.

In addition to her participation in parental networks, Clara Park became a major force in reclaiming the experience of parents as valuable expertise through her writings. Over the years, *The Siege* became essential reading for generations of parents with autistic children. It was translated into German, Spanish, Japanese and other languages. The book's success helped to establish Park's reputation not only as a gifted writer but also as a mother confident in her own knowledge of her daughter and a master of relevant scientific literature. Encouraged by other parents, Park left the domestic space to venture into scientific meetings and conferences.

Parents' experience as valuable expertise

Sociologists of science Chloe Silverman and Gil Eyal and colleagues have shown that in the United States parents were instrumental in the struggles to obtain education, healthcare, social acceptance and disability rights as well as in shaping directions for autism research.[34] In this context, Clara Park was a prominent figure, expanding on her early ideas and continuing to fight to have parents' experience recognized as valuable expertise.

As a leading member of the National Society for Autistic Children (NSAC), which a group of parents established in 1968, Park encouraged the collaboration between scientists and parents. Park was a member of NSAC's board of directors from 1970 to 1972 and she edited its official Newsletter from 1968 to 1972. The aim of NSAC's first national congress, held in Washington, DC in July 1969, was to foster the interaction of parents and professionals 'on an equal footing, with the parents acting as coordinators and the professionals as resources'. One of the first tasks taken up by NSAC's leadership was to make their presence felt in scientific conferences. The society's first elected president, Ruth Sullivan, and Clara Park attended the Joint Conference on Childhood Mental Illness, sponsored by the National Association for Mental Health and the congressional Joint Commission on Childhood Mental Illness, held in New York City in February 1968. In her report of the conference, Park emphasized that they attended the conference as 'parent-experts'.[35]

In an important way, psychiatrists and psychologists working on children's conditions had always relied on parental expertise, as these professionals obtained much of their information about the children from their parents. For example, Kanner's famous 1943 first paper identifying infantile autism as a unique condition relied extensively on information provided by the parents, especially the mothers. He also quoted from letters or reports sent to him by the parents over the years. Kanner encouraged the mothers to make detailed observations of their children's behaviour. When he published in 1971 a paper reporting his follow-up of those children, Kanner also relied on the parents' reports about their children's development. On some occasions, Kanner praised the detailed and reliable observations made by mothers. He was not alone in this practice. In this sense, the scientific experts used the mothers' knowledge to build their own expertise on autism and other conditions. However, the mothers remained mere 'informants' whose own expertise was often unacknowledged, denied or contested.

Yet, the mothers not only knew their own children better than anybody else, but they also knew a lot about their friends' children since they tried to meet each other and develop networks for exchanging information and support. Park and other mothers of autistic children, such as Sullivan and

Margaret Dewey, maintained an extensive correspondence, discussing new developments in the field, always 'testing' them against their own lived experience. They attended scientific meetings, published reviews of scientific publications and shared with many other families their experiences and knowledge about each other's children. For example, one day, Ruth Sullivan's autistic son Joseph became so upset about something that his mother could not communicate with him. She decided to put a question to him in writing. He responded by writing the answer. Sullivan wrote about this newfound method of communication to Clara, wondering if her son could not process verbal input during excited periods.[36] Clara decided to try this method with Jessy. She liked it and it became a good way of finding out about Jessy's own views on many issues.

Clara Park and other parents also started including the voices of autistic individuals in the National Society for Autistic Children's newsletter, first through stories Clara told in a column entitled 'How they grow'. Eventually, many autistic individuals wrote their own columns. Working in collaboration with each other and with many supportive researchers and therapists, these mothers helped to broaden the educational opportunities, social experiences and life options available for autistic people.

Park contributed as well to the visibility of parents' expertise and to building the autism community through her tireless reviewing activity. An excellent writer and a kind reader, editors sought her opinions about the books written by parents raising children with developmental difficulties, a genre that grew tremendously in the 1970s. In the NSAC's newsletter, Park also reviewed numerous scientific publications, thus bridging the scientific and autism communities.

Park started giving speeches too, first about her experiences with Jessica, then expanding her range to talk about the role of parents and family. In 1974, Albert Solnit, from the Yale Study Center at Yale, invited her to participate in a session about 'The vulnerable child' at the International Child Psychiatry and Allied Professions Congress. At the conference Clara spoke her mind, starting with: 'Ten years ago I would not have dared to speak to you on "The Positive Value of Parents" – nor would I have been invited.' Acknowledging her child was autistic 'would have been to display myself as a "refrigerator parent," whose emotional emptiness was demonstrated by the indubitable fact that I observed my child's abnormal behaviour and tried to do so objectively'. But now, she noted, 'The fact of this panel suggests that there is a *frail* new willingness to recognize the capacity of parents for doing things right.' Park argued that 'Professionals have the advantage of their expertise, but they should recognize that parents have theirs', based on 'total familiarity with the child's past and present, his emotional and physical environment and his language, verbal and non-verbal', and a conviction that they can help their children develop.[37] Invitations to conferences were one sign of progress regarding the experts' willingness to hear and recognize the value of the parents' voices.

There were other signs of progress, though not without some controversy. In 1974, *The Journal of Autism and Child Schizophrenia*, edited by psychologist Eric Schopler, started publishing a new column entitled 'Parents speak', envisioned as a forum for dialogue between parents and professionals. In a section called 'Questions, answers, and comments', – launched in the same volume – the query to the editor came from an editorial board member, E. James Anthony. A British child psychiatrist and psychoanalyst, Anthony wondered about the new policy of including parents as contributors. He was worried that 'the scientific level might well be lowered as a result'. In reply, Schopler said that the new section was 'nothing more than an opportunity for parents to respond and ask questions of the author' since researchers often used scientific jargon. 'Such dialogue', he continued, 'also opens the way for clinicians and parents to ask about research implications of practical experience and for investigators to question implications of one study for another'. In Schopler's view, 'an attempt at reasonable dialogue over scientific and clinical material will enhance the scientific level rather than diminish it'.[38]

Exemplifying the benefits of collaboration, in 1976, Park published *You Are Not Alone: Understanding and Dealing with Mental Illness. A Guide for Patients, Families, Doctors and Other Professionals*, written with Leon N. Shapiro, professor of psychiatry at Cornell University Medical School. The authors aimed to reach both professionals and parents. Unlike *The Siege*, which focused on the personal experience of raising her autistic child, this book was about the practical needs of any family with a child who had developmental difficulties or a mental illness. The book proposed establishing a common front of scientists and families, of experts and amateurs, for progressive change. Park called for scientists to collaborate with parents, for parents to collaborate with therapists and for those who had suffered because they were not considered 'normal' to come forward and fight for recognition, respect and services together.

In those collaborations, Park wanted to provide her expertise as a mother. The description of a course on 'The Development of Autistic Children', which she gave at the University of California in 1976, noted that she had laboured in an extended 'internship with her own autistic child'.[39] Park wanted others to recognize that experience as a valid form of expertise. Always relying on her first-hand experience with Jessica, Park and other mothers emphasized the need for effective techniques to help autistic children develop their skills and interests. In May 1979, she lectured about these topics to the Association of Child Care Workers. In September that year, she was the 'featured luncheon speaker' at the annual meeting of the Michigan Association for Emotionally Disturbed Children. Park was also a featured speaker at the 1979 annual NSAC conference in San Jose, California. Here, she was still the only non-scientist on the programme, which identified her as author, teacher and parent.[40]

An approachable, engaging and articulate speaker, Park was invited to many scientific conferences and public forums over the following decade.[41] She gave talks about her experiences with her daughter and presented updates on Jessica's progress. Speaking to many different professional groups, Park took off 'that armor parents wear when they go among professionals, whether they know it or not', and made her plea for a healthy partnership between parents and professionals.[42]

At a 1979 American Orthopsychiatric Association panel, Park suggested that there were two reasons why she was now being called on to participate in professional conferences. One was her experience living and working with Jessica. The second was her books about mental health from the perspective of a parent. But the first reason was more important:

> If you have something to learn from me, it's because I and my family have had significant experiences, not because I can write about them. There are thousands of families who have had experiences like ours, who, sometimes with help, often without, have worked and coped their way to some sort of equilibrium, even to improvement and growth. You could learn from any one of them what I have to tell you today. Our experience, our search for help, and our needs are astonishingly the same.[43]

Thus, it was not that she was an expert parent, meaning a parent with scientific expertise, but that she was a parent expert, that is, somebody with the expertise of a parent. In *The Autism Matrix*, Eyal and his colleagues have argued that NSAC developed a new type of actor, the 'parent-activist-therapist-researcher', which helped constitute 'a new modality of expertise, anchored in networks of knowledge exchange among parents, researchers, and therapists'.[44] Indeed, this applies to many NSAC parents. But Park always emphasized that her contribution to knowledge came from her lived experience as a mother. By implication other parents could make significant contributions as well because their experiences added to what scientists could find out about autism.

As she had done in *The Siege*, in many of her presentations Park made a case for recognizing that parents are the best ethnographers of their children's lives. They had knowledge of the daily life and changes over time of the children in their natural habitat, where their actions have meanings that are often not visible in the lab, the observation room or the clinic. It did not matter that the ethnographer – often the mother – was not a detached observer. On the contrary, not by going native, but by being native already, so to speak, the mother could provide a uniquely valuable and useful account of the lived experience of the non-standard world of an autistic person. By providing an account of their autistic children in their natural environments – the house, the school, the playground – parents could add to the multiple perspectives needed to understand children's perceptions of the world. The parents could provide narratives that complemented the knowledge reached through scientific means. With that knowledge, researchers and clinicians could better understand the needs of autistic people.

However, the fight for recognition was neither easy nor straightforward. Park did not think parents' contributions received appropriate recognition though, because their narratives still 'existed in a kind of limbo, lacking the legitimation either of science or of literature'. But she continued to draw attention to the potential contribution 'to clinical understanding'. This is not because parents could supplant clinicians or therapists. Rather, parents could add a dimension that others could not: 'the feel of lived experience. For it is the experience – not the mere fact – of disability that is what all our work and study, parental and professional, is about.' Furthermore, a good parent narrative presented in a holistic manner 'what the analyzing mind necessarily puts asunder. It offers not only the detail of behavior exactly observed, the complex emotion accurately expressed, but behavior and emotion together, to be apprehended as lived, holistically, in their total web of being.' For Park, bringing these together could 'tell the larger truth of indivisible experience.'[45]

In this way, Park once more called for overcoming dichotomies. Earlier she had argued that separating the emotional and the cognitive had hindered the understanding of autism. The separation between natural feelings and the intellect had also led to an impoverished view of good mothering. And the opposition between subjective and objective in science had banished the contributions of mothers in the field of child development. Now she added that a parent's narrative could offer a holistic vision, one that integrated experiences, behaviours and emotions in understanding autism.

Emphasizing the need for a full picture of the autistic experience, as their autistic children grew up, parents also invited them to become participants in conferences and offer their own insights about their own personal experiences. As Park had put it earlier, it was the experience – not the mere fact – of disability that they needed to understand.

In my view, Park was calling for a new type of knowledge, a new hybrid scientific epistemology. It would include experimental work or observations in laboratories and clinics. But not exclusively so because those sources of knowledge should be integrated with the lived experiences and multiple perspectives provided by narrative accounts such as memoirs written by parents and autistic people. This new epistemology would see the need for understanding behaviour in its developmental context and, thus, would recognize the significance of intimate knowledge.

Conclusion

In fighting to have a mother's knowledge recognized as a legitimate source of expertise in autism research, Clara Park challenged a deeply entrenched dichotomy between emotion and cognition that has been pervasive in views about mothering and scientific knowledge. In doing that, Park and other parents called for society to value not only different ways of mothering but also new ways of constructing scientific knowledge.

In the field of autism, Park was calling for a new type of science that valued lived experience, multiple perspectives and the historical knowledge expressed in parents' narrative accounts. Her position addressed central issues about scientific epistemology. Objectivity has been the hallmark of modern science. But what is objectivity? In the development of modern Western science, the meaning of objectivity has changed over time, always in close relation to the meaning of its counterpart, subjectivity. For many scientists, objectivity required erasing any trace of the knowing subject, thus obliterating the subjectivity of the scientist.[46] Though this view has also been contested, in many scientific areas much further work is still needed to incorporate experiential knowledge along with experimental studies carried out in the laboratory and the field.

Clara Park's story shows that to fully recover the hidden contributions of women in science, historians need to look beyond the traditional spaces of knowledge-making and beyond the scholarly elites. In the case of autism, by going beyond the lab and the clinic and examining the domestic space, we can see that women developed vernacular knowledge that helped transform practices of care and knowledge-making. While Clara Park's story is extraordinary at some levels because of her tremendous intellect and literary gifts, it is not unique in other crucial respects. Several sociological and ethnographic studies have revealed that mothers in many places have translated scientific knowledge and adapted therapies in pragmatic ways that helped their disabled children.[47] In doing so, they made valuable contributions to the understanding and acceptance of autism and other developmental conditions. Recovering their contributions will compel us to redraw traditional boundaries between experience and expertise and to expand our views about legitimate sites of knowledge production.

Notes

1. On women's 'invisible' work in science, see M. W. Rossiter, *Women Scientists in America: Struggles and Strategies to 1940* (Baltimore, MD: The Johns Hopkins University Press, 1982). On domestic sites and gendering of knowledge, see P. G. Abir-Am and D. Outram (eds), *Uneasy Careers and Intimate Lives: Women in Science, 1789–1979* (New Brunswick, NJ: Rutgers University Press, 1987); H. M. Pycior, N. G. Slack and P. G. Abir-Am (eds), *Creative Couples in the Sciences* (New Brunswick, NJ: Rutgers University Press, 1996); A. Lykknes, D. L. Opitz and B. Van Tiggelen (eds), *For Better or For Worse? Collaborative Couples in the Sciences* (Basel: Birkhäuser, 2012); C. von Oertzen, M. Rentetzi and E. S. Watkins (eds), *Beyond the Academy: Histories of Gender and Knowledge*, Special Issue of *Centaurus*, 55: 2 (2013), DOI:10.1111/1600-0498.12018; D. R. Coen, 'The common world: Histories of science and domestic intimacy', *Modern Intellectual History*, 11: 2 (2014), 417–38; D. L. Opitz, 'Domestic space', in B. Lightman (ed.), *A Companion to the History of Science* (Chichester: John Wiley & Sons Ltd., 2016), pp. 252–67; D. L. Opitz, S. Bergwik and B. Van Tiggelen (eds), *Domesticity in the Making of Modern Science* (New York: Palgrave Macmillan, 2015).
2. On the history of views about maternal affects, see E. Badinter, *Mother Love: Myth and Reality* (New York: Macmillan, 1981); C. E. Russett, *Sexual Science: The Victorian Construction of Womanhood* (Cambridge, MA: Harvard University Press, 1989); S. A. Shields, 'To pet, coddle, and "do for": Caretaking and the concept of maternal instinct', in M. Lewin (ed.), *In the Shadow of the Past: Psychology Portrays the Sexes* (New York: Columbia University Press, 1984), pp. 256–73; M. Vicedo, 'Mother love and human nature: A history of the maternal instinct' (PhD dissertation, Harvard University, 2005). On the history of maternal care, see D. Riley, *War in the Nursery: Theories of Mother and Child* (London: Virago, 1983); R. J. Plant, *Mom: The Transformation of Motherhood in Modern America* (Chicago, IL: University of Chicago Press, 2010).
3. R. Schwartz Cowan, *More Work for Mother: The Ironies of Household Technology from the Open Hearth to the Microwave* (New York: Basic Books, 1985); A. Stanley, *Mothers and Daughters of Invention: Notes for a Revised History of Technology* (New Brunswick, NJ: Rutgers University Press, 1995).
4. On blaming mothers, see M. Ladd-Taylor and L. Umansky (eds), *'Bad' Mothers: The Politics of Blame in Twentieth-Century America* (New York: New York University Press, 1998); D. E. Eyer, *Motherguilt: How Our Culture Blames Mothers for What's Wrong with Society* (New York: Random House, 1996); L. C. Fentiman, *Blaming Mothers: American Law and the Risks to Children's Health* (New York: New York University Press, 2019). On mother-blame in psychiatry and psychoanalysis, see S. Chess, 'The "blame the mother" ideology', *International Journal of Mental Health*, 11 (1982), 95–107; E. Dolnick, *Madness on the Couch: Blaming the Victim in the Heyday of Psychoanalysis* (New York: Simon and Schuster, 1998); D. Weinstein, *The*

Pathological Family: Postwar America and the Rise of Family Therapy (Ithaca, NY: Cornell University Press, 2013); A. Harrington, 'Mother love and mental illness: An emotional story', *Osiris*, 31 (2016), 94–115; S. S. Richardson, *The Maternal Imprint: The Contested Science of Maternal-Fetal Effects* (Chicago, IL: University of Chicago Press, 2021).

5 On how American mothers were increasingly influenced by scientific advice in the twentieth century, see J. Grant, *Raising Baby by the Book: The Education of American Mothers* (New Haven, CT: Yale University Press, 1998); R. Apple, *Perfect Motherhood: Science and Childrearing in America* (New Brunswick, NJ: Rutgers University Press, 2006).

6 S. Shuttleworth, *The Mind of the Child: Child Development in Literature, Science, and Medicine 1840–1900* (Oxford: Oxford University Press, 2013); C. von Oertzen, 'Science in the cradle: Milicent Shinn and her home-based network of baby observers, 1890–1910', *Centaurus*, 55 (2013), 175–95. See also D. Hoogland Noon, 'Situating gender and professional identity in American child study, 1880–1910', *History of Psychology*, 7 (2004), 107–29.

7 H. Rose, 'Hand, brain, and heart: A feminist epistemology for the natural sciences', *Signs*, 9 (1983), 73–90; E. F. Keller, *Reflections on Gender and Science* (New Haven, CT: Yale University Press, 1985); E. F. Keller, *A Feeling for the Organism: The Life and Work of Barbara McClintock* (New York: Times Books, 1984). On the history of scientific objectivity, see L. Daston and P. Galison, *Objectivity* (New York: Zone Books, 2017).

8 Clara Park personal archive, Williamstown, MA. This essay expands on themes that I presented in my book: M. Vicedo, *Intelligent Love: The Story of Clara Park, Her Autistic Daughter Jessica, and the Myth of the Refrigerator Mother* (Boston, MA: Beacon, 2021). See this book for references to the history of autism.

9 For biographical details on Clara Park, see Vicedo, *Intelligent Love*, chapter 1.

10 L. Kanner, 'Autistic disturbances of affective contact', *Nervous Child*, 2 (1943), 217–50; L. Kanner, 'Early infantile autism', *Journal of Pediatrics*, 25 (1944), 211–17. Around the same time, Hans Asperger in Vienna described similar children and introduced the concept of 'autistic psychopathy'. Nevertheless, Asperger's work was not discussed in the United States until much later.

11 R. A. Spitz, 'Hospitalism: An inquiry into the genesis of psychiatric conditions in early childhood', *Psychoanalytic Study of the Child*, 1 (1945), 53–74; R. A. Spitz, 'Anaclitic depression: An inquiry into the genesis of psychiatric conditions in early childhood, II', *Psychoanalytic Study of the Child*, 2 (1946), 313–42; R. A. Spitz, 'The importance of mother-child relationship during the first year of life', *Mental Health Today*, 7 (1948), 7–13; J. Bowlby, 'Maternal care and mental health', *Bulletin of the World Health Organization*, 3 (1951), 355–534, reissued as J. Bowlby, *Maternal Care and Mental Health*, 2nd ed. (Geneva: World Health Organization, 1952). On maternal deprivation and its effects on children, see E. Duniec and M. Raz, 'Vitamins for the soul: John Bowlby's thesis of maternal deprivation, biomedical metaphors and the deficiency model of disease', *History Psychiatry*, 22 (2011), 93–107; M. Vicedo,

'The social nature of the mother's tie to her child: John Bowlby's theory of attachment in post-war America', *British Journal of the History of Science*, 44 (2011), 401–26; and M. Vicedo, *The Nature and Nurture of Love: From Imprinting to Attachment in Cold War America* (Chicago, IL: University of Chicago Press, 2013).

12 'Frosted children', *Time* (26 April 1948), p. 81.
13 B. Bettelheim, *The Empty Fortress: Infantile Autism and the Birth of the Self* (New York: Free Press, 1967).
14 See Vicedo, *Intelligent Love*, chapter 3.
15 B. Rimland, *Infantile Autism: The Syndrome and Its Implications for a Neural Theory of Behavior* (New York: Appleton-Century-Crofts, 1964).
16 Clara Park to Bernard Rimland, 18 April 1964, Bernard Rimland personal archive. Autism Research Institute, San Diego, CA.
17 C. C. Park, *The Siege: A Family's Journey into the World of an Autistic Child*, orig. 1967 (Boston, MA: Little, Brown, 1982). For studies of *The Siege*, see J. T. McDonell, 'Mothering an autistic child: Reclaiming the voice of the mother', in B. O. Daly and M. T. Reddy (eds), *Narrating Mothers: Theorizing Maternal Subjectivities* (Knoxville: The University of Tennessee Press, 1991), pp. 58–75; D. L. Cumberland, 'Crossing over: Writing the autistic memoir', pp. 183–96 and S. Stevenson, '(M)Othering and autism: Maternal rhetorics of self-revision', pp. 197–211, both in M. Osteen (ed.), *Autism and Representation* (New York: Routledge, 2008).
18 Park, *The Siege*, p. 195.
19 E. H. Erikson, *Childhood and Society* (New York: W. W. Norton, 1950); M. Ribble, *The Rights of Infants: Early Psychological Needs and Their Satisfaction* (New York: Columbia University Press, 1943); Spitz, 'The importance of mother-child relationship during the first year of life'; Bettelheim, *The Empty Fortress*; Vicedo, *Nature and Nurture of Love*.
20 C. Smythe, 'The siege', *Christchurch Press* (New Zealand) (3 May 1969), copy in Clara Park personal archive.
21 See Daston and Galison, *Objectivity*; Rose, 'Hand, brain, and heart'; Keller, *Reflections on Gender and Science*.
22 Shuttleworth, *The Mind of the Child*.
23 On Darwin not using his wife's observations, see Shuttleworth, *The Mind of the Child*. Charles Darwin to Thomas Henry Huxley, 30 January 1868, in Darwin Correspondence Project, 'Letter No. 5817', www.darwinproject.ac.uk/DCP-LETT-5817 (accessed 30 June 2024). C. Darwin, 'A biographical sketch of an infant', *Mind*, 2 (1877), 285–94. On the complex and evolving role of the emotions in Darwin's views about himself as scientist and father, see P. White, 'Darwin's emotions: The scientific self and the sentiment of objectivity', *Isis*, 100 (2009), 811–26.
24 See von Oertzen, 'Science in the cradle'.
25 E. Herman, *The Romance of American Psychology: Political Culture in the Age of Experts* (Berkeley: University of California Press, 1995).

26 M. Solovey, *Social Science for What? Battles over Public Funding for the 'Other Sciences' at the National Science Foundation* (Cambridge, MA: The MIT Press, 2020).
27 Park, *The Siege*, pp. 180–83.
28 Park, *The Siege*, p. 181.
29 Park, *The Siege*, p. 306.
30 On therapeutic approaches to autism, see G. Eyal, B. Hart, E. Oncular, N. Oren and N. Rossi, *The Autism Matrix: The Social Origins of the Autism Epidemic* (Cambridge: Polity, 2010); Vicedo, *Intelligent Love*, chapter 7.
31 O. I. Lovaas, G. Freitag, V. J. Gold and I. C. Kassorla, 'Experimental studies in childhood schizophrenia: Analysis of self-destructive behavior', *Journal of Experimental Child Psychology*, 2 (1965), 67–84; O. I. Lovaas, with A. Ackerman, D. Alexander, P. Firestone, M. Perkins and D. B. Young, *Teaching Developmentally Disabled Children: The Me Book* (Baltimore, MD: University Park Press, 1981).
32 Vicedo, *Intelligent Love*, chapter 6.
33 C. C. Park, *Exiting Nirvana: A Daughter's Life with Autism* (Boston, MA: Little, Brown, 2001).
34 C. Silverman, *Understanding Autism: Parents, Doctors, and the History of a Disorder* (Princeton, NJ: Princeton University Press, 2012); G. Eyal et al., *The Autism Matrix*; B. Hart, 'Autism parents and neurodiversity: Radical translation, joint embodiment and the prosthetic environment', *BioSocieties*, 9 (2014), 284–303.
35 'Report: Joint Conference on Childhood Mental Illness, New York City, 1968'. In Clara Park personal archive.
36 Ruth Sullivan to Clara, 4 August 1979, Clara Park personal archive.
37 C. Park, 'The positive value of parents', 8th International Congress of the International Association for Child Psychiatry and Allied Professions, Philadelphia, 28 July–2 August 1974; typescript in Clara Park personal archive.
38 E. Schopler, 'Editorial: New publisher, new editor, expanded editorial policy – Goal: An improved journal', *Journal of Autism and Childhood Schizophrenia*, 4 (1974), 91–92; E. J. Anthony, 'To the editor', *Journal of Autism and Childhood Schizophrenia*, 4 (1974), 93; see also Schopler's 'Response', p. 94.
39 University of California Extension, brief course description for 'The development of autistic children', 1976. Clara Park personal archive.
40 '1979 NSAC Annual Meeting and Conference', *Journal of Autism and Developmental Disorders*, 9 (1979), 137–38.
41 For instance, she presented at 'Autism: The Emotional and Social Dimensions', 20th anniversary symposium sponsored by the League School, Boston, MA, 1–2 May 1987, and the next year was a featured speaker at the 6th International Conference on 'Issues in Autism' held at Rutgers University, 25–26 March 1988. Clara Park personal archive.
42 Conference on 'Parent-Professional Communication on Behalf of Children with Special Needs', Wheelock College, Boston, MA, 14 October 1978; workshop on 'Autism: A Communication Disorder', 7th Annual Spring Workshop

of the Association of Child Care Workers, Albany, NY, 19 May 1979; Annual Meeting of the Michigan Association for Emotionally Disturbed Children, Ann Arbor, MI, 29 September 1979. Brochures and drafts of Park's speeches in Clara Park personal archive.

43 C. Park, 'Working together: Deinstitutionalization, professionals, and the family', Annual Meeting of the American Orthopsychiatric Association, Washington, DC, March–April 1979. Clara Park personal archive.

44 Eyal et al., *The Autism Matrix*, pp. 171–72.

45 C. Park, 'Review of Dorothy Johnson Beavers, *Autism: Nightmare Without End* (New York: Ashley Books, 1982)', *Journal of Autism and Developmental Disorders*, 15 (1985), pp. 113, 116, 117. See also Park, 'Growth in language: The parent's part', *Topics in Language Disorders*, 3 (1982), 50–57.

46 For example, see the debates about objectivity and subjectivity in the study of animal behaviour: M. Vicedo, 'Epistemological discipline in animal behavior studies: Konrad Lorenz and Daniel Lehrman on intuition and empathy', *History and Philosophy of the Life Sciences*, 45: 1 (28 February 2023), DOI:10.1007/s40656-023-00558-7.

47 See: G. H. Landsman, *Reconstructing Motherhood and Disability in the Age of 'Perfect' Babies* (New York: Routledge, 2009); B. Hart, 'Autism parents and neurodiversity'. On mothers, maternal care and advocacy for disabled children, see also G. H. Landsman, 'Mothers and models of disability', *Journal of Medical Humanities* 26: 2/3 (2005), 121–39; E. F. Kittay, *Love's Labor: Essays on Women, Equality, and Dependency* (New York: Routledge, 1999); S. Ryan and K. Runswick Cole, 'From advocate to activist? Mapping the experiences of mothers of children on the autism spectrum', *Journal of Applied Research on Intellectual Disabilities*, 22: 1 (2009), 43–53; P. Lalvani (ed.), *Constructing the (M)other. Narratives of Disability, Motherhood, and the Politics of Normal* (New York: Peter Lang, 2019).

V

Towards visible change? Publics, pedagogies and politics of science

12

The valuable 's': Publics and counterpublics of abortion and contraception in late-twentieth-century Greece

Evangelia Chordaki

Constructing the public arena, creating the audiences of science

The relationship between science and the public is a well-studied issue in the history of science and the field of science communication. It explains the public presence of science, which embodies a certain peculiarity. The two notions are mutually co-constructed: society produces science and science 'needs' the public arena for its legitimization and realization. Scientists are part of the public and those who participate in the public may hold different types of expertise or be experts in different fields. The public presence of science presupposes the circulation of scientific knowledge. Thus, the conceptualization of the notion of the public sphere becomes crucial and is linked to questions related to the practices, places and temporalities of science communication and the interests, agendas and aims of those who communicate science. To examine who, where, when, why and how scientific knowledge is produced and circulated,[1] one should study the social and gender power relations, as well as how they are embodied in the content of communication practices. In that sense, the concept of the public sphere intertwines with social and epistemic justice and cannot be examined without its political dimensions. While public and private are relational concepts, their division implies that the constitution of the former (as the most visible and powerful) presupposes the latter's formation. This means that issues of marginalization, exclusion, discrimination and, by extension, (in) accessibility and democracy should be at the centre of attention.

Women's health and, more specifically, gynaecological issues are excellent examples to understand how public spheres and the audiences of science are constructed. A closer look at the history of the public presence of science in Greece (and globally) can shed light on those who control the production and circulation of knowledge, the gendered aspects of science communication and how this shapes specific audiences and makes them relevant, marginalizing those considered irrelevant. Reading the 1987 issue of a Greek medical textbook, one discerns a concerted effort to exclude women from medical science. The (male) author comments:

> We started from the primitive human and now we discuss family planning and oocyte cryopreservation. Similarly, we had a wooden birth chair and now we have maternity hospitals and ultrasounds, while midwives who were passively observing labour have been replaced by prenatal care and monitoring. It was only when our specialization passed from the midwives to men's hands during the seventeenth century that such a development and progress could be achieved.[2]

Women's exclusion is also apparent in an international conference on gynaecology and obstetrics held in the 1980s and attended by male doctors, with the exception of one woman: a secretary who was not a doctor.[3] Closer to our days, we witnessed the US president Donald Trump signing an anti-abortion executive order surrounded by men[4] and the cover of the Greek newspaper *Sportime* – whose audience is dominated by men – campaigning against abortion and promoting the anti-choice organization 'Let Me Live!'[5]

My point here is not to depict an enemy. Instead, I want to explore a mechanism of gender discrimination and women's exclusion from a historical perspective, while at the same time locating women in their marginalized social positions. I have discussed elsewhere the relationship between the practices of production and circulation of knowledge and the methodological schemes of science communication,[6] but here I explore the publics and counterpublics of birth control in Greece in the late twentieth century, focusing on their multiplicity, tensions, their meanings and narratives. Discussing specific theoretical concerns regarding the gender connotations entailed by the public sphere as a homogeneous concept, I argue that the way that notion is conceptualized affects the in/visibility of marginalized groups. In other words, I suggest that to examine women's herstories of science communication, one should follow feminist and queer approaches that advocate the multiplicity of public spheres. Maintaining the image of the public sphere as a homogeneous category means that women will remain absent even in the communication of medical issues that pertain to them. Here, herstories refer to women's histories or histories written by women, while at the same time the category 'women' does not reflect an essentialist gaze but a category of resistance, collective action and politicization.

Re-conceptualizing the public sphere, deconstructing the male gaze

Debates about the nature of the public sphere have a long history that goes back to the second half of the twentieth century and cannot be understood separately from the emergence of the second wave of feminism. More specifically, under the powerful slogan 'the personal is political', women questioned the public/private dichotomy and its impact on everyday political life[7] and the naturalization of the relationship between gender and the public sphere,[8] introducing-in-action the transformation of the public sphere through their activism.[9]

As the most effective response to the Habermasian theory and its focus on the bourgeois public sphere,[10] the feminist critique against the established notion of the public sphere aimed at revealing the hidden inequalities,[11] promoting the democratization of the concept on the basis of diversity and inclusion.[12] Identifying the public sphere as 'a bourgeois, masculinist, white supremacist public sphere',[13] scholars have discussed issues of idealization and exclusion of other forms of counterpublicity or competing public spheres.[14]

Simon Susen's points against the established conceptualization of the public sphere are crucial to understanding such critiques. In his insightful work, he discusses the problem of singularity, accompanied by repeated efforts to silence any other forms of collective, thus simplifying a complex social phenomenon.[15] He also refers to the problems of idealization and gender, highlighting the hegemonic dominance of rich, powerful, white males.[16] Finally, he discusses universality, emphasizing the pitfalls of conceiving the public good and interest as singular and universal entities.[17] Mojca Pajnik similarly criticizes the established monolithic approaches to the notion of accessibility concerning the public sphere. Approaching accessibility through the ethics of care, she discusses the relationship between discourse, action and responsibility, enriching the 'right to speak' with the equally important 'right to be listened to'.[18]

Contrary to the approaches that oversimplify the public sphere, by widening the concept, we will be able to see the presence of different social groups 'far beyond the borders'[19] and move forward towards equality and a more democratic society.[20] Nancy Fraser proposes the acknowledgment of different public arenas, called *counterpublics*, where 'subordinated social groups – women, workers, people of color, gays and lesbians – have repeatedly found it advantageous to constitute alternative publics'. For her, subaltern counterpublics are 'parallel discursive arenas, where members of subordinated social groups invent and circulate counterdiscourses, which in turn permit them to formulate oppositional interpretations of their identities, interests, and needs ... They create new meanings, terms, and practices',[21] which emerge from people's need to 'be seen and heard'.[22] Additionally, Michael Warner argues that counterpublics is 'a space for argument and exchange which is separated from authority and critical to power',[23] 'a space of mediation of the most private and intimate meanings of gender and sexuality ... which creates and transforms cultures and social relations ... publicizes subjectivity and brings it outside the domestic sphere'.[24] Melanie Loehwing and Jeff Motter offer a similar definition as a space of production of discourses and a place of representation of identities, needs and interests.[25]

Situating my case study in the theoretical exploration of the multiplicity of public spheres, it is apparent that the democratization of the concept affects what the research conceives and acknowledges as: the relevant actors regarding birth control; the gender power relations among them; the extent of their involvement in the production and circulation of knowledge; their role in the construction of meanings; their impact on the in/visibility of knowledge; and, undoubtedly, the construction of the relationship between science and society. More specifically, the following examination of the publics and counterpublics of birth control in Greece will allow me to locate women in the dominant and so-called public sphere and explore their practices of science communication in their own counterpublics. Nevertheless, the analysis will also provide a framework regarding the established and knowledge-related regime that women had to fight against by creating an alternative social realm.

Reconfiguring the 'public' of the public debate about birth control

Before discussing the publics and counterpublics of birth control in Greece, it is crucial to return to the title of the current chapter to justify two critical points. The first one is related to the proposal regarding the multiplicity of the public landscape – what I have called the valuable 's'. In the previously analysed theoretical framework, I have shown that the feminist approaches towards the notion of the public sphere reveal multiple public realms by acknowledging the presence and actions of marginalized and socially isolated or ignored social groups. Such an approach implied that one needs to shift methodologically and theoretically from stable categories towards malleable and transformable entities to promote diversity and inclusion regarding the reconstruction of the past or the vision of the future. However, the dynamic character of such approaches is not limited to the destabilization of the umbrella-notion of the public sphere. It moves forward to the enrichment of other related analytical categories, such as the notion of the counterpublics. This means that the exploration of the women's public sphere – the counterpublics of my case study – should not be conceived as a homogeneous category. Instead, by adding the 's' and acknowledging the presence of multiple counterpublics, we will be able to see additional differentiations embodied in the numerous manifestations of the identities of 'women'.

The second point I would like to justify is the division between publics and counterpublics. Earlier, I have referred to the gendered aspect of the public/private dichotomy: the meaning of the private sphere was constructed around women's isolation and exclusion from the public and resulted both in the isolation of the private from the political and, consequently, in the underestimation of the private sphere.[26] What I want to emphasize is the differentiation between the two types of dichotomy: while the latter is generated by hierarchies and boundaries, the former emerges as an act of resistance.

The public discussion about birth control in Greece started after the fall of the dictatorial regime[27] through the emergence of the feminist birth control movement and peaked in the late 1980s, more specifically, with the decriminalization of abortions in 1986.[28] The analysis of the primary archival material unveiled the two suggested significant categories, the publics and counterpublics, that communicated the meanings, ideas and knowledge related to abortion and contraception.

In the first category – the *publics* of abortion and contraception – the archival examination was staged in four different levels and included the analysis of the following types of material (1970s–1980s): articles published in Greek newspapers and magazines of general interest, medical textbooks, the related laws and the articles published in more conservative and/or religious press and para-ecclesiastical organizations' publications. In other words, *the publics* included the popular press and mass media, the medical community, the political and governmental sphere and the religious or religion-friendly and anti-choice supporters. However, it is essential to emphasize the multiple social groups involved in the public debate about birth control in the popular press: the journalists and authors, women and feminists, physicians, politicians, anti-choice supporters and lawyers. Raising such differentiations even in the debate performed in the popular press, one can draw attention to issues of accessibility and content of the communicated ideas and narratives to examine more thoroughly how the relationship between science and society is constructed.

As the counterpublics of abortion and contraception, I approach the spheres of women and feminists, which are often considered private. Against such approaches, the analysis of the archival material (consisting mostly of magazines, publications, brochures, leaflets, etc.) demonstrated the construction of a competing public arena that emerged through numerous collective communication practices. In this free and safe space, women self-organized the circulation of relevant knowledge, expressing their needs, desires and interests freely. Here again, the counterpublics were examined by carefully locating distinct groups. This category included the sphere of autonomous women (groups and collectives) and the state feminists (organizations and associations) that were often linked to political parties. Keeping in mind this multiplicity of the public spheres, I will now summarize the meanings and narratives produced and circulated in the multiple public spheres.

Exploring the meanings and narratives of abortion and contraception in publics and counterpublics

The publics: Popular press

The debate about abortions in the press was mainly framed in terms of *support or rejection* of their decriminalization.[29] Promoting their decriminalization, journalists conceived it as a 'human right',[30] while doctors regarded it as an 'act of modernity', establishing themselves as those responsible for sharing information about the medical procedure of abortion.[31] Moreover, women and feminists communicated abortion and contraception as a unified phenomenon, emphasizing their 'right to control their bodies, sexuality and reproduction'.[32] Additionally, politicians emphasized the negative consequences of the illegal abortion regime (with a special focus on the risks to women's health and the conditions under which illegal abortions took place)[33] and pro-choice lawyers conceived criminalization as a 'means of women's oppression'.[34] On the contrary, opposing decriminalization, journalists communicated it as an 'act against the fertility of the Greek population',[35] developing a narrative related to population decline.[36] Similarly, doctors argued that it would 'establish an inappropriate morality in Greek society' and argued for protecting women's health from the risks of abortions.[37] Politicians also correlated abortions to the population decline, questioning women's right to control their bodies.[38] Anti-choice supporters described it as a 'crime, bloodletting', 'slaughter' and a social, political, moral and religious issue, instead of a scientific one.[39] Regarding contraception, the need for widespread contraception and the decrease of abortion rates were a common topic for the involved groups. Journalists discussed contraception, focusing on the side effects and contraindications of the pill and presenting experimental contraceptive methods.[40] Furthermore, references that communicated different contraceptive methods were limited, while sterilization, male contraception and male sterilization were rarely discussed publicly.[41]

Feminists and women did not communicate abortion and contraception as two separate issues but as single issue; thus, there were no references in the press presenting their individual opinions on contraception. Interestingly, contraception was not discussed as an independent issue by the opponents of decriminalization. However, this was an act of silencing this issue, which was also devaluated and presented as irrelevant. Moreover, contraception was not part of the discussion with lawyers and politicians, who focused on the (de)criminalization of abortions. The medical community (besides journalists) appeared to be most connected to references to contraception and sterilization. Again, the pill – its use, efficacy, indications, contraindications and side effects – was the most discussed method, while articles with a summary of contraceptive methods were limited and sterilization was presented only in passing.[42]

The publics: The medical community

The opinion of the medical community was ascertained by analysing the medical textbooks. More specifically, in these type of publications, contraception was described as 'the temporary measures for preventing pregnancy, along with sterilization, which is permanent, abortion, which means the process of ending an unwanted pregnancy and family planning'.[43] In these publications, doctors were responsible for 'find[ing] the most appropriate contraceptive method';[44] at the same time, they explained women's hesitation towards the use of contraception in terms of their social backgrounds and not as an outcome of a limited circulation of knowledge.[45] For the medical community, performing an abortion was related to the termination of pregnancy.[46] However, the alternative subcategories of the term defined the legal status of the act (i.e., medical, artificial, criminal). This differentiation in terminology derives from the linguistic division of the term in the Greek language – έκτρωση/*ektrosi* indicates the termination of pregnancy for medical reasons, while άμβλωση/*amvlosi* refers to the termination of an unwanted pregnancy.[47]

The publics: The political sphere

While abortions have been criminalized since the first Greek Penal Code (1834), the first alterations appeared in the law passed in 1950,[48] when the political regime had been described as a crowned democracy.[49] In this law, the article about abortions is part of the section on 'Crimes against life', and it prohibits, with some exceptions, women from getting abortions.

Additionally, the public circulation of any information or means that lead to abortions is punishable. The next relevant law appeared in 1978[50] under the governance of New Democracy, a liberal conservative political party. The regulation 'For the removal or transplant of human biological substances' includes a paragraph related to abortions, according to which abortions are allowed until the twentieth week in cases of abnormalities in the foetus and until the twelfth week when the mother faces mental health problems that a psychiatrist has validated. Finally, in 1986[51] and under the governance of the Panhellenic Socialist Movement, a social democratic political party in Greece,[52] a new law was passed regarding the 'Technical termination of pregnancy and protection of women's health and other regulations', which ensured and legalized women's access to medical clinics for the technical termination of pregnancies. In parallel, the public circulation of information related to abortion remained prohibited, unless it was performed by family planning centres and doctors, or when it was published in medical or pharmaceutical journals.

The publics: Religious/religion-friendly and para-ecclesiastical press and publications

In the religious press,[53] abortion and contraception were contextualized as arguments related to science. Several articles argued that abortion was *not* a scientific issue and that its supporters followed anti-scientific arguments, while, paradoxically, they used science to justify their own position, arguing that the 'human foetus is soul and body since the first moment of conception'. For example, the controversial documentary *The Silent Scream* appeared as the most accurate scientific approach regarding the harmful and unethical procedure of abortion.[54] In the religious press, abortions were associated with the homicide of Greek children and, by extension, the Greek nation. At the same time, the press correlated them to the birth deficit and the issue of the Greek supremacy over the Turkish one, linking them to nationalistic ideologies and regenerating the famous conservative slogan that met with great popularity during the dictatorial regime (Junta of the Colonels): 'Fatherland, Religion, Family'.[55]

The counterpublics

Before presenting the essential meanings and narratives related to birth control that emerged in women's counterpublics, I will draw attention to the difference between autonomous women and state feminists, and the common ground of their fight. The historiography of women's studies and the content of the archival material suggest that these spheres were separated by a particular line.⁵⁶ The crucial difference between them was their political and ideological character: state feminists such as the Union of Greek Women, the Association of Greek Women Scientists or the Federation of Greek Women were often linked to political parties and formed their collective actions into associations and organizations that held political dependencies and hierarchical structures in one way or another. On the contrary, autonomous groups such as the Free Women's Movement, the Movement of the Liberation of Women or the Women's Group of Self-Examination were independent of political parties and formed based on horizontal relations and self-organization, embodying a plurality of ideological directions and agendas. The distinction between them is also apparent in the content, aims and analysis of birth control and the birth control movement.

Schematically, we can say that the former focused on their active participation in decision-making centres, prioritizing their cooperation with the state and the medical community, aiming for the state to manage birth control. By contrast, the latter emphasized self-education, self-help and self-consciousness, prioritizing the infusion of the institutional and scientific knowledge and culture with their own experiences, developing a unique system of knowledge production and circulation, wherein experts and the state would hold a secondary role.

While this chapter emphasizes the meaning and narratives of abortion and contraception in current research, it is essential to discuss the common context of women's counterpublics. Despite the differences in means, practices, organizational forms, ideologies or the related analysis, women fought against an established regime that was characterized by the profitable market of illegal abortions, women's financial exploitation, mortality and health complications, the limited circulation of contraception, the problematic and limited operation of family planning centres, the lack of sex education programmes, the established sexual taboos and gender roles, gender inequality, the lack of research related to male contraception and women's invisibility and inaccessibility from the public and political realms.⁵⁷

Additionally, the communication of birth control in women's and feminists' counterpublics was contextualized by the parallel discussion of the following issues: abortion, contraception, sexuality, family planning, sex education, local and international practices of networking, development of specific practices of communication, deliberation regarding the form of organization of their collective action that would ensure better communication of the issue of birth control (i.e., closed groups, open groups, organizations etc.), the analysis of feminist theories and the development of theoretical/political and ideological approaches and women's participation in political life. Last, it is important to note that both state feminists and autonomous women's groups communicated birth control in their publications by circulating medical knowledge about multiple contraceptive methods, the surgical procedure of abortion and male and female sterilization.[58]

The counterpublics: Autonomous groups

In the publications and magazines of the autonomous groups such as that of the Movement for the Liberation of Women, the *Anarchist Informational Bulletin*, *Broom* or *Katina*, women and feminists discussed multiple aspects of abortions and the impact of their decriminalization on numerous levels. For them, abortions could not be understood without examining the dependence of their legal status on social conditions and dominant ideologies, while at the same time highlighting their complex character against their perception as a simple 'medical procedure'. Such an analysis also challenged the dominant regime, which condemned women to remain 'passive receivers of the medical authority', while at the same time conceiving the fight for the decriminalization of abortions as 'a means of reshaping the relationship between the body, science and doctors' that was identified to women's sexual liberation and their struggle against gender inferiority.[59]

In their struggle to make a 'painful experience' visible,[60] they expanded the notion of motherhood, demanding the visibility and protection of single mothers.[61] Additionally, women's analysis concerning abortions highlighted the right of choice regarding motherhood as a 'fundamental right of every woman' and criticized the 'social construction of a dilemma' – the dipole pro or against abortion – for oversimplifying a complex social and gendered issue. Their narratives revealed and complicated women's complete epistemological dependency on doctors, emphasizing their right to control reproductivity and sexuality and reclaim knowledge.[62] Furthermore, women stood against scientists, politicians and the Church for treating abortions as a technical problem that could be solved 'scientifically without considering women's complex realities'.[63] In such a conceptualization of abortions, 'women's needs and desires rather than the family's or the state's interests' were prioritized and became the starting point of analysis. In that regard, abortions were linked to women's right to know their bodies, their functions and their organs, exculpating them and their sexual pleasure and destabilizing the social, scientific and political 'silence and hypocrisy'.[64]

In the same vein, the meaning of contraception was shaped by critically analysing the construction of the notion of family, of the institution of marriage and the dominant perception of contraception. As they stated, contraception is communicated by the Church and by the majority of the medical community as a 'risky set of methods' – narratives that 'distort the truth, maintaining women's passive role and challenging their right to sex, pleasure and independence'.[65] For them, the issue of contraception was linked to the autonomy of the female body, the conceptualization of which demanded a 'responsible education' that moved beyond the communication of risks, side effects and counterindications and enriched women's knowledge of their bodies.[66] Moreover, the public discussion of contraception in women's counterpublics was correlated with the deconstruction of the dominant and gendered conceptualization of female sexuality – the identification of sexuality with reproduction and its perception as a means for men's pleasure – and the struggle to associate reclaiming control over the body with control over knowledge.[67]

The counterpublics: State feminists

Exploring women's public sphere and, more specifically, magazines and publications related to state feminists such as the *Bulletin of the Democratic Women's Movement*, *Modern Woman* or the *Bulletin of the Association of Greek Women Scientists*, one sees the construction of alternative meanings and narratives concerning birth control. In such articles, the struggle for the decriminalization of abortions was often analysed as part of the promotion of birth control. This exact identification was contextualized with narratives about protecting women's health and interpreted as 'acts against women's exploitation by different experts'.[68] Additionally, birth control and contraception were crucial practices for decreasing high abortion rates (200,000–500,000 illegal abortions per year),[69] the reconstruction of the relationship between men and women and the reformation of sexual health.[70] In parallel, women highlighted the need for the circulation of knowledge related to birth control, embodying in their analysis the crucial role of family planning in the protection of motherhood. Here, abortion was conceived as a 'medical, social and economic issue'.[71] In their critique against the illegal abortion regime, they focused on gender inequality, a phenomenon resulting from the 'lack of related institutions and the limited circulation of knowledge', while at the same time arguing that decriminalization would 'transform abortions into an emergency solution', instead of a contraceptive method.[72]

Women's struggle to decriminalize abortion also included their participation in governmental or experts' committees, where cooperation between different actors was suggested primarily.[73] Within their practices, abortion was correlated with women's right to control their bodies, the deconstruction of the dominant conceptualization of motherhood, the transformation of the notion of family into a choice and the distinction between sexuality and reproduction.[74]

Similarly, contraception was communicated through collective actions for sharing experiences and knowledge, as a continuous mutual learning process.[75] In such articles, contraception as part of birth control was correlated with women's and social liberation, which contradicted the established 'sexophobic ideology'. The contextualization of the issue included the study of women's reproductive organs and menstrual cycle, while at the same time it was conceived as a step towards the reformation of the women's relationship with the doctors based on equality.[76] Last, related analysis perceived contraception as a 'women's right and state's duty' linked to the 'improvement of women's sexual life'.[77]

Conclusion

In this chapter, I examined the relationship between science and society in the second half of the twentieth century in Greece by focusing on the public presence of abortion and contraception, right after the first feminist attempts to secure reproductive justice. More specifically, I documented the publics and counterpublics of birth control, focusing on the meanings embodied in the discourses produced by different social groups. I have emphasized the role of gender power relations in science communication practices and argued that this methodological path could operate as a *lens* that enables us to depict the overlooked role of women in the reconstruction of the relationship between science and society. To that end, I have used the feminist and queer approaches regarding the notion of the public sphere and showed that approaching the concept as a complex and fluid entity is necessary for examining herstories of science communication and, consequently, women's involvement in science. It is exactly this fluidity that I suggest shows the multilayered phenomenon of the relationship between science and society and reveals women's engagement with different forms of knowledge.

In examining the publics of birth control, I have focused on the press – the public debate, the medical, political and religious/religion-friendly spheres. Regarding the press, I demonstrated how polarization became the primary practice of the communication of birth control and resulted in the de-scientization of abortion and contraception in the public discourses. In other words, focusing on presenting opinions of different social groups for or against the decriminalization of abortions or the use of contraception, the circulation of medical knowledge and the management of birth control (as appeared in the published articles) was limited to the experts' territory, establishing a specific boundary between science and society.

Concerning the medical community's public sphere, I have shown experts' efforts to legitimize themselves as the ones responsible for circulating knowledge related to birth control, choosing based on their own criteria which contraceptive method might be the most appropriate for each woman and defining the criteria by which an abortion could be medical, artificial, criminal and so on. Interestingly, and in line with the regulations, abortion here (except for the criminal abortion) was not correlated with an unwanted pregnancy, but only to a pregnancy that should be terminated for medical reasons. Additionally, the historical study of related regulation, up to their decriminalization, has demonstrated the state's efforts to discipline the female body, devalue women's demand for decriminalization and the importance of controlling knowledge and managing, alongside the experts, birth control. Last, the study of the religious (and religion-friendly) sphere has shown how the moralization of science can be a practice of science communication even without reproducing scientific content.

By contrast, exploring women's counterpublics has shown a multileveled body of science communication practices through which women reclaimed their bodies and knowledge. More specifically, women promoted their needs, desires and experiences both by demanding their cooperation with the experts and the state and by infusing medical knowledge with their own experiences, a situation that brought a female gaze to science. Indeed, while birth control was presented in the dominant discourses as a political and moral issue whose management should be male-centred, women struggled to reveal its scientific character to circulate relevant scientific knowledge. However, they approached science as a social activity that embodies ideological narratives and gendered power relationships.

To that end, women's polymorphous involvement in science and active engagement with the production and circulation of knowledge related to birth control operated as a mechanism that shifted the boundaries between science and society. Their practices of science communication were an act of resistance and an act against the given biopolitical order, which was correlated to hegemonic ways of knowing. Indeed, women ruptured the dominant public spheres to democratize knowledge, science and society. The counter-practices, counterdiscourses and countermeanings that women developed in their counterpublics were aimed at epistemic, social and reproductive justice. Thus, the feminist and queer approaches to the concept of the public sphere not only reveal women's hidden, complex and often neglected relationship with science, but also draw attention to the political dimensions of the relationship between science and society and the steps required towards the prioritization of diversity and inclusion in science studies.

Notes

1 The chapter is based on the author's dissertation 'Science communication in late twentieth-century Greece: Public intersections of gender and knowledge circulation in the feminist birth control movements'. Research was supported by the Hellenic Foundation for Research and Innovation under the HFRI PhD

Fellowship Grant (Fellowship No. 873) and is available online at: www.didaktorika.gr/eadd/handle/10442/52660. Former versions of the paper were presented at the Thirteenth International Graduate Student Conference in Modern Greek Studies (Princeton University, Seeger Center for Hellenic Studies), the First Flying Colloquium of the History of Science in Central, Eastern and Southeastern Europe, and the virtual conference 'Hidden Histories: Women and Science in the Twentieth Century' (Heidelberg Centre for Transcultural Studies and the University of Bucharest). J. Östling, E. Sandmo, D. L. Heidenblad, A. N. Hammar, K. Nordberg (eds), *Circulation of Knowledge: Explorations in the History of Knowledge* (Lund: Nordic Academic Press, 2018), p. 12.

2   N. A. Papanikolaou, *Μαιευτική* [*Obstetrics*] (Thessaloniki: Library of the Hellenic Society of Obstetrics and Gynecology, 1987), pp. 11–17.

3   OB/GYN Clinic of the University of Patras, Greek Society for the Study of Reproduction, Greek Society of IVF and Fetus Transportation, 'Πανελλήνιο συνέδριο για την ανθρώπινη αναπαραγωγή. Ελληνική Εταιρεία Εξωσωματικής Γονιμοποίησης και Εμβρυομεταφοράς' [Panhellenic Conference on Human Reproduction], 11–13 March 1988 (Patras: Library of the Hellenic Society of Obstetrics and Gynecology, 1988), p. 201.

4   Anon., 'Trump's order on abortion policy: What does it mean?', BBC News (24 January 2017).

5   S. Andrikakis, 'Αφήστε με να ζήσω' [Let me live], *Sportime* (29 December 2019), www.sportime.gr/entypi-ekdosi/diavaste-simera-sto-sportime-afiste-mena-ziso/ (accessed 25 November 2022).

6   See also E. Chordaki and A. Lazopoulou, 'Reclaiming our health: Greek feminist birth control movements as a form of women's engagement with science', in C. C. Harry and G. N. Vlahakis (eds), *Exploring the Contributions of Women in the History of Philosophy, Science, and Literature, Throughout Time: Women in the History of Philosophy and Sciences*, vol. 20 (Cham: Springer, 2023), pp. 179–98, https://doi.org/10.1007/978-3-031-39630-4_12.

7   E. Geoff, 'Politics, culture and the public sphere', *Positions*, 10: 1 (2002), 230.

8   M. Hawkesworth, 'Gender and the "public": A theoretical overview', paper presented at the 20th Congress of the International Political Science Association, 9–13 July 2006.

9   P. Werbner, 'Political motherhood and the feminization of citizenship: Women's activism and the transformation of the public sphere', in N. Yuval-Davis and P. Werbner (eds), *Women, Citizenship and Difference (Postcolonial Encounters)* (London: Zed Books, 1999), p. 221.

10 Geoff, 'Politics, culture and the public sphere', p. 230.

11 M. M. Ferree, W. A. Gamson, J. Gerhards and D. Rucht, 'Four models of the public sphere in modern democracies', *Theory and Society*, 31 (2002), 307.

12 R. Benson, 'Shaping the public sphere: Habermas and beyond', *American Sociologist*, 40 (2009), 178.

13 M. Loehwing and J. Motter, 'Publics, counterpublics and the promise of democracy', *Philosophy & Rhetoric*, 42: 3 (2009), 220–41.

14 Loehwing and Motter, 'Public, counterpublics', p. 221.

15 S. Susen, 'Critical notes on Habermas's theory of the public sphere', *Sociological Analysis*, 5: 1 (2011), 52.
16 Susen, 'Critical notes', p. 53.
17 Susen, 'Critical notes', p. 55.
18 M. Pajnik, 'Feminist reflections on Habermas's communicative action: The need for an inclusive political theory', *European Journal of Social Theory*, 9: 3 (2006), 391.
19 J. Rendal, 'Women and the public sphere', *Gender & History*, 11: 3 (1999), 482.
20 N. Fraser, 'Transnationalizing the public sphere: On the legitimacy and efficacy of public opinion in a post-Westphalian world', *Theory Culture Society*, 24 (2007), 7–30.
21 N. Fraser, 'Rethinking the public sphere: A contribution to the critique of actually existing democracy', *Social Text*, 25: 26 (1990), 67–68.
22 N. Fraser, 'Politics, culture, and the public sphere: Toward a postmodern conception', in L. Nicholson and S. Seidman (eds), *Social Postmodernism: Beyond Identity Politics* (Cambridge: Cambridge University Press 1995), p. 291.
23 M. Warner, *Publics and Counterpublics* (London: Zone Books, 2005), p. 56.
24 Warner, *Publics and Counterpublics*, p. 58.
25 Loehwing and Motter, 'Publics and counterpublics', p. 223.
26 R. Cohen and S. O'Byrne, ' "Can you hear me now…Good!" Feminism(s), the public/private divide, and Citizens United v. FEC', *UCLA Journal of Gender and Law*, 20: 1 (2013), 39.
27 The Junta of the Colonels (1967–74). The fall of the dictatorial regime and the military government was defined by the Polytechnic Uprising, a dynamic rebellion that started with the occupation of the National Technical University of Athens and ended with the tank invasion and bloodshed of 17 November 1973. See A. Liakos, *Ο ελληνικός 20ος αιώνας* [*The Greek 20th Century*] (Athens: Polis Publications 2020), pp. 392–462.
28 Law 1609/1986. Greek literature related to the issue of birth control is mostly focused on the emergence of the feminist movements of the period (1970s–1980s), to which the feminist birth control movement belonged. See Catalog, Ο φεμινισμός στα χρόνια της μεταπολίτευσης 1974–1990: Ιδέες, συλλογικότητες, διεκδικήσεις [Feminism during the era of the democratic transition 1974–1990: Ideas, collectivities, claims] (Athens: The Hellenic Parliament Foundation for Parliamentarism and Democracy, n.d.); Nt. Vaiou and Ag. Psarra, 'Εισαγωγικό σημείωμα επιμελητριών' [Editors' introductory note] in Nt. Vaiou and A. Psarra (eds), *Εννοιολογήσεις και πρακτικές του φεμινισμού: Μεταπολίτευση και «μετά»* [*Meanings and Practices of Feminism: Democratic Transition and Onwards*] (Athens: The Hellenic Parliament Foundation for Parliamentarism and Democracy, 2018), pp. ix–1. An introduction to the feminist birth control movement in Greece can be found in the feminist magazine *Dini* published right after the decriminalization of abortion in 1986: E. Avdela, M. Papagiannaki and K. Sklaveniti, Έκτρωση 1976–1986: Το χρονικό μιας διεκδίκησης [Abortion 1976–1986: The timeline of a claim], *Dini Feminist Magazine*, 1 (1986), 4–29.

Abortion and contraception in Greece are an understudied area. However, crucial aspects of those phenomena and practices have been examined from a sociological, statistical, anthropological, biopolitical and ethnographical point of view. See A. Barmpouti, 'Issues of biopolitics of reproduction in postwar Greece', *Studies in History and Philosophy of Biological and Biomedical Sciences*, 83 (2020); A. Chalkia, *Το άδειο λίκνο της δημοκρατίας. Σεξ, έκτρωση και εθνικισμός στην σύγχρονη Ελλάδα* [*The empty cradle of democracy: Sex, abortion and nationalism in modern Greece*] (Athens: Alexandreia Publications, 2007); E. Georges, *Bodies of Knowledge: The Medicalization of Reproduction in Greece* (Nashville, TN: Vanderbilt University Press, 2008); V. Hionidou, *Abortion and Contraception in Modern Greece, 1830–1967: Medicine, Sexuality and Popular Culture* (Cham: Palgrave Macmillan, 2020); E. Zaragkali, *Οι βουβές πληγές. Η αντι-σύλληψη και η έκτρωση ως βίωμα και ως πράξη* [*Silent Wounds: Contraception and Abortion as an Experience and Praxis*] (Athens: Nisos Academic Publishing, 2010).

29 The analysis of the archival material related to the popular press included magazines and newspapers of general interest for the period between 1976 and 1986. Newspapers included *Mesimvrini* [*Midday Press*], a daily political newspaper, *Avgi* [*Dawn*], the daily newspaper of the KKE (Eurocommunist), *Eleftherotypia* [*Press Freedom*], a daily broadsheet and political newspaper, *Ta Nea* [*News*], a daily broadsheet and political newspaper, *To Vima* [*The Step*], a weekly broadsheet and political newspaper, *Vradini* [*Nightly Press*], a weekly political newspaper, *To Vima tis Kyriakis* [*Sunday's Step*], a weekly newspaper, *Eleftheros Typos* [*Free Press*], a daily broadsheet and political newspaper, and *Ntomino* [*Domino*], a newspaper addressed to women. Magazines included *Fylladio* [*Flyer*], a magazine related to art and literature published every three months, *Pantheon*, a magazine of general interest addressed to women, and *Ena* [*One*], a magazine of general interest.

30 Anon., 'Θύματα οι γυναίκες' [Women are the victims], *Mesimvrini* (10 December 1983), Delfis Archival Center (hereafter, DAC); L. Mpenou, 'Μ' αφορμή το νομοσχέδιο για τις εκτρώσεις' [Occasioned by the abortion bill], *Filladio* (October 1985), DAC; Anon., 'Αμβλώσεις' [Abortions], *Ethnos* (26 January 1986), DAC.

31 Anon., 'Συνέντευξη -αστραπή Ήταν χρέος μας...μας μιλάει ο Πρόεδρος του Ι.Σ.Α. κ. Φραγκλίνος Παπαδέλης' ['It was our duty': The president of the Medical Association of Athens, Fr. Papadelis, talks to us], short interview, *Pantheon* (22 November 1983), DAC.

32 Anon., 'Ελεύθερες εκτρώσεις υπό όρους ζητά η ΚΔΓ και ενημέρωση για την αντισύλληψη' [The Democratic Women's Movement asks for free abortions under conditions and information regarding contraception], *Αυγή* [*Avgi*] (1 April 1983), DAC; Chr. Papastathopoulou, '1390 γιατροί μοιράζονται 4,5 δισ. από τις αμβλώσεις [Εκστρατεία γυναικών για δικαίωμα σε έκτρωση – αντισύλληψη – σεξουαλικότητα' [1,390 doctors share 4.5 billion from abortions: Women's campaign for the right to abortion-contraception-sexuality], *Ελευθεροτυπία* [*Eleftherotipia*] (9 December 1983), DAC.

33 L. Papadopoulos, 'Μια μεγάλη έρευνα για ένα μεγάλο πρόβλημα: Όχι στις αμβλώσεις αλλά πρέπει να νομιμοποιηθούν' [A major research for a major problem: No to abortions but they have to be legal], *Ta Nea* (24 January 1976), DAC.
34 O. Ioannou, 'Οι δικηγορίνες των αμβλώσεων' [The women lawyers of abortions], *Pantheon* (12 February 1985), DAC.
35 Anon., 'Λιγοστεύουμε οι Έλληνες' [The Greeks become fewer], *To Vima* (5 March 1985), DAC.
36 A. Athanasiou, 'Bloodlines: Performing the body of the "demos," reckoning the time of the "ethnos"', *Journal of Modern Greek Studies*, 24: 2 (2006), 229–56; A. Athanasiou, 'Το εθνικό σώμα σε κατάσταση έκτακτης ανάγκης: Η δημογραφική πολιτική της ζωής και τα όρια του πολιτικού' [The national body in a state of emergency: Demographic biopolitics and the limits of the political], in P. Ethymios (ed.), *Πολιτικές της καθημερινότητας Σύνορο, σώμα και ιδιότητα του πολίτη στην Ελλάδα* [*Everyday Politics: Border, Body and Citizenship in Greece*] (Athens: Alexandreia, 2014).
37 Th. Roumpanis and A. Papaioannou, 'Κόντρα της εκκλησίας για τις αμβλώσεις' [The Church's wedge regarding abortions], *Vradini* (28 April 1986), DAC.
38 N. Psaroudakis, 'Αμβλώσεις: το ψέμα και η αλήθεια' [Abortions: The lie and the truth], *Eleftherotipia* (3 March 1986), DAC.
39 Papadopoulos, 'A major research for a major problem'; Anon., 'Ιερείς κατά των αμβλώσεων' [Monks against abortions], *Eleftherotipia* (7 February 1984), DAC; Anon., 'Υστερία από παρεκκλησιαστικές οργανώσεις κατά των αμβλώσεων' [Hysteria from the para-Christian organizations against abortions], *Avgi* (25 February 1986), DAC.
40 Anon., 'Αντισύλληψη Καταφάσεις και αντιφάσεις' [Contraception: Affirmatives and contradictions], *Pantheon* (22 March 1988), DAC; E. Bioubi, '16 χρόνια από την κυκλοφορία του χαπιού: Θαύμα ή κατάρα' [Sixteen years since the circulation of the pill: Miracle or curse?] *To Vima tis kiriakis* (23 May 1976), DAC; Anon., 'Και αντισυλληπτικό σπόγγος' [Contraceptive sponge], *To Vima* (25 March 1984), DAC; Anon., 'Χάπι Μετά' [The morning-after pill], *Ena* (7 November 1985), DAC.
41 Anon., 'Στις εκτρώσεις/Αντισυλληπτικά μέτρα' [In abortions: Measures related to contraception], *Eleftheros Tipos* (11 February 1985), DAC.
42 M. Mpredakis, 'Ο γυναικολόγος κοντά σας/ μέθοδοι αντισύλληψης' [A gynecologist close to women: Contraceptive methods], *Ntomino* (18 April 1986), DAC.
43 N. A. Papanikolaou, *Γυναικολογία* [*Gynecology*] (Thessaloniki: Library of the Hellenic Society of Obstetrics and Gynecology, 1986). See also D. I. Aravandinos, *Φυσιολογία της γυναίκας* [*Physiology of Woman*] (Scientific Publication Gr. Parisianos, n.d.), p. 192.
44 Papanikolaou, *Gynecology*.
45 Aravandinos, *Physiology of Woman*, p. 194.
46 N. A. Papanikolaou, *Μαιευτική* [*Obstetrics*] (Thessaloniki: Library of the Hellenic Society of Obstetrics and Gynecology, 1987); D. I. Aravandinos, *Μαιευτική* [*Obstetrics*] (Scientific Publication Gr. Parisianos, 1989), p. 380.

47 For the linguistic division regarding the term 'abortion' in the Greek language, see Chordaki, 'Science communication in late twentieth-century Greece'.
48 Law 1492/1950, 'Περί κυρώσεως του Ποινικού Κώδικα' [For the constitutional validity of the Greek Penal Code], Library of the Hellenic Parliament.
49 Th. Veremis and S. I. Koliopoulos, *Νεότερη Ελλάδα: Μια ιστορία από το 1821* [Modern Greece: A history since 1821] (Athens: Patakis Publications, 2013).
50 Law 821/1978, 'Περί αφαιρέσεων και μεταμοσχεύσεων βιολογικών ουσιών ανθρώπινης προελεύσεως' [For the removal or transplant of human biological substances], Library of the Hellenic Parliament.
51 Law 1609/1986, 'Τεχνητή διακοπή της εγκυμοσύνης και προστασία της γυναίκας και άλλες διατάξεις' [Technical termination of pregnancy and protection of women's health and other regulations], Library of the Hellenic Parliament.
52 Liakos, *The Greek 20th Century*, p. 419.
53 The analysis of the religious press included the publication of the Panhellenic Union of Friends of Multi-Child Parents *Oi Filoi ton Polyteknon* [*Friends of Multi-Child Parents*], the magazine *Oikogeneia* [*Family*], published every three months by the Panhellenic Union of Friends of Multi-Child Parents and the conservative daily newspaper *Akropolis* [*Acropolis*].
54 Anon., 'Μελέτη -έρευνα/ Προς νομιμοποίηση των εκτρώσεων!' [Study-research: Towards the decriminalization of abortions?!], *Oi Filoi ton Polyteknon* (March 1983), DAC; Anon., 'Σπάνια απεικόνιση -δογματική της θείας συλλήψεως – εναρθρωπήσεως του θεού Λόγου ως θεανθρώπου Χριστού με πρόσληψη ψυχής και σώματος εξ άκρας συλλήψεως' [Rare visualization of the holly conception with soul and body from the moment of conception], *Oikogeneia* (1985), DAC; Anon., 'Απλώνεται η σταυροφορία κατά των εκτρώσεων' [The crusade of abortions is spreading], *Akropolis* (1986), DAC.
55 Gr. Mihalopoulos, 'Εθνικά μίλησε ο κ. Σαρτζετάκης' [Sartzetakis spoke for the fatherland], *Unknown Newspaper* (1985), DAC. For the history of the slogan in the Greek area, see E. Gazi, *Πατρίς, θρησκεία, οικογένεια. Ιστορία ενός συνθήματος 1880–1930* [*Fatherland, religion, family: Exploring the history of a slogan in Greece, 1880–1930*] (Athens: Polis Publications, 2011).
56 See Avdela, Papagiannaki and Sklaveniti, *Abortion 1976–1986*.
57 Catalog, *Feminism in the Post-Colonial Years*.
58 E. Roboti, 'Αντισύλληψη: Κάτι που θα έπρεπε να ξέρουμε' [Contraception: Something that we should have known about], *Σύγχρονη Γυναίκα* [*Modern Woman*], 26 (n.d.), DAC; Anon., 'Η αυτόματη άμβλωση' [The 'spontaneous' abortion], *Modern Woman*, 59 (1988), DAC.
59 Anon., unknown title, *Κίνηση για την Απελευθέρωση των γυναικών* [*Publication of the Movement for Liberation of Women*], 1 (1978), DAC.
60 A. Ntaifa, 'Η έκτρωση δεν είναι αντισύλληψη' [Abortion is not contraception], *Πόλη των γυναικών* [*City of Women*], 7 (1982), DAC.
61 Anon., 'Εκτρώσεις! Τακτική 'στρουθοκαμηλισμού"' [Abortions: The tactic of hypocrisy], *Κίνηση για την απελευθέρωση των γυναικών* [*Publication of the Movement for Liberation of Women*], 4 (1978), DAC.

62 Anon., 'Έλεγχος γεννήσεων: εκτρώσεις' [Birth control: Abortions], *Αναρχικό Ενημερωτικό Δελτίο* [*Anarchist Informational Bulletin*] (1984), DAC.
63 Editorial Board, 'Ποιός θέλεις παιδιά' [Who are those who want to have children: Special Issue for abortions], *Σκούπα* [*Broom*], 1 (1979), DAC.
64 Anon., 'Αμβλώσεις' [Abortions], *Αδέσμευτη Κίνηση Γυναικών* [*Bulletin of the Free Women's Movement*] (1986), DAC; Anon., 'Αν οι γυναίκες είχαν το δικαίωμα της επιλογής δεν θα υπήρχε δημογραφικό πρόβλημα' [If women had the right to choose, the demographic problem would not have existed], *Bulletin of the Free Women's Movement* (1994), DAC; D. Naziri, 'Έκτρωση: Αναπόφευκτη εμπειρία στην ζωή της Ελληνίδας' [Abortion: Is it an inevitable experience in Greek women's lives?] *Δίνη* [*Swirling*], 7 (1994), DAC.
65 Anon., 'Αντισύλληψη ή το δικαίωμα να ελέγχουμε οι ίδιοι το σώμα και τη ζωή μας' [Contraception or the right to control ourselves, our body, and our lives], *Περιοδικό Ομάδας Παρέμβασης Κουφαλίων* [*Magazine of the Group of Intervention in Koufalia*] (1984), DAC.
66 Anon., 'Αμβλώσεις αντισύλληψη: Μια ακόμη παράμετρος του γυναικείου ζήτήματος' [Abortion, contraception: Another parameter in the female issue], *Κατίνα* [*Katina*], 1 (n.d.), DAC.
67 'Κάλεσμα υπογεγραμμένο από διαφορετικές ομάδες γυναικών' [Call for action signed by different women's groups, National Day of Contraception], *Broom*, 4 (1980), DAC.
68 R. Kaklamanaki, 'Αμβλώσεις: Λίγο πιο νόμιμες από πριν' [Abortions: A bit more legal than before], *Δελτίο Κίνησης Δημοκρατικών Γυναικών* [*Bulletin of the Democratic Women's Movement*], 3 (1979), DAC.
69 Anon., 'Αμβλωσεις' [Abortions], *Ethnos* (26 January 1986), DAC.
70 Anon., 'Σκεφτείτε πριν φτάσετε στην έκτρωση' [Think before you end up in abortion], *Bulletin of the Democratic Women's Movement*, 4 (1976), DAC.
71 Anon., 'Αμβλώσεις. Τι συμβαίνει' [Abortions: What is happening], *Modern Woman* 47 (1989), DAC.
72 Roboti, 'Contraception'.
73 Anon., 'Σκεφτείτε πριν φτάσετε στην έκτρωση' [Think before you end up in abortion], *Bulletin of the Democratic Women's Movement*, 4 (1976), DAC; S. Mezili, 'Αγώνας για την νομιμοποίηση των αμβλώσεων μέχρι την κατάργηση τους: Παρά τις λυσσαλέες αντιδράσεις εκκλησίας και δεξιάς, οι αμβλώσεις πρέπει να νομιμοποιηθούν' [Fight for the decriminalization of abortions till their abolition: Despite the reactions of the right wing and the Church, abortions must be decriminalized], *Modern Woman*, 33 (n.d.), DAC.
74 K. T., 'Αμβλώσεις. Νομιμοποίηση τώρα' [Abortions: Decriminalization now], *Modern Woman*, 38 (undated), DAC; Anon., 'Αμβλώσεις: Τι συμβαίνει' [Abortions: What is happening], *Modern Woman*, 47 (1989), DAC.
75 Anon., 'Αντισύλληψη' [Contraception], *Bulletin of the Democratic Women's Movement*, 19 (1981), DAC.
76 A. Douka, 'Αντισύλληψη, μέρος πρώτο: Δικαίωμα της γυναίκας, υποχρέωση της πολιτείας' [Contraception, part a: Women's right and state's duty], *Ανοιχτό Παράθυρο* [*Open Window*], 20 (n.d.), DAC.

77 A. Douka, 'Αντισύλληψη, μέρος δεύτερο. Δικαίωμα της γυναίκας, υποχρέωση της πολιτείας' [Contraception, part b: Women's right and state's duty], *Open Window*, 21 (1983), DAC.

13

The power of autobiography: Documenting women scientists through a lecture series at the University of Illinois

Bethany G. Anderson and Kristen Allen Wilson

Introduction

Archives and museums have long been sites of study for the history of science and for science education. Yet these institutions have not, for the most part, sufficiently documented or told a fuller story of science. Indeed, most museums and archives[1] largely chronicle the lives and contributions of scientists who are white and male and are thus replete with silences that obscure the contributions of women and other under-represented and marginalized communities in science. In the case of archives, even when records exist for these communities, because they are not always the creators of these records (or credited as creators), their experiences and contributions can be difficult to discern. These archival absences affect museums, which often rely on records to create exhibits. Who gets to create records and have those records preserved for posterity reflects who holds power and authority in society.[2] The history of women in science is typically shrouded by these absences[3] and the effects of male-dominated record creation, preservation and credit.

In addition to the ways that women have been disempowered in both science and its record-keeping, the larger under-representation of women scientists is also mirrored in archival repositories and museums. Despite a few specialized repositories and initiatives, women scientists are under-represented in the holdings of cultural heritage institutions.[4] Documenting women in science has been challenging given the historically fewer numbers of women in scientific fields, the ways in which women have been largely unacknowledged for their contributions and biases and discrimination they have faced in science, all of which are compounded by the politics of record-creation and passive collection development as well as archival appraisal approaches.[5] The lack of critical mass of records, objects and exhibitions has meant women scientists have not seen themselves reflected in the historical record of science. In addition to a lack of representation, women may be depicted in passive roles or in negative ways that perpetuate biases and stereotypes about them. For example, in a study of gender representation in museums, Gaby Porter notes that women are depicted as 'relatively passive, shallow, undeveloped, muted and closed'.[6]

In archives, representation matters not only in how the institution depicts women in collection descriptions, exhibits and so on, but also in terms of the types of documentation present. Women scientists may be present in administrative records, directories or publications, yet a dearth of materials exist in which they are the creators and tell their stories in their own words. Women scientists may thus be *in* the archives, but often without the agency as record creators. At the same time, it is important to acknowledge that archives are not neutral spaces, creating and perpetuating their own biases through acquisitions, metadata creation, digitization and other decisions that may impede access or create and widen gaps. These biases and gaps may not only go unquestioned, but they become further compounded when researchers bring their own biases and stereotypes (unconscious or not) to the archival materials they use to piece together histories about science. For example, one may conclude that the dearth of records about women scientists is because of the misconception that women do not have the aptitude for STEMM fields.[7] Such silences may also become self-perpetuating; women scientists may conclude that their stories and materials do not belong when their representation is scarce or their contributions misrepresented. Archives must understand these challenges to confront silences, absences and invisibilities to better document and represent a fuller and more accurate history of science. Finding ways to mediate archival engagement and provoke introspection and questions about gender stereotypes and identity can be a useful intervention.[8]

This chapter discusses an initiative by the University of Illinois Archives and the Illinois Distributed Museum to build archival holdings that enhance the historical representation of women scientists at the University of Illinois Urbana-Champaign through a lecture series that captures and preserves autobiographical accounts of their research and pathways into science. In an effort to better engage the scientific community, and the university community as a whole, the authors launched a Women in Science Lecture Series to highlight and capture the work and contributions of women scientists at the University of Illinois. By inviting women scientists to discuss their research and reflect on the importance of preserving a record of their own stories, we hope to take a step towards a more diverse and inclusive archival record and change perceptions about whose records and stories belong in archives and museums. We describe the ways in which women have been under-represented in cultural heritage institutions, as well as the successes and challenges of the lecture series, including lessons learned and an evaluation of the initiative through surveys. We use these takeaways as a point of departure for evaluating the utility of a lecture series as a documentation approach for beginning to redress archival and museum silences about women in science.

The University of Illinois Archives

Established in 1963, the University of Illinois Archives (hereafter, U of I Archives) is one of several special collections units that are part of the University of Illinois Library.[9] The U of I Archives broadly documents the history and activities of the University of Illinois Urbana-Champaign, aiming to holistically capture the experiences of students, the work and contributions of faculty, alumni and staff, and the activities and decisions of administrators. To document these aspects of university history, the U of I Archives has programmatic areas that focus on science and technology, the fine and applied arts, multicultural and under-represented communities, faculty and student life. In line with this work, the U of I Archives has sought to preserve and make available the archival records of the University of Illinois' science units and programmes. As one of many land-grant institutions that came out of the Morrill Land Grant Acts in the nineteenth century,[10] the University of Illinois' early history was very much grounded in the teaching of 'agriculture and mechanic arts',[11] similar to other land-grant colleges and universities of the time in the United States. To this day, the University of Illinois is still a very science-intensive university, albeit characterized by many interdisciplinary programmes.

While the U of I Archives has sought to comprehensively and broadly document the university's scientific enterprise,[12] its holdings largely represent the experiences and contributions of white men in science – many of whom have been historically highlighted and celebrated by the university itself.[13] This has led to gaps in knowledge and to an incomplete history of science at the University of Illinois. The Women in Science Lecture Series was launched to begin addressing these gaps and begin reshaping the collection from one that is predominantly white and male to one that is more diverse and inclusive and better represents current and past science.

The Illinois Distributed Museum (IDM) is an online museum that highlights innovations and the people behind them. It also has a component that leads visitors to related locations around campus where they can see papers, objects, pictures or buildings. 'Innovation' is defined broadly so that it encompasses all disciplines. An engineering alumnus and faculty members from the Department of Anthropology and from the School of Information Sciences originally developed the IDM. It became a project of the University of Illinois Archives in 2017, seven years after its creation. When the U of I

Archives received the project, it heavily focused on the sciences and only featured one woman. While there has been much work done to try to diversify innovators featured on the IDM, an imbalance in representation between men and women scientists has persisted. This was partly due to faculty and student volunteers helping to write content and picking from a giant list of topics. More active and intentional approaches were needed to help create better representation of different identities. Thus, one of the goals of the Women in Science Lecture Series is to help fill in the IDM's gaps by creating an archival record, through the lectures, as a source for future exhibits that highlight the speakers and connect visitors to their lectures.

Women in Science Lecture Series

The Women in Science Lecture Series was envisioned as both a documentation and preservation project as well as an outreach endeavour. We, the authors – one of whom is an archivist and one of whom is a museum specialist – have been thinking critically about how cultural heritage institutions can create more inclusive and diverse holdings that document women in this space. At the same time, we have been seeking to create collaborative and equitable relationships with the University of Illinois' communities, to steward and co-curate their records. The idea for this lecture series was inspired in part by a similar one held at the Field Museum in Chicago, Illinois, which we thought could serve as a model for a successful lecture series that engages the public in the work and contributions of women in science.[14] We drew additional inspiration from other initiatives and lecture series that centre women scientists, both past and present. For example, the Association for Women in Science (USA) maintains a speaker's bureau and the Cranbrook Institute of Science has a speaker series focused on women in science.[15] The *Lady Science* magazine and podcast and the *Lost Women of Science* podcast present additional exemplars for meaningfully engaging the public in the too-often-overlooked accomplishments of women in the history of science.[16] We sought to create a lecture series that takes inspiration from these examples of public engagement and scholarship, but felt it was vital to situate the lectures as autobiographical records that would be recorded and preserved in the University of Illinois Library's digital preservation repository.[17] As an archival record of these scientists' research and individual life stories, the lectures are akin to biographical 'vertical files' that one might find at archives. However, the lectures are not meant to replace the documentation that one might find in someone's personal papers; rather, they are meant to be an autobiographical archival record *created by* women scientists that also reshapes the contours of the Archives' holdings and who is represented in them.[18] As a means to highlight the research and contributions of women scientists at the university, we aimed to promote the U of I Archives as a place where these histories can be preserved and stewarded and where women scientists could tell their stories in their own words.

The lecture series was initially financially supported with the University of Illinois Library's innovation seed fund, which enabled us to hire a graduate student to assist with logistics for planning and promoting the lectures. The student also researched and developed digital exhibits that highlight materials from the U of I Archives on the speakers' areas of research or records documenting women scientists.[19] During our initial planning and launch, we recognized that the lecture series would have greater reach and impact if it were part of a larger network on campus. To form a network around the lecture series, we reached out to science units and to the Women and Gender in Global Perspectives programme on campus to invite them to be co-sponsors. We requested their help by suggesting speakers and promoting the talks through their email lists and social media.

We scheduled the lectures for the second Tuesday of the month over the noon lunch hour. Spanning an hour in length (that included time at the end for questions from the audience), we modelled the lecture series after a typical academic lecture, with which the university community would be familiar. While we had originally envisioned having the lectures onsite in the U of I Archives, the Covid-19 pandemic meant that we had to change the format to virtual. As we later realized, the change to the virtual format over Zoom proved fortuitous, enabling a broader audience to attend the lectures and engage with the speakers, including attendees from the general public. This ultimately facilitated greater accessibility to the lectures and a larger platform from which the speakers could share their work.

For the purposes of the lecture series, we deliberately decided to define 'science' broadly to cast a wide net that captures as many different scientific disciplines as possible and, at the same time, to acknowledge that women science faculty and staff conduct interdisciplinary research and are frequently affiliated with several campus units. We therefore sought speakers from the natural sciences, engineering, physical sciences and the social sciences. Some disciplines our speakers have represented include anthropology, psychology, epidemiology, chemistry, ecology, sociology, nutritional sciences, engineering and kinesiology. Another reason we decided to define science broadly is because women are under-represented across scientific disciplines in different ways.[20] Hosting a diverse representation of women from different sciences has helped to attract a broad audience and attendees from departments across the university. We hope this will in turn foster relationships between the U of I Archives and the university to help remediate the gaps in our holdings.

Capturing and preserving a diverse and inclusive record of women in science

In recognition that women are under-represented in different ways across scientific fields, as we began planning the lecture series it was also important for us to have speakers who have different identities. Women are not a monolithic category and experience biases and discrimination in science differently depending on their positionalities. We strived to have a diverse array of speakers, in particular by inviting women of colour as speakers. Because we are capturing women's experiences in an academic setting, we also sought women at different points in their careers, including at various stages in the tenure process, women who are administrators for their department or those who hold senior-level administrator positions at the university. Having women at different stages in their careers allows for role models who may not be well-known names to younger generations, which makes them more relatable and can be encouraging for students in STEMM.[21] We also strove to capture as many different experiences as possible. These initial eight lectures provided a springboard to continue the lecture series beyond the 2020–21 academic year and led to requests from other units on campus to join our list of co-sponsors.

As the lectures are more biographical than might be typical of an academic lecture, we asked the speakers to provide more details about why they decided to pursue a career in science, specifically in their field, and to discuss their research. In many cases, the speakers talked about how their identities informed their own research. This was also beneficial to our audience, as it allowed for women with many different interests and across different intersecting identities to serve as role models for current and future STEMM students. Role models are more effective when students can connect their interests and do not follow stereotypes of scientists, such as scientists being 'nerdy' and 'introverted'.[22] Our speakers' personalities and research interests both came through and many candidly discussed their struggles in academia, which may enable more opportunities for other women to relate to our speakers and see a place for themselves in STEMM fields.

Preserving their stories

Because one of the chief goals for the lecture series is to preserve women scientists' stories, we wanted these lectures to go beyond what one might learn about a scientist's work from just her published papers. We therefore provided the speakers with guiding questions which asked them to discuss how they became interested in science and their field, their current research and research process, how they became interested in those specific topics and with whom they have collaborated (if applicable). Because we hoped students would attend the lectures, we asked the speakers to offer advice to anyone who would also want to pursue similar research or enter their specific scientific field. Given the archival nature of the lecture series, we also asked the speakers to reflect on the historical significance of preserving their research and stories.

Before each lecture, we met with the speaker and discussed the format of the lecture, the departments they wanted us to contact to advertise their lecture and to see if they had any questions about the guiding questions we produced. Most speakers asked about the question on the historical significance of their work and experiences, as many had not thought about it before. Many of the speakers mentioned either during the talk or afterwards that they were grateful for the opportunity to reflect on this question. A few even situated their work within the history of their fields and past women scientists' work.[23]

Each speaker took different approaches to answering these questions. Some speakers answered all the guiding questions, while others chose to only address two or three. These questions helped our speakers discuss the process of conducting their research, their career trajectories and other women scientists who inspired them. For instance, Associate Professor of Epidemiology Rebecca Lee Smith talked about different ways she has been able to secure funding and advice for identifying funding sources for research, and the importance of mentoring, supporting and uplifting women entering scientific fields. She also described the challenges of grant proposal-writing and mentioned that sometimes one needs to apply multiple times to the same or different grant programmes. When asked to contemplate preserving her own story, Associate Professor of Nutrition Zeynep Madak-Erdogan mentioned that she hopes that one day in the future when her daughter watches the lecture, she and women scientists will be living in a more equitable world. These kinds of stories about scientists' challenges and hopes are difficult to glean from published research. The preserved lectures, along with papers that we hope our speakers will donate someday, will shed light on the stories behind the science, the challenges they faced, the hopes they held and the ways they persisted despite pervasive inequities.

Some of our speakers focused solely on their work and did not provide as much detail about how being a woman and other intersecting identities impact their work as scientists. However, a few of the speakers conduct research that draws on their identities and personal experiences. Our first lecture was by Kathryn B. H. Clancy, associate professor and director of graduate studies, anthropology, and the Beckman Institute. Clancy divided her talk into two sections, one about her work on systemic sexism and racism in the sciences as well as other systems in society[24] and the other part focused on her work on menstruation.[25] She has researched sexual harassment in science and testified before the US House of Representatives.[26] Her work brings to light all the ways in which women in science are harassed and discriminated against. In her lecture, she spoke about how, in thirty years of research, very little change has been enacted. Clancy hypothesized that because most of these cases are nonviolent, the criminal justice system cannot be as easily deployed for ending harassment, but that a victim-centred restorative justice would be a more beneficial approach. Clancy's research is intersectional, and she studies the ways in which racism in organizations and systems can affect women of colour in more and different ways than white women. She advocates for centring the voices of those who experience racism and sexual harassment, but everyone should be responsible for calling out harmful behaviours, reporting them and/or supporting victims. The rest of the talk focused on her work on menstruation. She provided a brief history of research on menstruation and discussed how most of our understanding over the last 400 years comes from the works of a monk, despite no evidence that he ever medically treated women or performed autopsies. Clancy noted that it has only been more recently that menstruation has been seriously studied and explained several theories on the evolution of menstruation.

Another speaker who discussed how her identities inform her work is Carla Desi-Ann Hunter, associate professor of psychology. Hunter's research focuses on differences of perceived racism by African Americans and Black immigrants.[27] She noticed that racial groups are often treated as a monolith and thus studies differences in experiences among different communities of African descent. As a Black immigrant herself, Hunter is curious about others' experiences. She calls this 'me-search', which was originally perceived negatively and not as real research. However, perceptions about me-search have recently shifted and it is now viewed as valuable and serious work. Her research has demonstrated that the culture in which people grow up does influence perceived racism. Hunter has studied African Americans and immigrants who have worldviews that emphasize individuality, self-reliance and equality but were learned in different social contexts.[28] She has found that immigrants minimized perceived racial discrimination while African Americans who shared these views did not alter their perceptions of racial discrimination. Hunter emphasized that cultural context matters and that we need to actively reject scientific approaches that perpetuate the myth of monolithic racial groups. There is a need for voices from ethnically diverse participants and ethnically diverse researchers; me-search can contribute substantive research in this area and beyond.

Susan Martinis, a biochemist and vice chancellor for research and innovation, was another speaker who discussed the intersection of one's identities and science. Martinis talked about her work with tRNA synthetases and how understanding genetic code has been exciting for her. During her time researching tRNA synthetase,[29] scientists have discovered that tRNA synthetase goes beyond protein synthesis and has been used in many ways throughout evolution. It can be used to show how life evolves and she believes there is much more to be discovered in this area. Martinis also discussed her career path and colleagues who have influenced her. She mentioned that her parents and partner all had a great impact on her career choices and encouraged her in her career (with her partner also being a scientist). Both of her grandfathers were immigrants, and many male members of her family were fishermen. She decided to pursue science because she had worked summer jobs in fishing as a teenager but felt the industry was male-dominated and she could not break in. She decided to study biochemistry instead. Martinis was also fortunate enough to have supportive mentors who inspired her and helped her to succeed and who encouraged her to apply for leadership positions that she would otherwise not have considered.[30] Martinis did apply for these positions and was successful in receiving them.

Zeynep Madak-Erdogan's career choices were in part inspired by her family history. Her paternal grandmother died in childbirth and her maternal grandmother lost her husband and had to raise three children as a single mother who always emphasized the importance of education. Madak-Erdogan thus decided to study women's health. In her talk, she described three projects she has recently been working on with her research team. One focuses on cancer health disparities and the factors that influence these disparities; her group has found socio-economic status has the largest effect.[31] A newer project looks at gestational diabetes that her graduate student was leading, which was in the data-gathering phase. The last project relates to finding diagnostic biomarkers for coronary heart disease.

Madak-Erdogan consistently referred to her team of graduate and undergraduate students by name during the presentation, setting a great example for recognition of all contributors. In her conclusions, she mentioned that there is a greater need for women in academia to work on women's health. She presented data that underscored this point: the ratio of researchers (men and women) to patients was 1:126 and researchers (women only) to patients was 1:262.[32] The statistics for researchers for reproductive health were even more dismal, with 1:3,333 for men and women researchers and 1:6,930 for women only.[33] With evidence that bias in grant funding for research grants leads to projects being funded by issues that matter to the reviewers, this means there are fewer women at the top looking at these grants and possibly making it more difficult for research on women's reproductive health to receive funding.[34]

These stories illustrate the ways that women's experiences and identities not only inform their research but are integral to it. It is important to acknowledge and highlight all the ways in which identities affect scientific research so that under-represented individuals can see both themselves and their communities represented in science. Science and scientific research become more equitable, inclusive and representative of broader society the more under-represented communities enter science and conduct research germane to their communities.[35] As Zeynep Madak-Erdogan noted in her talk, more women need to research and work on women's reproductive health in large part because women are not only under-represented as researchers, but also because women's concerns and bodies have been marginalized in scientific research. Thus, one important takeaway from these lectures is not only the importance of women's representation in science, but also that this representation has an impact on broader society and specifically the ways that women's concerns, lives and bodies gain better representation in the historical record of scientific research. These autobiographical accounts create a much-needed space to consider not only the politics of historical representation, but also the more immediate and even urgent ways that women's representation in this space matters.

Advertising and audience

Each month we sent out the announcement about the forthcoming lecture to our co-sponsors for promotion through their communication channels. We also sent announcements about the talks to any departments with which the speaker was affiliated. This assistance from the co-sponsors proved to be critical in helping to publicize the lecture series. Indeed, 47.6 per cent of the people who registered heard about the lecture series through our co-sponsors' emails.

We also advertised through university-wide newsletters that reached undergraduate and graduate students, and faculty members and staff. About 20 per cent of our audience resulted from these advertisements (see Table 13.1). Some other ways that attendees learned about the lecture series were through social media, professors sharing the announcements with their classes and university calendars. After several lectures, we also began asking registrants if they would be interested in subscribing to an email list for future events. We subsequently shared announcements through this list as well. In the registration form we asked a question about how people heard about the lecture series. We listed several options for hearing about the lecture through the organizers, but not the mailing list. It seems people chose

Table 13.1 Count of how registrants learned about the lecture series for 2020–21 academic year

Method of learning about lecture	Number of responses
Department email	485
EWeek, GradLinks or iNews	207
Social media	109
Organizers	108
Referral	41
Professor	29
Mailing list	13
Library	10
Calendar	9
Website	8

'organizers' if they had heard about the lecture from the mailing list, but a few did write in 'mailing list'. The number of registrants typically ranged from 53 to 185 and approximately 50 per cent of those who registered attended the lecture.

We made sure to convey to our speakers that the audience would comprise individuals who were not specialists and therefore unfamiliar with their discipline, and thus it was important for their talk to be accessible to a broad audience. Through the many different advertising approaches, we were able to have a broad mix of faculty, staff and students registering for the lectures (see Table 13.2). Staff was the highest percentage of registrants (330 attendees), followed by students (309 attendees) and faculty (213 attendees). We also had a large number of alumni (102 attendees) and attendees not affiliated with the university (110 attendees). These numbers are not reflective of unique registrants, as some individuals may have registered for multiple lectures. Alumni found out about the lectures from emails that they still received from their department and from social media. Attendees not affiliated with the university discovered the lectures through social media, email from colleagues and, in some cases, they were family members of the speakers. While we did not specifically target these last two groups, it was encouraging to see many of them attend. It helped us think about future lecture series advertising and the format, since many attendees would not have been able to attend if we hosted these lectures only in person.

Table 13.2 Number of participants who were staff, students, faculty, alumni and not affiliated with the University of Illinois for the 2020–21 academic year

Affiliation with the University of Illinois	Number of responses
Staff	330
Students	309
Faculty	213
Not affiliated	110
Alumni	102

Evaluation

In the summer of 2021, a survey was emailed to everyone who had agreed to be a part of the Women in Science mailing list. We received a total of twenty-two responses. One key takeaway from the survey was why people were coming to the lectures. Most attendees came because they wanted to learn about and support women scientists' work on campus. Many were not in the same discipline but attended because they wanted to support women scientists and thought it was important to document their stories. When asked what they learned from the lecture, many responses did not focus on the details of the research but rather the personal stories that shaped the speakers' research. They stated they learned about being flexible in research, finding good mentors, following your passions and other advice that could apply to any scientist or researcher no matter their discipline or gender identity. While most responses focused on these concepts, some did mention that they attended a lecture because it was in their field and they did share specific facts that they learned from the lecture, such as uses for gold in nanotechnology.

Another question from the survey asked: 'Do you think it is important to actively work on documenting the history of women scientists? Why or why not?' Nineteen people responded with a yes, and three people did not answer the question. Many respondents felt this series had inspired them and made them not feel as alone in the scientific fields as a woman. Others mentioned how important it is to document and share these stories of women scientists so that young girls would be inspired to enter science. Research has shown that having women role models can increase the likelihood of women choosing to enter STEMM fields.[36] Susana Gonzalez-Perez, Ruth Mateos de Cabo and Milagros Sainz demonstrated that 'the optimal way to encourage young girls to pursue emerging high-growth roles, particularly those requiring STEMM math skills, is to expose them to the professional and personal experiences of actual female role models'.[37] Women students' exposure to diverse women role models helps increase students' expectations of value and success with a STEMM career and reduces gender stereotypes typically associated with STEMM fields.[38] The Women in Science Lecture Series allows for women undergraduate students to be exposed to these types of role models and the recorded videos allow middle and high school students to also learn about successful women in STEMM.

Another way we sought to understand engagement was through access statistics for the recordings online. After uploading and publishing each lecture in our digital preservation repository, everyone who registered received a link to the recording in our digital collections portal. The lecture series' records received 116 views as of October 2021. For each lecture we did have several registrants ask if we would be recording the lecture since they were unable to make it live. It is encouraging that people are returning and watching these lectures – it illustrates that these recordings have continued value and interest.

Since writing this paper, we completed a fourth year of the Women in Science Lecture Series. In spring 2024, we sought to update our understanding of how the lecture series is impacting our audience and emailed another survey. This survey received 48 responses from the 900 people it was sent to. The survey showed similar positive results of participants attending to learn more about women scientists and their stories. Most participants were interested in supporting the work of women scientists and were not active research scientists themselves (but often had a background in a scientific discipline). Participants enjoyed that they were able to hear more autobiographical details through these lectures in addition to the research. We also found that 90 per cent of respondents felt this lecture series helped participants understand what people with different identities and backgrounds experienced in the scientific fields and 40 per cent of participants found attending lectures helped them understand others' experiences in their own fields. Finally, the survey also gave us some insight into logistical details of the lecture series that could help it improve, such as better audio quality and having the lectures in different places across campus so they are easier to attend in person.

Next steps and future goals

In addition to this evaluation, we plan to debrief with our co-sponsors and partners and get feedback and ideas for how we can continue building on this work and improving the lecture series. We also plan to share a four-year report with the co-sponsors as a means to not only illustrate the impact of the lecture series, but to reiterate its goals, gather additional feedback and outline a plan for making the lecture series a sustainable ongoing initiative. As of May 2024, we have had a total of 31 speakers, 2,733 registrants and 1,330 attendees, and our email list has grown to 901 subscribers. We have also moved to a hybrid approach so that attendees can join in person or online. In addition to continuing the traditional lecture format, we hosted a panel discussion in response to feedback from the 2021 survey.[39] Above all, we would like to continue creating a broad and diverse collection of women scientists' research. Through support from our co-sponsors and dedicated funding from the Archives' budget itself, we have been able to continue to hire graduate students to help with the logistics, preservation of the lectures and supplemental exhibits.

One of our larger goals is to more systematically understand the ways that women are under-represented in science across our holdings. While we have few personal papers overall, it is important to understand how women are under-represented across identities and how differences in identities can inform archival selection and collection development.[40] Collection development is an activity we hope to do in collaboration with our co-sponsors and the communities with whom we have been engaging as part of this lecture series. As a part of the preservation of the lectures themselves, we will also preserve records related to the management of the lecture series itself, including the digital flyer advertisements, presentations we have given on the lecture series, reports and the survey questionnaires.

A longer-term goal is to reach out to teachers of students in K-12 classes and local schools. This is especially important for girls around the age of twelve, which is when girls start to feel they would not be successful in STEMM fields.[41] We have developed a primary-source toolkit for students to engage with the lectures and digital exhibits to learn more about women in science. The toolkit includes lesson plans and questions to answer while watching the video, investigating digitized primary-source documents from our holdings and hands-on activities where students can be creative and try to understand our speakers' research processes more. We were able to implement these lesson plans in a workshop focused on exploring the natural sciences at Expanding Your Horizons Chicago 2023, an event geared towards middle school girls interested in STEMM.[42] We are also exploring creating 'teacher trunks' that will be available for instructors to check out from the U of I Archives so that they are able to have all the materials they need for these activities.

We will continue our relationships with our speakers, with the goal of being able to share their stories through these recorded lectures, as well as to remind them to think of the U of I Archives as a place to donate their papers so that their materials and stories are not lost. The lecture series is a useful tool to help engage with the community whose work and history we aim to preserve, but an hour-long lecture is not enough to capture the whole story. Papers related to their research and personal journeys are incredibly valuable and we want to make sure that our speakers recognize this and remember to donate their own materials in the future. While we have received donations of other women scientists since launching the lecture series,[43] we hope that this initiative continues to be impactful and to lead to a re-imaging of the U of I Archives as a place where women scientists' papers are welcomed and stewarded. Some speakers have mentioned how frustrating it is to not be able to find much information about important women scientists whose work may have been done quietly or have been largely unacknowledged or credited but was extremely impactful in their field. We want our speakers to recognize that their papers can help make a difference so that it is easier for future scientists and researchers to learn about them and inspire other women scientists to donate their papers.

Conclusion

Women scientists have remained under-represented in archives and museums in distinct and different ways depending on the scientific field of study and a variety of 'social, historical, and cultural reasons':[44] their contributions have sometimes been credited to male colleagues and spouses; sometimes they have been steered to the 'feminine sciences' (which were not always given the same scientific credibility as other sciences); sometimes they have been pushed to the margins of their own fields due to gender biases, discrimination and harassment; oftentimes, anti-nepotism rules in academia meant that women scientists were not given paid or tenure-track positions.[45] There has also been the matter of misperceptions about what kinds of records are valuable, which has led women scientists to discard important documents that provide insight into their experiences and research process and not just the final published work.[46]

Women of colour in particular have had to navigate additional barriers to participate in science,[47] which are important to understanding their impact on record-creation, record-keeping and absences in archives. Finding ways to proactively collect the stories of people who have been left out of the historical record, such as through this lecture series, will help in creating a more inclusive and representative record.[48]

The ways in which women scientists have been under-represented in archives and museums can be complicated and complex, but it is nonetheless important for archivists and museum professionals to begin unpacking these reasons and to be intentional about collection development policies and practices, the outreach programmes they create and who they highlight as being important in history and in science. Initiatives such as the Women in Science Lecture Series can be useful for addressing inequities in holdings, but they should be part of broader programmatic efforts and an institutional ethos that seeks to steward a diverse and inclusive history science. Such efforts can also assist with capturing current science – history as it is being made – and the powerful autobiographical accounts of the diverse scientists who advance science.

Notes

1 While beyond the scope of this chapter, which focuses mainly on archives and museums in the US, a global survey of science documentation in archives and museums would be valuable and provide more fulsome and nuanced insights into these silences.
2 Carrie C. Heitman, for example, discusses the dearth of women in photographic archives of archaeological digs. When women were photographed, they were represented in less-active and less-powerful positions. C. C. Heitman, 'The creation of gender bias in museum collections: Recontextualizing archaeological and archival collections from Chaco Canyon, New Mexico', *Museum Anthropology*, 40: 2 (2017), 128–48.
3 A. Reser and L. McNeill, *Forces of Nature: The Women Who Changed Science* (London: Frances Lincoln, 2021), pp. 7–13.
4 See T. Zanish-Belcher, 'The archives of women in science and engineering and future directions for oral history: Questions for women scientists', *Centaurus*, 54 (2012), 292; and L. Madsen-Brooks, 'Challenging science as usual: Women's participation in American Natural History Museum work, 1870–1950', *Journal of Women's History*, 21: 2 (2009), 11.
5 Archivists have long debated decisions about which appraisal and selection strategies they should employ. Many have questioned the merits of passive approaches to appraisal and collection development. See P. Hohmann, 'On impartiality and interrelatedness: Reactions to Jenkinsonian appraisal in the twentieth century', *American* Archivist, 79: 1 (2016), 14–25, https://doi.org/10.17723/0360-9081.79.1.14.
6 G. Porter, 'Seeing through solidity: A feminist perspective on museums', *Sociological Review*, 43: 1 (1995), 110, https://doi.org/10.1111/j.1467-954X.1995.tb03427.x.
7 Stereotypes that women face in science have been overwhelmingly documented, despite still persisting. These stereotypes impact hiring decisions, job performance evaluations, not to mention advancement in science more generally. See E. Reuben, P. Sapienza and L. Zingales, 'How stereotypes impair women's careers in science', *PNAS*, 111 (2014), 4403–8, https://doi.org/10.1073/pnas.1314788111.

8 A. Fisher and K. Henningsen, 'Women in science through an archival lens', *Transformations: The Journal of Inclusive Scholarship and Pedagogy*, 27 (2017), 158–79, https://doi.org/10.1353/tnf.2017.0015.

9 University of Illinois Archives, https://archives.library.illinois.edu/ (accessed 31 May 2022).

10 B. E. Seely, 'Engineering and the land-grant tradition at the University of Illinois, 1868–1950', in A. I. Marcus (ed.), *Science as Service: Establishing and Reformulating American Land-Grant Universities, 1865–1930* (Tuscaloosa: University of Alabama Press, 2015), pp. 269–96.

11 National Archives and Records Administration, Morrill Act (1862), www.archives.gov/milestone-documents/morrill-act (accessed 25 May 2022).

12 The University of Illinois Archives' first archivist conducted a comprehensive survey of the University of Illinois' scientific activities in seeking to create a set of guidelines for other college and university archives likewise documenting science and technology. See M. J. Brichford, *Scientific and Technological Documentation: Archival Evaluation and Processing of University Records Relating to Science and Technology* (Urbana: University of Illinois at Urbana-Champaign, 1969), https://archives.library.illinois.edu/workpap/Sci-Tech-Documentation.pdf (accessed 25 May 2022).

13 One only has to look at the University of Illinois Engineering's Hall of Fame to see the ways that mostly white men have dominated the narrative about who is important in the University's history of engineering: https://grainger.illinois.edu/alumni/hall-of-fame (accessed 25 August 2019). These portraits, which are also displayed in the Grainger Engineering Library Information Center, illustrate the all-too-pervasive phenomenon of the 'dude wall' in science. See N. Greenfieldboyce, 'Academic science rethinks all-too-white "dude walls" of honor', *NPR* (25 August 2019), www.npr.org/sections/health-shots/2019/08/25/749886989/academic-science-rethinks-all-too-white-dude-walls-of-honor.

14 Women in Science Lectures, Field Museum, www.fieldmuseum.org/our-events/women-science-lectures (accessed 9 September 2021).

15 Association for Women in Science, Speaker's Bureau, https://awis.org/speakers-bureau/ (accessed 24 May 2024); Cranbrook Institute of Science, Women in Science Speaker Series, https://science.cranbrook.edu/women-rock-science/speaker-series (accessed 24 May 2024).

16 *Lady Science*, www.ladyscience.com/ (accessed 26 May 2023); *Lost Women of Science*, www.lostwomenofscience.org/ (accessed 26 May 2024).

17 'Women in Science Lecture Series (Born Digital Records)', University of Illinois Library, University of Illinois Archives, https://digital.library.illinois.edu/collections/4fe01460-e960-0138-74c9-02d0d7bfd6e4-0 (accessed 31 May 2022).

18 One persistent issue in the history of women in science is the lack of credit and recognition they have received, thus making it all the more important to capture and preserve an archival record created by women who can specifically talk about their work. Women have historically (and still today) received less credit for their contributions and ideas, especially in collaborative scientific research. It is therefore important to have more autobiographically oriented

materials created by women. See, for example, M. B. Ross et al., 'Women are credited less in science than men', *Nature*, 608 (2022), 135–45, https://doi.org/10.1038/s41586-022-04966-w.

19 Women in Science Lectures, University of Illinois Archives digital exhibits, https://omeka-s.library.illinois.edu/s/archives/page/women-in-science (accessed 30 May 2022).

20 Michelle Rodrigues and Kathryn Clancy discuss ways women are under-represented across different disciplines and differing, intersecting identities. See M. Rodrigues and K. Clancy, 'Factors that drive the underrepresentation of women in scientific, engineering, and medical disciplines', in R. Colwell, A. Bear and A. Helman (eds), *Promising Practices for Addressing the Underrepresentation of Women in Science, Engineering, and Medicine* (Washington, DC: The National Academies Press, 2020), pp. 37–72, https://doi.org/10.17226/25585.

21 A. A. Philips, C. R. Walsh, K. A. Grayson, C. E. Penney, F. Husain and the Women Doing Science Team, 'Diversifying representations of female scientists on social media: A case study from the women doing science Instagram', *Social Media + Society*, 8: 3 (2022), https://doi.org/10.1177/20563051221113068.

22 I. R. Johnson, E. S. Pietri, F. Fullilove and S. Mowrer, 'Exploring identity-safety cues and allyship among Black women students in STEM environments', *Psychology of Women Quarterly*, 43: 2 (2019), 133, https://doi.org/10.1177/0361684319830926.

23 For example, Brenda Molano-Flores discussed the work of naturalist Mary Davis Treat (1830–1929), 12:32:55, and 12:57:15, https://digital.library.illinois.edu/items/caeefc70-1342-013a-7af9-02d0d7bfd6e4-9 (accessed 31 May 2022).

24 For an example of Dr Clancy's research on sexism and racism, see K. B. H. Clancy, L. M. Cortina, and A. R. Kirkland, 'Use science to stop sexual harassment in higher education', *PNAS*, 117: 37 (2020), 22614–18, https://doi.org/10.1073/pnas.2016164117.

25 For example, some of Dr Clancy's most recent work relating to menstruation has focused on the effects of Covid-19 vaccines on menstrual cycles and women's bodies. See K. M. N. Lee, E. J. Junkins, C. Luo, U. A. Fatima, M. L. Cox and K. B. H. Clancy, 'Investigating trends in those who experience menstrual bleeding changes after SAR-CoV-2 vaccination', *ScienceAdvances*, 8: 28 (2022), https://doi.org/10.1126/sciadv.abm7201.

26 S. J. Toel, 'Anthropology professor testifies before U.S. Congress', College of Liberal Arts & Sciences, University of Illinois at Urbana-Champaign, 7 March 2018, https://las.illinois.edu/news/2018-03-07/anthropology-professor-testifies-us-congress.

27 For example, see C. D. Hunter, A. D. Case, N. Joseph, Y. Mekawi and E. Bokhari, 'The roles of shared racial fate and a sense of belonging with African Americans in Black immigrants' race-related stress and depression', *Journal of Black Psychology*, 43: 2 (2017), 135–58, https://doi.org/10.1177/0095798415627114.

28 See, for example, A. D. Case and C. D. Hunter, 'Cultural racism-related stress in Black Caribbean immigrants: Examining the predictive roles of length of residence and racial identity', *Journal of Black Psychology*, 40: 5 (2014), 410–23, https://doi.org/10.1177/0095798413493926.
29 For an example of Martinis' research on tRNA synthetases, see R. S. Mursinna and S. A. Martinis, 'Rational design to block amino acid editing of a tRNA synthetase', *Journal of the American Chemical Society*, 124: 25 (2002), 7286–87, https://doi.org/10.1021/ja025879s.
30 More women have entered leadership positions in higher education, though they still face significant challenges. Encouragement is one important aspect in fostering women leaders. However, institutions 'need to more actively encourage and develop women leaders'. See C. D. Williams, 'Investing in and supporting women's leadership in higher education', *HigherEdJobs*, 28 May 2021, www.higheredjobs.com/Articles/articleDisplay.cfm?ID=2726&Title=Investing%20In%20and%20Supporting%20Women%E2%80%99s%20Leadership%20in%20Higher%20Education (accessed 28 May 2021).
31 See, for example, A. Santaliz Casiano, A. Lee, D. Teteh, Z. Madak Erdogan and L. Treviño, 'Endocrine-disrupting chemicals and breast cancer: Disparities in exposure and importance of research inclusivity', *Endocrinology*, 163 (2022), https://doi.org/10.1210/endocr/bqac034.
32 From L. Latimer, 'Calling all women: The disparity in researchers dedicated to women's health', *Ellevate*, www.ellevatenetwork.com/articles/8081-calling-all-women-the-disparity-in-researchers-dedicated-to-women-s-health (accessed 27 January 2023).
33 Latimer, 'Calling all women'.
34 Latimer, 'Calling all women'.
35 D. Kozlowski, V. Larivière, C. R. Sugimoto and T. Monroe-White, 'Intersectional inequalities in science', *PNAS*, 119: 2 (2022), https://doi.org/10.1073/pnas.211306711.
36 S. Gonzalez-Perez, R. Mateos de Cabo and M. Sainz, 'Girls in STEM: Is it a female role-model thing?' *Frontiers in Psychology*, 11, Article 2204 (2020), 2, https://doi.org/10.3389/fpsyg.2020.02204.
37 Gonzalez-Perez, Mateos de Cabo and Sainz, 'Girls in STEM', p. 15.
38 Gonzalez-Perez, Mateos de Cabo and Sainz, 'Girls in STEM', p. 2.
39 For example, we hosted a panel discussion on the film *Picture a Scientist* during Women's History Month 2022. See: www.pictureascientist.com/ (accessed 31 May 2022).
40 Elizabeth Novara discusses developing appraisal and collection development criteria for the papers of women politicians. Developing criteria for the selection and appraisal of women scientists' papers, especially to ensure that women from under-represented groups are documented, is also important. See E. Novara, 'Documenting Maryland women state legislators: The politics of collecting women's political papers', *American Archivist*, 76: 1 (2013), 196–214, https://doi.org/10.17723/aarc.76.1.u57m635512311v48.
41 Gonzalez-Perez, Mateos de Cabo and Sainz, 'Girls in STEM', p. 2.

42 Expanding Your Horizons Chicago, https://eyhchicago.wordpress.com/ (accessed 27 May 2024).
43 For example, recently donated materials from women scientists include the papers of Donna J. Cox, faculty member and an innovator in scientific visualization, https://archon.library.illinois.edu/archives/index.php?p=collections/controlcard&id=13152; Rosalyn Sussman Yalow, alumna and Nobel Prize-winner in physics (1977), https://archon.library.illinois.edu/archives/index.php?p=accessions/accession&id=1010.
44 T. Zanish-Belcher, 'Documenting the sometimes invisible: Working with women scientists', *Humanities Collections*, 1 (2001), 4.
45 Zanish-Belcher, 'Documenting the sometimes invisible', pp. 4–5.
46 Zanish-Belcher, 'Documenting the sometimes invisible', pp. 10–11.
47 S. J. Halsey, L. R. Strickland, M. Scott-Richardson, T. Perrin-Stowe and L. Masenburg, 'Elevate, don't assimilate, to revolutionize the experience of scientists who are Black, Indigenous and people of colour', *Nature Ecology and Evolution*, 4 (October 2020), 1291–93, www.nature.com/articles/s41559-020-01297-9.
48 The authors previously discussed the importance of proactively documenting women in science. See B. G. Anderson and K. A. Wilson, 'Remembering women scientists: The case for proactively documenting women in science', *Collections*, 18: 4 (2022), 453–78, https://doi.org/10.1177/15501906221129336.

14

How to do science as a woman and laugh? Insights and lessons from Hungary

Andrea Pető

It is difficult to write the closing chapter to this multifaceted yet coherent volume. It is multifaceted because the examples and case studies are drawn from many countries around the world and coherent because very similar patterns emerge about the position of women in science, engineering and medicine in the twentieth century. The contributions analyse trends in women's participation in STEMM, complicating our understanding of what it meant to be in/visible in these fields and documenting the challenges women encountered as well as the strategies they devised to overcome them. Many of these hurdles have changed little in character over the decades. In this final chapter, which also functions as an epilogue, I will add another case study to this impressive collection, that of Hungary, to illustrate continuities in the challenges women have faced and suggest ways forward, discussing and drawing inspiration from the strategies and tactics they have employed to overcome them.

Dilemmas of women's participation in science

I want to start with the photograph below (Figure 14.1), taken in a university lecture hall in Budapest, to argue that the frame within which we talk about women in science matters. When I first saw the image, I was struck by the year, 1929. There is much debate today about the Horthy era (1919–44), with some portraying it as a static, stagnant neo-baroque world built on discrimination against individual citizens. Others, like illiberal politicians, idealize it, imagining it as a time of social peace and progress.[1] Drawing on an excellent recent book by Balázs Sipos and Barbara Papp, which examines the history of women graduates during this period, we can say that a decisive change in the acceptance of modern womanhood took place in those years.[2] Let us take the example of Margit Techert, Zoltán Magyaryné (1900–45) who, as editor of *Magyar Női Szemle* (Hungarian Women's Review), played a decisive role in the widespread recognition of

Figure 14.1 Geographer Jenő Cholnoky (1870–1950) with his students in a classroom at Pázmány Péter University on Museum Boulevard, Budapest, 1929 (photograph donated by Tamás Cholnoky, Fortepan 29841, open access, https://fortepan.hu/hu/photos/?id=29841)

women as intellectuals and in paving the way for a change of the norm; and changing the norm undoubtedly achieved some degree of success.[3] In 1941, the University of Budapest supported her application for a private teaching post. At the time, debates centred on the question of whether women were suitable for university study at all. If so, what subjects should they study? Did women have aptitudes for study at all? If a woman's purpose and duty was to raise a family, was it not inconsistent to attend university? Could a degree and a family be reconciled? These questions may sound familiar to today's female students. My point is not 'how much would have to change here for nothing to change', as a well-known saying goes, but rather, that the questions that remain unanswered are still with us today.

To understand the origins of the dilemma, we must go back to the famous and much-quoted inaugural address of the great Hungarian writer Imre Madách (1823–64) to the Hungarian Academy of Sciences (*Magyar Tudományos Akadémia*, MTA), which illustrates the nineteenth-century point of view:

> The woman develops early, but never reaches full male maturity; she comprehends and learns more easily but lacks creative genius and does not rise to

the leading spirits of mankind. She is always a sufferer, never the penetrating element, and therefore, although she supplies the kindest contingent of dilettantism, she has never substantially advanced art and science. This irrefutable fact cannot be attributed to a contrary education. Most men who have been possessed by genius have made their way to their calling even by the most contrary paths and have triumphed, because the spirit is stronger than any earthly obstacle.[4]

But this difference between men and women can also be interpreted differently. Who has not heard the argument that we need more women in politics because women are more peaceful and more willing to compromise, and therefore their participation is good for the competitive work environment? Some argue, as Sandra Harding does, that women's particular epistemological perspective – their greater empathy and intuitive insights – enrich the process of scientific cognition.[5]

To further probe this paradox of equality or difference, I would like to quote Ida Bobula, a forgotten pioneer of Hungarian women's historiography. Unfortunately, only her works on Sumerian–Hungarian kinship are cited, although Bobula did much to promote women's education. Following the family tradition, Bobula (1900–81), one of the first Hungarian women scholars, went to the United States in the 1920s, where she studied sociology for two years. In 1924, she was awarded a doctorate in history in Budapest and in 1939, she became a lecturer at Tisza István University in Debrecen. Between 1920 and 1924, she was president of the Women's Section of the National Association of Hungarian University and College Students, and later a member of the staff of the *Hungarian Women's Review*, a periodical of the Association of Hungarian Women Graduates of University and College. She did not have a research position in Hungary, but she worked for the Ministry of Religion and Public Education, and then directed a girls' college in Budapest until 1944. However, her work, including her monograph *Women in the Hungarian Society of the 18th Century* (1933), continues to be recognized in Hungarian historiography. In her words, we can already trace the change in the perception of the role of women in science and the fact that Madách's views were no longer tenable: 'I advise anyone who has a serious and sacrificial vocation to take up his cross and go to college, even if it is a woman. Being a woman is a difficult handicap in today's life – you have to know that! – but that's no reason to give up the fight.'[6] What is this 'handicap'? And what can be done about it? In this chapter, I argue that a triple start strategy is one way to try to overcome this obstacle: increasing the number of women, changing the institutions and changing knowledge.

To return to the picture at the beginning of this chapter, taken in 1929: if you look closely, there was an impressive number of women in the crowded lecture hall in Professor Cholnoky's Earth science class. They are clearly visible in the photograph, even if they have not necessarily been so in historiography. This immediately raises questions, especially considering the data on women's participation in science today. First, how were women admitted to the Faculty of Geography in the first place? And second, what happened to them later?

In Hungary, certain areas of higher education were opened to women in 1895. Even before that, there were women scientists who were allowed to study because of their family situation or because they were 'girls brought up as boys', meaning that if the desired male child in the family turned out to be a girl, she was given the opportunity for an education, which was essential for social mobility. Some women had the opportunity to study science alongside their husbands. But women were not involved in the creation of institutional and institutionalized knowledge. This changed briefly with the establishment of the Hungarian People's Republic in 1918, which stipulated that the faculties of all universities (except the theological ones) had to admit deserving women. This seemed to settle the question of women's suitability for intellectual professions in general and for fields that required 'male intelligence' (skills) in particular: the advocates of the essentialist view – who, like Madách, believed that the biological-spiritual differences between women and men determined women's place in society – were 'defeated' by the supporters of the constructivist view, such as Ida Bobula a good half-century later. However, the term 'defeated' does not accurately describe the struggle that went on for many decades.

Various groups in the emancipation movements demanded equal rights for women – in politics, education and culture – in the name of equity. Not different, but equal rights. The question of women's education and access to science is a question of democracy. But if you look at the number of women in scientific research and higher education, women are not visible or do invisible work. They are assistants, secretaries and wives. For example, after Professor Cholnoky retired, his successor as the professorial chair in the department was also a man. Moreover, Barbara Papp and Balázs Sipos argue convincingly that the question of 'modern' women cannot be narrowed down only to those with a university degree, because then we would overlook the fundamental change that took place in Hungarian society in the period between the two world wars regarding the role of women.[7]

The canonical narrative of the history of women in science in Hungary states that women were allowed to study from 1895, many of them graduated and then, due to World War I, more and more professional opportunities were open to them. With the annexation of Bosnia, for example, Hungarian women doctors were given jobs there even before the war, as only women were allowed to examine the local female population. During World War I, however, not only women doctors, such as Vilma Hugonnai

(1847–1922), the first female doctor in Hungary, but also volunteer nurses learned the horrors of war and the material and psychological benefits of independence. Unfortunately, this was followed by the *numerus clausus* in 1921, the decree originally aimed at reducing the number of women in public life and in higher education, which later became the first political antisemitic law in Europe, limiting the number of Jewish students admitted to the university. As a result of this law, women were not admitted to universities for years. They were allowed only as private lecturers. Those women who had participated in the emancipation movement and thus played an active role in 1918 and 1919, such as Laura Polányi (1882–1959) or Irén Götz (1889–1941) – the first Hungarian chemistry professor and colleague of Marie Curie – left the country. None of them, however, obtained formal academic positions in universities in the countries to which they emigrated, the United States and the Soviet Union, respectively.

In 1946, the political space opened for women. Erzsébet Andics (1902–86) became the first woman to be elected a member of the Hungarian Academy of Sciences in 1950 and the massification of higher education began, followed only gradually by the rise of women to leadership positions. But from then on, women could also become civil servants. After 1989, not much changed; on the contrary, things turned worse.[8] The first reports of the Helsinki Group showed that, in the former socialist countries, more women were represented in the sciences due to equality policies imposed from above. Yet, they were less likely to be promoted to senior positions. After 1989, when the state no longer intervened, the number of women declined sharply, especially in senior positions. At the same time, there has been a steady drain of funds from higher education and women have been socially disadvantaged in the pursuit of excellence. As a member of the Horizon 2020 Advisory Board, I have repeatedly pointed out, to no avail, that EU research policies and frameworks structurally discriminate against Eastern European academics and researchers. Now that gender equality in society has unfortunately become a political battleground and many use this issue as a political hate card, the situation is even more complicated.[9]

The same sad chronology of the status of women in science can also be told as a success story. That is precisely what I want to propose in this epilogue. The lives of the women smiling at us in the 1929 photograph have certainly changed, even if they have not become deans or professors. I am sure many of them experienced bitterness because they could not join the university and pursue academic work. Is it possible to change that state of affairs, to eliminate the bitterness or 'handicap'? I believe so. I will discuss this in the last part of this chapter.

Changes in the narrative that are significant

Change is not easy when it comes to the position of women in STEMM. In the Horizon 2020 proposals, which I am most familiar with, gender equality is considered by focusing on three areas: the composition of research teams, the composition of decision-making teams (40 per cent from the underrepresented gender should be present) and the integration of gender into the content of research. This is a new integrative approach that, as I mentioned earlier, can serve as a recipe for improving the situation. In short, we need to increase the number of women, change the institutions and change the knowledge itself. If we want to improve the current situation, we need concerted change in all these three areas. This reform process is not something we have just invented: it has a long history that goes back to figures like Vilma Hugonnai and Ida Bobula. This history also includes the women who smiled into the camera in the auditorium on Museum Boulevard in Budapest in 1929 and are now part of the Fortepan collection of photographs.[10]

Of course, this triple change – of numbers, institutions and knowledge – is not without its challenges. In 2004, for example, the Directorate General for Research and Innovation of the European Commission published a report on women and science in former socialist countries. The Hungarian contribution by Dóra Groó describes how, as a participant in the distribution of research funds, she increasingly noticed that male researchers were applying for grants, while the women who wrote the proposals were, at best, working their way into the successful male scientist's grant. This example illustrates why science is a particularly difficult area in which to implement gender equality in society. After all, the female researchers who wrote the successful proposals instead of their male bosses should be grateful: had they applied under their own names, the chances of securing the grants would have been smaller. Discrimination in science is both elusive, perhaps even intangible, and internalized.[11]

But why should it be a problem that there are very few women in Hungarian science, especially in STEMM, not to mention in leading positions in academia? Because gender equality is a core value of the European Union; this should be cause for cautious optimism for the future. The EU's core value is equality, which means that discrimination, whether direct or indirect, on the basis of gender is a violation of human rights. Excellence, which is another basis of EU science policy, cannot mean excellence of 'men' only.

There are three levels of faculty development where the problem arises: undergraduate education, PhD training and career development. Even though more women are entering universities today, fewer and fewer are getting opportunities throughout their careers. Earning a doctorate and then starting a first job comes at a time when women are often left to fend for themselves under the pressure of family responsibilities. Another major watershed is the appointment as a university professor. How can we change the ever-shrinking pyramid-like academic selection system? When asked this question in an interview, Maria Ormos (1930–2019), a member of the Hungarian Academy of Sciences, stated half-jokingly that it is clear why fewer and fewer women are being appointed to socially prestigious and secure, but increasingly lower-paid, university positions: because women become frustrated over time. Or, irony aside, to continue Ida Bobula's thought, they cannot overcome this particular handicap. But let us call it what it is: a structural handicap.

There is another framework within which we can and should talk about the role of women in science, namely that of efficiency. Another fundamental principle of EU science policy, efficiency underlines that no economy can afford the luxury of not making the best use of women's skills for its economic interests. Put differently, the female geology students seen smiling in the Fortepan picture in 1929 should have been allowed to work in the profession they mastered as a result of long years of university education. I do not necessarily agree with the neoliberal argument that evaluates science based on 'what can be measured is what matters', but this can become an argument for those who seek to identify statistically demonstrable inequalities and change the situation. A regulated, quantified assessment system helps rather than hinders women's progress. At the same time, the quantified system of requirements for obtaining a PhD helps women and even encourages them to apply for it. Without such a clear set of requirements, there would be much more room for informal relationships, and this, we know from research, encourages male advocacy.

Research at Yale University, for example, found that university lecturers – both men and women – who served on an application committee consistently rated applications submitted by men higher than those submitted by women.[12] Researchers at Maastricht University found that female lecturers were rated three times lower by students than male lecturers.[13] They conclude, and I can only agree, that this is not conscious misogyny, probably not even conscious discrimination, but the result of stereotypes about the social roles expected of women, that is, the operation of the cultural unconscious, influenced by the hidden curricula of our socialization. Take, for example, meetings attended by both male and female academics. When it comes to deciding who takes the minutes – a person who is clearly unable to participate meaningfully in the meeting – or who pops out to make the coffee or looks up where the coffee is, there are often puzzled male glances in the direction of the attending women. From this perspective, however, it is perhaps less clear that academic careers and promotions are based solely on principles of meritocracy.

Valéria Csépe is a former deputy secretary general of the Hungarian Academy of Sciences, an academic and former leading Hungarian delegate to the expert forum of the Helsinki Group on Gender in Research and Innovation (now Standing Working Group on Gender in Research and Innovation, SWG GRI). A Helsinki Group's report shows that:

> the proportion of women in senior positions in scientific institutions in Hungary has almost doubled in four years, from 9 percent in 2010 to 17 percent in 2014. The number of research institutions in Hungary that have a gender equality plan is growing faster than the EU average and the proportion of women among PhDs is also increasing faster than that of men.[14]

Valeria Csépe played a key role in creating the Framework Program for Equal Opportunities, which was launched by the Hungarian Academy of Sciences in 2009. The programme, based on the German model, extends the age limit for female researchers with children under the age of ten by two years per child for applications tied to age or date of scientific graduation. This was an important step in addressing the structural disadvantage of women through a key instrument.

The handicap mentioned by Ida Bobula is not only structural. There is also a supply without demand, meaning that more women want to go into academia, but do not have the opportunity. That is something this programme can improve. In addition, there is another factor to this handicap: it is internalized. To understand and improve this situation, we need to analyse women's life paths. Vilma Hugonnai, for example, had all the privileges a woman could have at the end of the nineteenth century: she came from a wealthy family, was healthy and had a supportive husband, if only initially. She pioneered the idea that when we say 'doctor', we do not just think of men. On the other hand, if we look at the photo from 1929, Professor Viktor Cholnoky fits the stereotype of the professor: he is old, wears a white beard and exudes confident scientific knowledge. While the image of the older man culturally suggests wisdom, knowledge and authority, the image of the older woman has no such associations – hopefully only for the moment. Wisdom, knowledge and authority are essential elements of the stereotypical image of the professor and it is this stereotype that still makes it difficult to recognize and accept older women professors.

This is not easy to live with or even survive, as I hope to show by drawing on my own personal experience. I worked for years as a mentor for the German Catholic foundation Cusanuswerk. The situation of women in German higher education is much worse than in Hungary, although the prestige and income of those working in higher education are considerably higher. For this reason, the Cusanuswerk launched a programme inviting PhD candidates to apply each year and bring their doctoral supervisor with them. Every six months, the foundation organized training sessions for the doctoral students, where they learned how to deal with difficult issues and decisions, for example, if they felt they were not getting the institutional support they deserved at their university or if they had a different perspective on a professional topic than their mentor. As subject leaders, we engaged in communication exercises, such as how to communicate effectively with students when their research was not progressing well.

If we look at the smiling women in the 1929 photograph, we may understand that it is the joy, security and integrity of the individual in intellectual work that attracts women to study. When that is lost because a male professor, students or administrators do not evaluate them fairly, supportive institutions are needed. Long-term changes in gender equality in society can only be expected from the implementation of EU policies, although it is in this particular area that EU regulation has little impact. Optimists are right, however, when they say that, thanks to EU norms, we at least have statistics.

I would now like to return to the strategy for change I mentioned above, which could take three forms – all top-down, all requiring strong political intervention. It should also be noted that all three are subject to opposition and criticism, proving that there is no magic formula for changing inequalities and that solutions are always context-specific.

The first tool for change is the principle of equal treatment, which should be institutionalized to ensure that men and women are treated equally. This applies to direct discrimination, which has become rare thanks to the many achievements of women's movements over the last 150 years. It might be tempting to say that policymakers can sit back with satisfaction and claim that there is not much left to be done. However, the education system, teaching materials and textbooks play an extremely important role in instilling gender stereotypes in society. As long as textbooks feature girls cooking and boys reading, we cannot expect much to change. If mathematics teachers explain equations differently for girls than boys, implying that it is harder for girls to understand, nothing will really change.

The second tool, equality of opportunity, can exist in science only if the distribution of scarce resources, goods and social burdens – which are always, conceptually, necessarily scarce – is guided by considerations that are relevant in terms of the competition for the goods in question. Thus, there are relevant elements such as race, religion, class, gender, disability, sexual orientation, ethnicity or other extrinsic factors that may impair the chances of success of certain competitors. Achieving equal opportunity would require 'positive discrimination' or, to use the more common term today, 'affirmative action'. These are measures to prevent or eliminate discrimination, or to compensate for disadvantages, targeting stigmatized groups at risk of discrimination. But the introduction of affirmative action is not accepted by either the conservative or the liberal part of the public, whether in Hungary or elsewhere. Who has not heard, 'I do not want to be a quota woman because I work hard' or 'We cannot do that to X and appoint her on the basis of a quota, they are all behind her saying she is a quota woman'?

But quotas only – and I stress this – give a chance to the candidate who is under-represented in that category if they have the same merits. There is also resistance because some people believe that the inequalities rooted in society cannot be eliminated by legal measures. Anti-discrimination laws, passed under pressure from the EU, declare that differences in opportunity at birth cannot be eliminated by legal means. This liberal interpretation of the law, by its very wording, calls into question the legitimacy of the entire law and precludes the use of supportive instruments, which historically have been shown to be the most effective method of addressing social inequalities. Equality in science policy is also opposed by those who view social inequality as merely a legal issue. These are those who consider the problem of inequality to be solved because it is already de facto regulated at the legal level and who do not recognize that 'women's rights' are distinct from human rights. However, legal measures alone cannot be expected to work miracles; they are no substitute for decisions at the institutional level, the level that most generates and legitimizes inequalities in public perception. This situation can be improved primarily through a gender-sensitive redesign of educational content. The main goal of such a redesign is gender equality in the content areas of education and research.

Another argument against the application of equal opportunity policies, especially in science, is that there are natural inequalities in talents and abilities among people, so the principle of equal opportunity, which builds on natural inequalities, does not eliminate them but increases them. Moreover, opponents of equal opportunity policies argue, support measures also imply that people in need of support are unable to perform well without help. This argument is also used by women researchers who express a visceral aversion to the so-called 'feminist' approach because they feel that their professional and scientific performance is compromised, and they are by no means supportive along the lines of women's solidarity. But 'natural' differences are relative, not self-evident, and therefore all inequalities arise from social institutions and practices, not 'by themselves', naturally. To take up the example of Dóra Groó mentioned earlier, the feudal, patriarchal nature of the institutional system of Hungarian research means that women researchers have a better chance of long-term scientific success if they choose the invisible status of 'servant' rather than seeking individual, that is, institutionally recognized, laurels.

A third means of achieving gender equality in society is gender mainstreaming, which places gender equality at the heart of organizations, institutions, programmes, policies and practices. There is a separate body of literature on the critique of gender mainstreaming. Part of the critique is that those who want to transform everything are in fact transforming nothing. Thus, the desire to change is expressed only in rhetorical phrases. The other criticism concerns the extent to which the Western European policy strategy of gender mainstreaming can be used to change the post-communist situation in Eastern Europe. With its membership in the EU, Hungary, like the other former socialist countries, had to adopt the most advanced policy mechanism currently available for implementing gender equality. However, its effectiveness is undermined, limited and alienated by the way it is implemented, which is often top-down, without consulting civil society organizations or even considering the views of experts in the field.

As long as science policymakers, most of whom are men, perceive the 'female' presence as a loss and a threat to themselves, little change is likely. Even our great-grandmothers would not have been able to win women's suffrage, not so long ago, if it had not been for supportive men like those from the Men's League for Women's Suffrage. Vilma Hugonnai could not have gone to school if her husband had not let her.

Conclusions

Hungary's membership in the European Union only provides an opportunity to improve the disadvantaged position of women in the field of science, because in post-communist countries there is a strong culture of formal, alibi-type implementation at the level of rhetoric. Implementation of international values at the national level is only possible through networking, continuous lobbying and open advocacy of values. It seems that, in the short term, this can only be achieved through the pragmatic argument that science that does not discriminate based on gender is more effective and generates more money. In 2022, the Hungarian National Audit Office (Állami Számvevőszék) published the research report 'Pink Education', which showed the number of women higher-education graduates and concluded that women's higher education is a waste of money. We should ask ourselves: why are these nineteenth-century issues returning? Moreover, the recent initiative to introduce an oral entrance exam for medical schools is in fact an instrument of indirect discrimination, aimed at reducing the number of women admitted. In future, only a sense of shared profit and strong economic arguments may be the key to change among scientific decision-makers, as a more democratic and efficient scientific life is in the long-term interest of all. Perhaps, then, women researchers will dare to apply for funding under their own names and in their own right, in the hope of success. The woman scientist will not be an eccentric freak or an exotic 'Other', but a simple working woman. Of course, that does not solve the problems, it just relegates them to a different place. The cheerfulness and sense of community that emanates from the 1929 photo with which I began this chapter should give us hope.

Notes

1 A. Pető, 'The illiberal memory politics in Hungary', *Journal of Genocide Research*, 24: 2 (2022), 241–49.
2 B. Papp and B. Sipos (eds), *A modern diplomás nő a Horthy korban* [*The Modern Graduate Women in the Horthy Era*] (Budapest: Napvilág, 2017).
3 Margit Techert was her name before marriage, to which her husband's name was then attached, in this case Zoltán Magyary.
4 I. Madách, 'A nőről, különösen esztétikai szempontból (Székfoglaló értekezés az Akadémián)' [About women, especially from an aesthetic point of view (Essay on the occasion of becoming a member of the Academy)], in I. Madách, *Válogatott művei* [*Selected Works*] (Budapest: Szépirodalmi Kiadó, 1958).
5 S. Harding, *Whose Science? Whose Knowledge?* (Ithaca, NY: Cornell University Press, 1991).
6 I. Bobula, 'A nő és a hivatal' [The woman and the office], *Új Idők* [*New Times*], 4 (1937), 123.
7 Papp and Sipos, *The Modern Graduate Women in the Horthy Era*.
8 V. Paksi, 'Miért kevés a női hallgató a természet- és műszaki tudományi képzésekben? Nemzetközi kitekintés a "leaky pipeline" metaforájára' [Why are there few female students in natural and technical science courses? An international perspective on the "leaky pipeline" metaphor], *Replika*, 85–86 (2014), 1–2, 193–214; B. Nagy and V. Paksi, 'A munka-magánélet összehangolásának kérdései a magasan képzett nők körében' [Issues of work-life balance among highly educated women], in Z. Spéder (ed.), *A család vonzásában: Tanulmányok Pongrácz Tiborné tiszteletére* [In the gravitational pull of the family: Studies in honour of Tiborné Pongrácz] (Budapest: KSH Population Studies Research Institute, 2014), pp. 159–75.
9 E. Kováts and A. Pető, 'Anti-gender movements in Hungary: A discourse without a movement?' in R. Kuhar and D. Paternotte (eds), *Anti-Gender Campaign in Europe* (Lanham, MD: Rowman and Littlefield, 2017), pp. 117–33.
10 Fortepan is an open-access photo collection started by Szepessy Ákos and Tamási Miklós in 2010.

11 See also A. Pető, 'Miért marad továbbra is alacsony a nők aránya a magyar tudományban az EU csatlakozás után is?' [Why does the percentage of women in Hungarian science remain low even after joining the EU?], *Magyar Tudomány*, 8 (2006), 1014–17.
12 M. Cummings, 'Study shows gender bias persists in performance-driven industry', *Yale News* (24 October 2017), https://news.yale.edu/2017/10/24/study-shows-gender-bias-persists-performance-driven-industry (accessed 21 January 2025).
13 'Research suggests students are biased against female lecturers', *The Economist* (11 September 2017), www.economist.com/news/science-and-technology/21729426-how-long-does-prejudice-last-research-suggests-students-are-biased-against.
14 A. Lipinsky, *Gender Equality Policies in Public Research: Based on a Survey among Members of the Helsinki Group on Gender in Research and Innovation, 2013* (Luxembourg: Publications Office of the European Union, 2014), https://data.europa.eu/doi/10.2777/65956.

Select bibliography

Archives

Archive of the Office of the President of the Republic, Prague, Czech Republic
Archives of Wuwei County, Anhui Province, China
Bentley Historical Library, University of Michigan, Ann Arbor, MI, USA
Clara Park personal archive, Williamstown, MA, USA
Delfis Archival Center, Athens, Greece
Illinois Distributed Museum, Urbana, IL, USA
The Institution of Engineering and Technology Archives, London, UK
Masaryk Institute and Archives of the Czech Academy of Sciences, Archive of the Institute of Tomáš Garrigue Masaryk, Prague, Czech Republic
Museum of Czech Literature, Prague, Czech Republic
National Archives of the Czech Republic, Prague, Czech Republic
National Archives of Romania, Bucharest, Romania
Shanghai Municipal Archives, Shanghai, China
Sichuan Provincial Archives, Chengdu, China
State Regional Archives, Písek, Czech Republic
University of Illinois Archives

Oral history interviews

Anne Hardgrove, oral history interview with Lady Irwin graduate, New Delhi, 1996.
Anne Hardgrove, oral history interview with Flower Silliman, Zoom, 5 April 2022.

Literature

Abir-Am, P. G. and D. Outram, eds. *Uneasy Careers and Intimate Lives: Women in Science, 1789–1979*. New Brunswick, NJ: Rutgers University Press, 1987.
American Men of Science, A Biographical Dictionary, The Physical and Biological Sciences (L-O). New York and London: R. R. Bowker Company, 1966.
Anderson, B. G. and K. A. Wilson. 'Remembering women scientists: The case for proactively documenting women in science.' *Collections*, 18: 4 (2022), 453–78.
Andrews, B. *The Making of Modern Chinese Medicine, 1850–1960*. Vancouver: University of British Columbia Press, 2014.
Apple, R. *Perfect Motherhood: Science and Childrearing in America*. New Brunswick, NJ: Rutgers University Press, 2006.

Appel, T. 'Physiology in American women's colleges: The rise and decline of a female subculture', *Isis*, 85: 1 (1994), 26–56.
Appel, T. 'Writing women into medical history in the 1930s.' *Bulletin of the History of Medicine*, 88: 3 (2014), 457–92.
Archer, L., E. Dawson, J. Dewitt, A. Seakins and B. Wong. ' "Science capital": A conceptual, methodological, and empirical argument for extending bourdieusian notions of capital beyond the arts.' *Journal of Research in Science Teaching*, 52: 7 (2015), 922–48.
Arikawa, Y., ed. *Joshi rigaku kyōiku o rīdo shita josei kagakushatachi: rekimeiki, Meijiki kōhan kara no kiseki* [Women Scientists Leading Science Education among Women: From the Dawn of Women's Participation, or the Late Meiji Period]. Tokyo: Akashi-shoten, 2013.
Arnold, D. *Everyday Technology: Machines and the Making of India's Modernity*. Chicago, IL: University of Chicago Press, 2013.
Athanasiou, A. 'Bloodlines: Performing the body of the "demos," reckoning the time of the "ethnos." ' *Journal of Modern Greek Studies*, 24 (2006), 229–56.
Attewell, G. N. A. *Refiguring Unani Tibb: Plural Healing in Late Colonial India*. Hyderabad: Orient Longman, 2007.
Avdela, E., M. Papagiannaki and K. Sklaveniti. Έκτρωση 1976–1986: Το χρονικό μιας διεκδίκησης' [Abortion 1976–1986: The timeline of a claim]. *Dini Feminist Magazine*, 1 (1986), 4–29.
Bama. *Karukku*. New Delhi: Oxford University Press, 2014.
Babiracki, P. and K. Zimmer, eds. *Cold War Crossings: International Travel and Exchange across the Soviet Bloc 1940s–1960s*. College Station, TX: A&M University Press, 2014.
Babiracki, P. and A. Jersild. *Socialist Internationalism in the Cold War: Exploring the Second World*. London: Palgrave Macmillan, 2016.
Badinter, E. *Mother Love: Myth and Reality*. New York: Macmillan, 1981.
Bailey, P. J. *Gender and Education in China: Gender Discourses and Women's Schooling in the Early Twentieth Century*. New York: Routledge, 2007.
Banerjee, M. *Power, Knowledge, Medicine: Ayurvedic Pharmaceuticals at Home and in the World*. Hyderabad: Orient Blackswan, 2009.
Banerjee, S. *Men, Women, and Domestics: Articulating Middle-Class Identity in Colonial Bengal*. New Delhi: Oxford University Press, 2004.
Bangham, J., X. Chacko and J. Kaplan, eds. *Invisible Labour in Modern Science*. London: Rowman and Littlefield Publishers, 2022.
Barahona, A. 'Transnational science and collaborative networks: The case of genetics and radiobiology in Mexico, 1950–1970.' *Dynamis*, 35: 2 (2015), 333–58.
Barahona, A. 'Local, global and transnational perspectives on the history of biology.' In M. R. Dietrich, M. E. Borrello and O. S. Harman (eds), *Handbook of the Historiography of Biology, Historiographies of Science 1*, pp. 371–91. Cham: Springer, 2021. https:/doi.org/10.1007/978-3-319-74456-8_17-1.
Barnes, N. E. *Intimate Communities: Wartime Healthcare and the Birth of Modern China, 1937–1945*. Berkeley: University of California Press, 2018.
Barrow, M. V., Jr. 'Carson in cartoon: A new window onto the noisy reception to Silent Spring.' *Endeavour*, 36: 4 (2012), 156–64.
Băluță, O. *Gen și putere: Partea leului în politica românească* [Gender and Power: The Lion's Share in Romanian Politics]. Iași: Polirom, 2006.
Behringer, M. P. 'Women's role and status in the sciences: An historical perspective.' In R. Berger (ed.), *Ayurveda Made Modern: Political Histories of Indigenous*

Medicine in North India, 1900–1955, pp. 4–26. Hampshire: Palgrave MacMillan, 2013.

Berger, R. 'Between digestion and desire: Genealogies of food in nationalist North India.' *Modern Asian Studies*, 47: 5 (2013), 1622–43.

Bettelheim, B. *The Empty Fortress: Infantile Autism and the Birth of the Self*. New York: Free Press, 1967.

Bix, A. S. *Girls Coming to Tech! A History of American Engineering Education for Women*. Cambridge, MA: The MIT Press, 2014.

Blackwell, J. *No Peace Without Freedom: Race and the Women's International League for Peace and Freedom, 1915–1975*. Carbondale: Southern Illinois University Press, 2004.

Blaga, D. *Tatăl meu, Lucian Blaga* [*My Father, Lucian Blaga*]. Bucharest: Humanitas, 2015.

Bobula, I. 'A nő és a hivatal' [The woman and the office]. *Új Idők* [*New Times*], 4 (1937), 123–25.

Bonea, A. '"Contagion by telephone": Print media and knowledge about infectious diseases in Britain, 1880s–1914.' *Technology and Culture*, 62: 4 (2021), 1063–86.

Bonneuil, C. 'La Cinquième République des sciences: Transformations des savoirs et des formes d'engagement des scientifiques' [The Fifth Republic of Sciences: Transformations of knowledge and forms of scientific engagement]. In C. Charles and L. Jeanpierre (eds), *La Vie intellectuelle en France* [*Intellectual Life in France*], vol. 2, pp. 515–36. Paris: Seuil, 2016.

Borská, I. *Doktorka z domu Trubačů* [*The Doctor from the Trumpeters' House*]. Prague: Mladá Fronta, 1978.

Borthwick, M. *The Changing Role of Women in Bengal*. Princeton, NJ: Princeton University Press, 1984.

Bretelle-Establet, F. 'Chinese biographies of experts in medicine: What uses can we make of them?' *East Asian Science, Technology and Society*, 3: 4 (2009), 421–51.

Bretelle-Establet, F. *La Santé en Chine du Sud, 1898–1928* [*Health in Southern China, 1898–1928*]. Paris: CNRS, 2002.

Brichford, M. J. *Scientific and Technological Documentation: Archival Evaluation and Processing of University Records Relating to Science and Technology*. Urbana: University of Illinois at Urbana-Champaign, 1969.

Brown, A. *J. D. Bernal: Sage of Science*. Oxford: Oxford University Press, 2006.

Bucur, M. *Gendering Modernism: A Historical Reappraisal of the Canon*. New York: Bloomsbury, 2017.

Bucur, M. 'From invisibility to marginality: Women's history in Romania.' *Women's History Review*, 27: 1 (2018), 48–57.

Butler Kahle, J., ed. *Women in Science: A Report from the Field*. Philadelphia, PA: The Falmer Press, 1985.

Canel, A., R. Oldenziel and K. Zachmann, eds. *Crossing Boundaries, Building Bridges: Comparing the History of Women Engineers, 1870s–1990s*. Amsterdam: Harwood Academic, 2005.

Cao, S. and B. Yang. 'Grain, local politics, and the making of Mao's famine in Wuwei, 1958–1961.' *Modern Asian Studies*, 49: 6 (2015), 1675–1703.

Carpenter, M. W. *Health, Medicine and Society in Victorian England*. Santa Barbara, CA and Oxford: Praeger, ABC Clio, 2009.

Case, A. D. and C. D. Hunter. 'Cultural racism-related stress in Black Caribbean immigrants: Examining the predictive roles of length of residence and racial identity.' *Journal of Black Psychology*, 40: 5 (2014), 410–23.
Chand, R. P. *Plague Darpan*. Patna City: Satya Sudhakar Press, 1916.
Chatterjee, P. 'Colonialism, nationalism, and colonized women: The contest in India.' *American Ethnologist*, 16: 4 (1989), 609–21.
Chatterjee, P. *The Nation and Its Fragment: Colonial and Postcolonial Histories*. Princeton, NJ: Princeton University Press, 1993.
Chimba, M. and J. Kitzinger. 'Bimbo or boffin? Women in science: An analysis of media representations and how female scientists negotiate cultural contradictions.' *Public Understanding of Science*, 19: 5 (2010), 609–24.
Cho, S., K. W. Crenshaw and L. McCall. 'Toward a field of intersectionality studies: Theory, applications, and praxis.' *Signs: Journal of Women in Culture and Society*, 38: 4 (2013), 785–810.
Chrisopol, I. *Curs de igienă pentru școlile profesionale și cele de ucenici din atelierele și depourile căilor ferate* [Manual of Hygiene for Professional Schools and Apprentice Schools Associated with Railway Workshops and Depots]. Bucharest: Tipografia Cultura, 1925.
Coen, D. R. 'The common world: Histories of science and domestic intimacy.' *Modern Intellectual History*, 11: 2 (2014), 417–38.
Cohen, J. 'The culture of credit.' *Science*, 268 (1995), 1706–11.
Cohen, R. and S. O'Byrne. ' "Can you hear me now… Good!" Feminism(s), the public/private divide and citizens united v. FEC.' *UCLA Women's Law Journal*, 20: 1 (2013), 39–70.
Comfort, N. *The Tangled Field: Barbara McClintock's Search for the Patterns of Genetic Control*. Cambridge, MA: Harvard University Press, 2003.
Confortini, C. C. 'Doing feminist peace.' *International Feminist Journal of Politics*, 13: 3 (2011), 349–70.
Cowan, R. S. *More Work for Mother: The Ironies of Household Technology from The Open Hearth to The Microwave*. New York: Basic Books, 1985.
Crenshaw, K. 'Demarginalizing the intersection of race and sex: A Black feminist critique of antidiscrimination doctrine, feminist theory and antiracist politics.' *University of Chicago Legal Forum*, 140 (1989), 139–67. https://chicagounbound.uchicago.edu/uclf/vol1989/iss1/8
Cristian, R. M. and A. Kérchy, eds. *Pioneering Hungarian Women in Science and Education*. Budapest: Akadémiai Kiadó, 2022.
Damodaran, V. 'Gender, race and science in twentieth-century India: E. K. Janaki Ammal and the history of science.' *History of Science*, 51: 3 (2013), 283–307.
Damodaran, V. 'Janaki Ammal, C. D. Darlington and J. B. S. Haldane: Scientific encounters at the end of Empire.' *Journal of Genetics*, 96: 5 (2017), 827–36.
Das, R. R. and B. Ray. *Teaching of Home Science*. New Delhi: Sterling Publishers Private Ltd., 1983.
Daston, L. and P. Galison. *Objectivity*. New York: Zone Books, 2017.
Deshpande, U. and T. P. Barbosa. *Iru: The Remarkable Life of Irawati Karve*. New Delhi: Speaking Tiger, 2024.
Devi, Y. *Shishu Raksha Vidhan arthat Balrog Chikitsa* [Manual for the Protection of Children]. Allahabad, 1912.
Devi, Y. *Ghar ka Vaidya* [Home Doctor]. Allahabad, 1912.
Devi, Y. *Santan Palan* [The Rearing of Progeny]. Allahabad, 1913.

Devi, Y. *Samsar ka Nari Itihas (Bharat ka Nari Itihas)* [*Women's History of the World (Women's History of India)*]. Vol. 2. Allahabad, 1922.

Devi, Y. *Dampati Arogyata Jeevanshashtra* [*The Science of a Healthy Conjugal Life*]. Allahabad, 1927.

Devi, Y. *Dampatya Prem aur Ratikriya ka Gupt Rahasya* [*Conjugal Love and Secrets of Sexual Intercourse*]. Allahabad, 1933.

Devi, Y. *Nari Sharir Vigyan Stri Chikitsa Sagar: Sambhog Vigyan* [*Women's Physiology and Medical Treatment: Science of Intercourse*]. Allahabad, 1938.

Devi, Y. *Vivah Vigyan Kamshastra: Anand Mandir* [*Science of Marriage and Sex*]. Allahabad, n.d.

Devika, J. *En-gendering Individuals: The Language of Re-forming in Early Twentieth-Century Keralam*. New Delhi: Orient Longman, 2007.

Dikötter, F. *The Discourse of Race in Modern China*. Oxford: Oxford University Press, 2015.

Domenach, J.-L. *The Origins of the Great Leap Forward: The Case of One Chinese Province*, trans. A. M. Berrett. Boulder, CO: Westview Press, 1995.

Dreilinger, D. *The Secret History of Home Economics*. New York: Norton, 2021.

Drucker, A. R. 'The role of the YWCA in the development of the Chinese Women's Movement, 1890–1927.' *Social Service Review*, 53: 3 (1979), 421–40.

Dyhouse, C. 'The British Federation of University Women and the status of women in universities, 1907–1939.' *Women's History Review*, 4: 4 (1995), 465–85.

Dyhouse, C. *No Distinction of Sex: Women in British Universities, 1870–1939*. London: UCL Press, 1995.

Dyson, F. *The Scientist as Rebel*. New York: New York Review of Books, 2006.

Elliott, C. A. 'The tercentenary of Harvard University in 1936: The scientific dimension.' *Osiris*, special issue, *Commemorative Practices in Science: Historical Perspectives on the Politics of Collective Memory*, 14 (1999), 153–75.

Eyal, G., B. Hart, E. Onculer, N. Oren and N. Rossi. *The Autism Matrix: The Social Origins of the Autism Epidemic*. Cambridge: Polity, 2010.

Faber, K. *Report in Medical Schools in China*. Geneva: League of Nations Health Organization, 1931.

Famintsyn, A. S. 'Die Symbiose als Mittel der Synthese von Organismen.' *Biologisches Centralblatt*, 27 (1907), 253–64.

Fan, F.-T. 'Science in cultural borderlands: Methodological reflections on the study of science, European imperialism, and cultural encounter.' *East Asian Science, Technology and Society: An International Journal*, 1: 2 (2007), 213–31.

Fang, X. *Barefoot Doctors and Western Medicine in China*. Rochester, NY: University of Rochester Press, 2012.

Fang, X. *China and the Cholera Pandemic: Restructuring Society under Mao*. Pittsburgh, PA: University of Pittsburgh Press, 2021.

Fara, P. *A Lab of One's Own: Science and Suffrage in the First World War*. Oxford: Oxford University Press, 2018.

Fisher, A. and K. Henningsen. 'Women in science through an archival lens.' *Transformations: The Journal of Inclusive Scholarship and Pedagogy*, 27 (2017), 158–79.

Fong, M. *One Child: The Story of China's Most Radical Experiment*. Boston, MA: Houghton Mifflin, 2016.

Franchini, C. 'Women pioneers in civil engineering and architecture in Italy: Emma Strada and Ada Bursi.' In C. Franchini, M. Groot, H. Seražin, E. Garda and

A. Di Battista (eds), *MoMoWo: Women Designers, Craftswomen, Architects and Engineers*, pp. 82–101. Ljubljana: Založba ZRC, 2017.
Fraser, N. 'Rethinking the public sphere: A contribution to the critique of actually existing democracy.' *Social Text*, 25: 26 (1990), 56–80.
Fraser, N. 'Politics, culture, and the public sphere: Toward a postmodern conception.' In L. Nicholson and S. Seidman (eds), *Social Postmodernism: Beyond Identity Politics*, pp. 287–312. Cambridge: Cambridge University Press, 1995.
Fraser, N. 'Transnationalizing the public sphere: On the legitimacy and efficacy of public opinion in a post-Westphalian world.' *Theory Culture Society*, 24: 4 (2007), 7–30.
Freidenreich, H. P. 'Jewish women physicians in Central Europe in the early twentieth century.' *Contemporary Jewry*, 17: 1 (1996), 79–105.
Frisch, R. E. 'Demographic implications of the biological determinants of female fecundity.' *Social Biology*, 22: 1 (1975), 17–22.
Frisch, R. E. *Female Fertility and the Body Fat Connection*. Chicago, IL: University of Chicago Press, 2002.
Fremlin, J. and Fremlin, M. 2004. *There Isn't a Snake in the Cupboard: A Review of the Life of J. H. Fremlin*. https://margaret.fremlin.org/book.html.
Funk, N. and M. Mueller, eds. *Gender Politics and Post-Communism: Reflections from Eastern Europe and the Former Soviet Union*. New York and London: Routledge, 1993.
Furth, C. *A Flourishing Yin: Gender in China's Medical History, 960–1665*. Berkeley, CA: University of California Press, 1999.
Furukawa, Y. *Tsuda Umeko: Kagaku he no michi, daigaku no yume* [*Umeko Tsuda: The Road to Science and the Dream of Founding a University*]. Tokyo: Tokyo University Press, 2022.
Geoff, E. 'Politics, culture and the public sphere.' *Positions*, 10: 1 (2002), 219–36.
Ghosh, A. *Making It Count: Statistics and Statecraft in the Early People's Republic of China*. Princeton, NJ: Princeton University Press, 2020.
Gilbert, P. K. *The Citizen's Body: Desire, Health and the Social in Victorian England*. Columbus: Ohio State University Press, 2007.
Gonzalez-Perez, S., R. Mateos de Cabo and M. Sainz. 'Girls in STEM: Is it a female role-model thing?' *Frontiers in Psychology*, 11: Article 2204 (2020), 1–21.
Gooday, G. and E. Rees Koerner. 'Formulating a transnational history of women in engineering and applied science.' *Women's History Today*, 3: 4 (2022), 4–13.
Gornick, V. *Women in Science: Then and Now*. New York: The Feminist Press at CUNY, 2013.
Govoni, P. and Z. A. Franceschi, eds. *Writing about Lives in Science: (Auto) Biography, Gender, and Genre*. Goettingen: Vandenhoeck & Ruprecht, 2014.
Grant, J. *Raising Baby by the Book: The Education of American Mothers*. New Haven, CT: Yale University Press, 1998.
Grecu, E. *Azilul Elena Doamna și ajutorul domnesc dat orfanilor* [*The Elena Doamna Asylum and the Prince's Assistance to the Orphans*]. Bucharest: Editura Casa Școalelor, 1944.
Green, N. *How Asia Found Herself: A Story of Intercultural Understanding*. New Haven, CT: Yale University Press, 2022.
Gross, M. *Farewell to the God of Plague: Chairman Mao's Campaign to Deworm China*. Berkeley, CA: University of California Press, 2016.
Guglielmo, L. 'Introduction: Re-collection as feminist rhetorical practice.' In L. L. Gaillet and H. G. Bailey (eds), *Remembering Women Differently:*

Refiguring Rhetorical Work, pp. 1–20. Columbia: University of South Carolina Press, 2019.

Guha, S. 'From dais to doctors: The medicalisation of childbirth in colonial India.' In L. Lingam (ed.), Understanding Women's Health Issues: A Reader, pp. 145–60. New Delhi: Kali for Women, 1998.

Gupta, C. Sexuality, Obscenity, Community: Women, Muslims, and the Hindu Public in Colonial India. Delhi: Permanent Black, 2001.

Gupta, C. 'Procreation and pleasure: Writings of a woman Ayurvedic practitioner in colonial North India.' Studies in History, 21: 1 (2005), 17–44.

Gupta, C. 'Vernacular sexology from the margins: A woman and a Shudra.' South Asia: Journal of South Asian Studies, 43: 6 (2020), 1105–27.

Gynecology Teaching and Research Group at the Chengdu College of Chinese Medicine. Zhongyi xueyuan shiyong jiaocai: Zhongyi fuke xue jiangyi [Trial Textbooks for Chinese Medicine Colleges: Lecture Notes in Chinese Gynecology]. Beijing: Renmin weisheng chubanshe, 1960.

Haines, C. M. C. and H. M. Stevens. International Women in Science: A Biographical Dictionary to 1950. Santa Barbara, CA and Oxford: ABC-Clio, 2001.

Halsey, S. J., L. R. Strickland, M. Scott-Richardson, T. Perrin-Stowe, L. Massenburg. 'Elevate, don't assimilate, to revolutionize the experience of scientists who are Black, Indigenous and people of colour.' Nature Ecology and Evolution, 4 (2020), 1291–93.

Hannah-Jones, N. 1619 Project: A New Origin Story. New York: One World, 2021.

Haraway, D. Primate Visions: Gender, Race, and Nature in the World of Modern Science. New York and London: Routledge, 1989.

Harding, S. Sciences from Below: Feminisms, Postcolonialisms, and Modernities. Durham, NC: Duke University Press, 2008.

Harding, S. Whose Science? Whose Knowledge? Ithaca, NY: Cornell University Press, 1991.

Hargittai, M. Women Scientists: Reflections, Challenges, and Breaking Boundaries. New York: Oxford University Press, 2015.

Hart, B. 'Autism parents and neurodiversity: Radical translation, joint embodiment and the prosthetic environment.' BioSocieties, 9 (2014), 284–303.

Hecht, D. K. 'The atomic hero: Robert Oppenheimer and the making of scientific icons in the early Cold War.' Technology and Culture, 49: 4 (2008), 943–66.

Hecht, D. K. 'Constructing a scientist: Expert authority and public images of Rachel Carson.' Historical Studies in the Natural Sciences, 41: 3 (2011), 277–302.

Heitman, C. C. 'The creation of gender bias in museum collections: Recontextualizing archaeological and archival collections from Chaco Canyon, New Mexico.' Museum Anthropology, 40: 2 (2017), 128–42.

Herman, E. The Romance of American Psychology: Political Culture in the Age of Experts. Berkeley: University of California Press, 1995.

Hershatter, G. The Gender of Memory: Rural Women and China's Collective Past. Berkeley, CA: University of California Press, 2014.

Hicks, M. Programmed Inequality: How Britain Discarded Women Technologists and Lost Its Edge in Computing. Cambridge, MA: The MIT Press, 2017.

Horrocks, S. 'Promising pioneer profession? Women in industrial chemistry in inter-war Britain.' British Journal of the History of Science, 33: 3 (2000), 351–67.

Horrocks, S. 'World War II, Post-war reconstruction and British women chemists.' Ambix, 58: 2 (2011), 150–70.

Horrocks, S. 'The women who cracked the glass ceiling.' *Nature* (7 November 2019), 243–46.
Howes, R. *Their Day in the Sun: Women of the Manhattan Project*. Philadelphia, PA: Temple University Press, 1999.
Hu, A. *The Great Leap Forward: 1957–1965*. Translated by G. Hu and V. C. W. Hui. Singapore: Enrich Professional Publishing, 2014.
Hui, W. 'The fate of "Mr. Science" in China: The concept of science and its application in modern Chinese thought.' *Positions: Asia Critique*, 3: 1 (1995), 1–68.
Hu, Y. 'Minguo shiqi yisheng zhi zhenxun he pinghe' [The screening and evaluation of doctors in the Republican Period]. *Zhejiang xue kan*, 5 (2008), 88–94.
Iacob, B. C. 'Malariology and decolonization: Eastern European experts from the League of Nations to the World Health Organization.' *Journal of Global History*, 17: 2 (2022), 233–53.
Institute for Statistics. 'Women in science.' Fact Sheet No. 55, June 2019, FS/2019/SCI/55, http://uis.unesco.org/sites/default/files/documents/fs55-women-in-science-2019-en.pdf.
Jack, J. *Science on the Home Front: American Women Scientists in World War II*. Champaign, IL: University of Illinois Press, 2009.
Jensen, K. 'The "open way of opportunity": Colorado women physicians and World War I.' *Western Historical Quarterly*, 27: 3 (1996), 327–48.
Jinga, L. M. *Gen și reprezentare în România comunistă, 1944–1989: Femeile în cadrul Partidului Communist Român* [*Gender and Representation in Communist Romania, 1944–1989: Women and the Romanian Communist Party*]. Bucharest: Polirom, 2015.
Johnson, T. P. *Childbirth in Republican China: Delivering Modernity*. Lanham, MD: Lexington Books, 2011.
Johnson, T. P. 'Yang Chongrui and the first National Midwifery School: Childbirth reform in early twentieth-century China.' *Asian Medicine*, 4: 2 (2008), 280–302.
Jones, C. *Femininity, Mathematics and Science, 1880–1914*. Cham: Springer, 2009.
Jones, C. G., A. E. Martin and A. Wolf. *The Palgrave Handbook of Women and Science since 1660*. London: Palgrave Macmillan, 2022.
Jöns, H. 'Feminizing the university: The mobilities, careers, and contributions of early female academics in the University of Cambridge, 1926–1955.' *The Professional Geographer*, 69: 4 (2017), 670–82.
Judge, J. *Republican Lens: Gender, Visuality, and Experience in the Early Chinese Periodical Press*. Oakland: University of California Press, 2015.
Judge, J. *The Precious Raft of History: The Past, the West, and the Woman Question in China*. Stanford, CA: Stanford University Press, 2008.
Kálalová Di-Lotti, V. *Přes Bospor k Tigridu* [*Across the Bosphorus to the Tigris*]. Unpublished manuscript.
Kálalová Di-Lottiova, V. 'O chorobách a zdravotních poměrech v Íráku' [On diseases and health conditions in Iraq]. *Časopis lékařů českých*, 1933, 72: 14 (1933), 430–33; 72: 15 (1933), 460–63; 72: 16 (1933), 495–98.
Kálalová Di-Lottiova, V. *Cařihradské a bagdadské kapitoly* [*The Constantinople and Baghdad Chapters*]. Praktický lékař, 1933.
Kanner, L. 'Autistic disturbances of affective contact.' *Nervous Child*, 2 (1943), 217–50.
Kanner, L. 'Early infantile autism.' *Journal of Pediatrics*, 25 (1944), 211–17.
Kapp, R. A. *Szechwan and the Chinese Republic: Provincial Militarism and Central Power, 1911–1938*. New Haven, CT: Yale University Press, 1973.

Källstrand, G. 'Warburg's dogs: Nobel laureates and scientific celebrity.' *Celebrity Studies*, 13: 1 (2022), 56–72.

Kawano, G. and M. Ogawa, eds. *Josei kenkyūsha shienseisaku no kokusaihikaku: Nihon no genjō to kadai* [*International Comparison of Support Policies for Women Researchers: Current Scenario and Issues in Japan*]. Tokyo: Akashi Shoten, 2021.

Kedharnath, S. 'Edavaleth Kakkat Janaki Ammal (1897–1984).' *Biographical Memoirs of Fellows of the Indian National Science Academy* 13 (1980), 90–101.

Keller, E. F. *A Feeling for the Organism: The Life and Work of Barbara McClintock*. New York: Times Books, 1984.

Keller, E. F. *Reflections on Gender and Science*. New Haven, CT: Yale University Press, 1985.

Keller, E. F. 'The gender/science system: Or, is sex to gender as nature is to science?' *Hypatia*, 2: 3 (1987), 37–49.

King, A. *The Bungalow*. London: Routledge & Kegan Paul, 1984.

King, R. 'Romancing the Leap: Euphoria in the moment before disaster.' In K. E. Manning and F. Wemheuer (eds), *Eating Bitterness: New Perspectives on China's Great Leap Forward and Famine*, pp. 51–71. Vancouver: University of British Columbia Press, 2011.

Kligman, G. *Politicile de gen în perioada postsocialistă* [*Gender Policies in the Post-Socialist Period*]. Iaşi: Polirom, 2006.

Knibiehler, Y. and C. Fouquet. *La femme et les médecins: Analyse historique* [*The Woman and the Doctors: Historical Analysis*]. Paris: Hachette, 1983.

Koblitz, A. H. 'Gender and science where science is on the margins.' *Bulletin of Science, Technology & Society*, 25: 2 (2005), 107–14.

Koblitz, A. H. 'Life in the fast lane: Arab women in science and technology.' *Bulletin of Science, Technology & Society*, 36: 2 (2016), 107–17.

Kodate, N. and K. Kodate. *Japanese Women in Science and Engineering: History and Policy Change*. New York: Routledge, 2016.

Kohlstedt, S. G. 'Sustaining gains: Reflections on women in science and technology in the 20th century.' *NWSA Journal*, 16: 1 (Re) Gendering Science Fields (Spring 2006), 1–26.

Kováts, E. and A. Pető. 'Anti-gender movements in Hungary. A Discourse without a movement?' In R. Kuhar and D. Paternotte (eds), *Anti-Gender Campaign in Europe*, pp. 117–33. London: Rowman and Littlefield, 2017.

Koyama, S. *Ryōsai Kenbo: Educational Ideal of 'Good Wife, Wise Mother' in Modern Japan*. Boston, MA: Brill, 2012.

Kozlowski, D., V. Larivière, C. R. Sugimoto and T. Monroe-White. 'Intersectional inequalities in science.' *PNAS*, 119: 2 (2022), e2113067119.

Kozo-Polyansky, B. M., V. Fet (ed. and trans.) and L. Margulis, ed. *Symbiogenesis: A New Principle of Evolution*. Cambridge, MA: Harvard University Press, 2010.

Kubica, G. *Maria Czaplicka: Gender, Shamanism, Race. An Anthropological Biography*, trans. B. Koschalka. Lincoln: University of Nebraska Press, 2020.

Kumar, K. *Political Agenda of Education: A Study of Colonialist and Nationalist Ideas*. New Delhi: Sage Publications, 1991.

Kumar, N. *Women and Science in India: A Reader*. New Delhi: Oxford University Press, 2009.

Kumar, N., ed. *Gender and Science: Studies across Cultures*. New Delhi: Cambridge University Press, 2012.

Kumar, N. 'Widows, education and social change in twentieth century Banaras.' *Economic and Political Weekly*, 26: 17 (1991), WS19–WS25.
Kwok, D. W. Y. *Scientism in Chinese Thought, 1900–1950*. New Haven, CT: Yale University Press, 1965.
Kwok, P.-L. *Chinese Women and Christianity, 1860–1927*. Atlanta, GA: Scholars Press, 1992.
Landsman, G. H. *Reconstructing Motherhood and Disability in the Age of 'Perfect' Babies*. New York: Routledge, 2009.
Lathers, M. ' "No official requirement": Women, history, time, and the U.S. Space Program.' *Feminist Studies*, 35: 1 (2009), 14–40.
Law 1492/1950, 'Περί κυρώσεως του Ποινικού Κώδικα' [For the Constitutional Validity of the Greek Penal Code], Library of the Hellenic Parliament.
Law 821/1978, 'Περί αφαιρέσεων και μεταμοσχεύσεων βιολογικών ουσιών ανθρώπινης προελεύσεως' [For the Removal or Transplant of Human Biological Substances], Library of the Hellenic Parliament.
Law 1609/1986, 'Τεχνητή διακοπή της εγκυμοσύνης και προστασία της γυναίκας και άλλες διατάξεις' [Technical Termination of Pregnancy and Protection of Women's Health and Other Regulations], Library of the Hellenic Parliament.
Laycock, J. and J. Johnson. 'Creating 'New Soviet Women' in Armenia? Gender and tradition in the early Soviet South Caucasus.' In C. Baker (ed.), *Gender in Twentieth-Century Eastern Europe and the USSR*, pp. 64–78. London: Palgrave Macmillan, 2017.
Lei, S. H.-L. *Neither Donkey nor Horse: Medicine in the Struggle Over China's Modernity*. Chicago, IL: University of Chicago Press, 2014.
Leung, A. K. C. 'Dignity of the nation, gender equality, or charity for all? Options for the first modern Chinese women doctors.' In S. Y. S. Chien and J. Fitzgerald (eds), *The Dignity of Nations: Equality, Competition, and Honor in East Asian Nationalisms*, pp. 71–91. Hong Kong: Hong Kong University Press, 2006.
Leung, A. K. C. 'Women practicing medicine in pre-modern China.' In H. T. Zurndorfer (ed.), *Chinese Women in the Imperial Past: New Perspectives*, pp. 101–34. Leiden: Brill, 1999.
Leslie, C., ed. *Asian Medical Systems: A Comparative Study*. Berkeley: University of California Press, 1976.
Lewis, W. H., A. A. Reznicek and R. K. Rabeler. 'Identifications and typifications of Rosa (Rosaceae) taxa in North America described or used by E. W. Erlanson, 1927–1934.' *Novon*, 22: 1 (2012), 41–42.
Liakos, A. *Ο ελληνικός 20ος αιώνας* [*The Greek 20th Century*]. Athens, GA: Polis Publications, 2022.
Light, J. 'When computers were women.' *Technology and Culture*, 40: 3 (1999), 455–83.
Lin, S.-T. 'The female hand: The making of Western medicine for women in China, 1880s–1920s.' PhD dissertation, Columbia University, New York, 2015.
Lincoln, A. E., S. Pincus, J. B. Koster and P. S. Leboy. 'The Matilda Effect in science: Awards and prizes in the US, 1990s and 2000s.' *Social Studies of Science*, 42 (2012), 307–20.
Livingstone, D. N. *Putting Science in Its Place: Geographies of Scientific Knowledge*. Chicago, IL and London: University of Chicago Press, 2003.
Loehwing, M. and J. Motter. 'Publics, counterpublics, and the promise of democracy.' *Philosophy & Rhetoric*, 42: 3 (2009), 220–41.
Longrigg, S. *Four Centuries of Modern Iraq*. Beirut: Librairie du Liban, 1968.

Lutkehaus, N. *Margaret Mead: The Making of an American Icon*. Princeton, NJ: Princeton University Press, 2008.

Lykknes, A., D. L. Opitz and B. Van Tiggelen, eds. *For Better or For Worse? Collaborative Couples in the Sciences*. Basel: Birkhäuser, 2012.

Macfarquhar, R. *The Origins of the Cultural Revolution*, Vol. 2, *The Great Leap Forward, 1958–1960*. New York: Columbia University Press, 1983.

Mack, P. E. 'What difference has feminism made to engineering in the twentieth century.' In A. N. H. Creager, E. Lunbeck, C. R. Stimpson and L. Schiebinger (eds), *Feminism in Twentieth Century Science, Technology and Medicine*. Chicago, IL: University of Chicago Press, 2001.

Maddox, B. *Rosalind Franklin: The Dark Lady of DNA*. New York: HarperCollins, 2002.

Mainz, V. V. and T. E. Strom, eds. *Ladies in Waiting for the Nobel Prize, The Posthumous Nobel Prize in Chemistry*. Vol. 2. Washington, DC: American Chemical Society, 2018.

Malhotra, A. *Gender, Caste, and Religious Identities: Restructuring Caste in Colonial Punjab*. New Delhi: Oxford University Press, 2002.

Mandelbaum, D. R. 'Women in medicine.' *Signs*, 4: 1 (1978), 136–45.

Manning, K. E. 'Communes, canteens, and creches: The gendered politics of remembering the Great Leap Forward.' In C. K. Lee and G. Yang (eds), *Re-Envisioning the Chinese Revolution: The Politics and Poetics of Collective Memories in Reform China*, pp. 93–118. Stanford, CA: Stanford University Press, 2007.

Manning, K. E. 'Making a Great Leap Forward? The politics of women's liberation in Maoist China.' *Gender & History*, 18: 3 (2006), 574–93.

Margulis, L. 'On the origin of mitosing cells.' *Journal of Theoretical Biology*, 14: 3 (1967), 225–74.

Margulis, L. 'Kingdom Animalia: The zoological malaise from a microbial perspective.' *American Zoologist*, 30: 4 (1990), 861–75.

Margulis, L., ed. *Lynn Margulis: The Life and Legacy of a Scientific Rebel*. White River Junction, VT: Chelsea Green Publishing, 2012.

Mark, J. and P. Betts. *Socialism Goes Global: The Soviet Union and Eastern Europe in the Age of Decolonization*. Oxford: Oxford University Press, 2022.

Martin, S. C., R. M. Arnold and R. Parker. 'Gender and medical socialization.' *Journal of Health and Social Behaviour*, 29: 4 (1988), 333–43.

Massino, J. *Ambiguous Transitions: Gender, the State, and Everyday Life in Socialist and Postsocialist Romania*. New York and Oxford: Berghahn, 2019.

Mayberry, M. and B. Subramaniam, eds. *Feminist Science Studies: A New Generation*. New York and London: Routledge, 2001.

McCormack, C. *Women in the Picture: What Culture Does with Female Bodies*. New York: Norton, 2021.

McDonell, J. T. 'Mothering an autistic child: Reclaiming the voice of the mother.' In B. O. Daly and M. T. Reddy (eds), *Narrating Mothers: Theorizing Maternal Subjectivities*, pp. 58–75. Knoxville: The University of Tennessee Press, 1991.

McGrayne, S. B. *Nobel Prize Women in Science: Their Lives, Struggles, and Momentous Discoveries*. New York: Basic Books, 1992.

McWilliams Tullberg, R. *Women at Cambridge*. Cambridge: Cambridge University Press, 1998.

Menyhért, A. *Women's Literary Tradition and Twentieth-Century Hungarian Writers: Renée Erdős, Ágnes Nemes Nagy, Minka Czóbel, Ilona Harmos Kosztolányi, Anna Lesznai*, trans. A. Bentley. Leiden: Brill NV, 2020.

Meng, X., N. Qian and P. Yared. 'The institutional causes of China's Great Famine, 1959–61.' *The Review of Economic Studies*, 82: 4 (2015), 1568–1611.

Midgley, C., A. Twells and J. Carlier. *Women in Transnational History: Connecting the Local and the Global*. London and New York: Routledge, 2016.

Mihăilescu, Ş. *Din istoria feminismului românesc. Studiu și antologie de texte (1929–1948)* [*From the History of Romanian Feminism: Study and Anthology of Texts (1929–1948)*]. Iași: Polirom, 2006.

Miller, A. H. and J. E. Adams, eds. *Sexualities in Victorian Britain*. Bloomington: Indiana University Press, 1996.

Minden, K. *Bamboo Stone: The Evolution of a Chinese Medical Elite*. Toronto: University of Toronto Press, 1994.

Modgil, S., R. Gill, V. L. Sharma, S. Velassery and A. Anand. 'Nobel nominations in science: Constraints of the fairer sex.' *Annals of Neurosciences*, 25 (2018), 63–78.

Mukharji, P. B. 'From serosocial to sanguinary identities: Caste, trans-national race science and the shifting metonymies of blood group B, India c.1918–1960.' *The Indian Economic and Social History Review*, 51: 2 (2014), 143–76.

Mukherjee, S. *Indian Suffragettes: Female Identities and Transnational Networks*. New Delhi: Oxford University Press, 2018.

Murgia, A. and B. Poggio, eds. *Gender and Precarious Research Careers: A Comparative Analysis*. London: Routledge, 2020.

Nagy, B. and V. Paksi. 'A munka-magánélet összehangolásának kérdései a magasan képzett nők körében' [Issues of work-life balance among highly educated women]. In Z. Spéder (ed.), *A család vonzásában: Tanulmányok Pongrácz Tiborné tiszteletére* [*In the Gravitational Pull of the Family: Studies in Honour of Tiborné Pongrácz*], pp. 159–75. Budapest: KSH Population Studies Research Institute, 2014.

Nahimas, N. 'Making science popular: Readers, nation, and the universe in Chinese popular science periodicals, 1933–1952.' PhD dissertation, York University, 2022.

Nair, S. P. *Chromosome Woman, Nomad Scientist: E. K. Janaki Ammal, A Life 1897–1984*. New York and London: Routledge, 2022.

Nastasă-Matei, I. 'Transnational far right and Nazi soft power in Eastern Europe: The Humboldt Fellowships for Romanians.' *East European Politics and Societies*, 35: 4 (2021), 899–923.

National Science Foundation. *Beyond Bias and Barriers: Fulfilling the Potential of Women in Academic Science and Engineering*. Arlington, VA: National Science Foundation, 2006.

Nazarska, G. 'Opportunities for an academic career of women scientists at the Bulgarian Academy of Sciences (mid-1940s–1980s).' *Balkanistic Forum*, 30: 1 (2021), 120–37.

Nazarska, G. 'An (un)established academic and scientific network: Branches of the International Federation of University Women on the Balkans (1920–1950s).' *Balkanistic Forum*, 31: 1 (2022), 32–58.

Noon, D. H. 'Situating gender and professional identity in American Child Study, 1880–1910.' *History of Psychology*, 7 (2004), 107–29.

Novara, E. 'Documenting Maryland women state legislators: The politics of collecting women's political papers.' *American Archivist*, 76: 1 (2013), 196–214.
Odeseanu, C. *Cartea femeii moderne. Ce trebuie să știe o femeie și chiar o fată* [*Modern Woman's Book: What Women and Even Girls Must Know*]. Bucharest: Editura Cartea Românească, 1934.
von Oertzen, C., M. Rentetzi and E. S. Watkins, eds. *Beyond the Academy: Histories of Gender and Knowledge*. Special Issue of *Centaurus*, 55: 2 (2013).
von Oertzen, C. *Science, Gender and Internationalism: Women's Academic Networks, 1917–1955*. London: Palgrave Macmillan, 2021.
von Oertzen, C. 'Science in the cradle: Milicent Shinn and her home-based network of baby observers, 1890–1910.' *Centaurus* 55 (2013), 175–95.
Ogawa, M. 'History of women's participation in STEM fields in Japan.' *Asian Women*, 33: 3 (2017), 65–85.
Ogawa, M. 'Nihon no STEMM bun'ya ni okeru josei jinzai no rekishi' [A history of women's resources in STEMM fields in Japan]. *Kagaku gijutsu shakairon kenkyū*, 19 (2021), 43–52.
Ogilvie, M. B. and J. D. Harvey, eds. *The Biographical Dictionary of Women in Science: Pioneering Lives from Ancient Times to the Mid-Twentieth Century*, 2 vols. New York and London: Routledge, 2000.
Opitz, D. L. 'Domestic space.' In B. Lightman (ed.), *A Companion to the History of Science*, pp. 252–67. Chichester: John Wiley & Sons Ltd, 2016.
Opitz, D. L., S. Bergwik and B. Van Tiggelen, eds. *Domesticity in the Making of Modern Science*. New York: Palgrave Macmillan, 2015.
Oreskes, N. 'Objectivity or heroism? On the invisibility of women in science.' *Osiris* 11 (1996), 87–113.
Organization for Economic Co-operation and Development. *Women in Scientific Careers: Unleashing the Potential*. Paris: OECD, 2006.
Östling, J., S. Erling, D. Larsson Heidenblad, A. Nilsson Hammar and K. Nordberg, eds. *Circulation of Knowledge: Explorations in the History of Knowledge*. Lund: Nordic Academic Press: Sweden, 2018.
Our West China Mission. Toronto: Missionary Society of the Methodist Church, 1920.
Pajnik, M. 'Feminist reflections on Habermas's communicative action: The need for an inclusive political theory.' *European Journal of Social Theory*, 9: 3 (2006), 385–404.
Paksi, V. 'Miért kevés a női hallgató a természet- és műszaki tudományi képzésekben? Nemzetközi kitekintés a "leaky pipeline" metaforájára' [Why are there few female students in natural and technical science courses? An international perspective on the "leaky pipeline" metaphor]. *Replika*, 85–86: 1–2 (2014).
Palasik, M. 'Women in technological higher education and in the sciences in 20th century Hungary.' *Hungarian Studies Review*, 29: 1–2 (2002).
Pandey, M. *Hamare Bachche* [*Our Children*]. Prayag, 1931.
Papanikolaou, A. N. Γυναικολογία [*Gynecology*]. Thessaloniki: Library of the Hellenic Society of Obstetrics and Gynecology, 1986.
Papanikolaou, A. N. Μαιευτική [*Obstetrics*]. Thessaloniki: Library of the Hellenic Society of Obstetrics and Gynecology, 1987.
Papp, B. and B. Sipos. *A modern diplomás nő a Horthy korban* [*The Modern Graduate Women in the Horthy Era*]. Budapest: Napvilág, 2017.
Park, C. C. *Exiting Nirvana: A Daughter's Life with Autism*. Boston, MA: Little, Brown, 2001.

Park, C. C. *The Siege: A Family's Journey into the World of an Autistic Child.* Boston, MA: Little, Brown, 1982 (1967).
Penn, S. and J. Massino, eds. *Gender Politics and Everyday Life in State Socialist Eastern and Central Europe.* New York: Palgrave MacMillan, 2009.
Péteri, G. 'Nylon Curtain: Transnational and transsystemic tendencies in the cultural life of state-socialist Russia and East-Central Europe.' *Slavonica*, 10: 2 (2004), 113–23.
Pető, A. 'Miért marad továbbra is alacsony a nők aránya a magyar tudományban az EU csatlakozás után is?' [Why does the percentage of women in Hungarian science remain low even after joining the EU?]. *Magyar Tudomány* 8 (2006), 1014–17.
Pető, A. 'From visibility to analysis: Gender and history.' In C. Salvaterra and B. Waaldijk (eds), *Paths to Gender: European Historical Perspectives on Women and Men*, pp. 1–9. Pisa: PLUS Pisa University Press, 2009.
Pető, A. 'The illiberal memory politics in Hungary.' *Journal of Genocide Research*, 24: 2 (2022), 241–49.
Philips, A. A., C. R. Walsh, K. A. Grayson, C. E. Penney, F. Husain and the Women Doing Science Team 'Diversifying representations of female scientists on social media: A case study from the Women Doing Science Instagram.' *Social Media and Society*, 8: 3 (2022), 1–17.
Pickles, K. 'Colonial counterparts: The first academic women in Anglo-Canada, New Zealand and Australia.' *Women's History Review*, 10: 2 (2001), 273–97.
Pietz, D. *The Yellow River: The Problem of Water in Modern China.* Cambridge, MA: Harvard University Press, 2015.
Porter, G. 'Seeing through solidity: A feminist perspective on museums.' *Sociological Review*, 43: 1 (1995), 105–26.
Prakash, G. *Another Reason: Science and Imagination of Modern India.* Princeton, NJ: Princeton University Press, 1999.
Pu, F. *Zhongyi dui jizhong funubing de zhiliao fa* [*Chinese Medicine Treatment Methods for Several Women's Illnesses*]. Beijing: Popular Science Press, 1958.
Puaca, L. M. 'Cold War women: Professional guidance, national defense, and the Society of Women of Engineers, 1950–60.' In A. M. Knupfer and C. Wayshner (eds), *The Educational Work of Women's Organizations, 1890–1960*, pp. 57–77. London: Palgrave Macmillan, 2008.
Puaca, L. M. *Searching for Scientific Womanpower: Technocratic Feminism and the Politics of National Security, 1940–1980.* Chapel Hill, NC: UNC Press, 2014.
Pycior, H. M., N. G. Slack and P. G. Abir-Am, eds. *Creative Couples in the Sciences.* New Brunswick, NJ: Rutgers University Press, 1996.
Rai, S. K. 'Gendering late colonial Ayurvedic discourse: United Provinces, c.1890–1937.' *History and Sociology of South Asia*, 10: 1 (2016), 21–34.
Rai, S. K. 'Gazing at the woman's body: Historicising lust and lechery in a patriarchal society.' *Social Scientist*, 47: 1–2 (2019), 49–62.
Rai, S. K. 'Invoking "Hindu" Ayurveda: Communalisation of the late colonial Ayurvedic discourse.' *Indian Economic and Social History Review*, 56: 3 (2019), 411–26.
Rai, S. K. 'Brahmanizing Ayurveda: Caste and class dimensions of late colonial Ayurvedic movement in Upper India.' *Summerhill: IIAS Review*, 25: 2 (2019): 4–9.

Rai, S. K. 'In search of indigenous medicine: Medical pluralism and the Ayurvedic movement in colonial India.' Occasional Paper, *History and Society (New Series) 104*, Nehru Memorial Museum and Library, 2020.

Ram, K. 'Anthropology as "Ananthropology": L. K. Ananthakrishna Iyer (1861–1937), colonial anthropology, and the "native anthropologist" as pioneer.' In P. Uberoi, N. Sundar and S. Deshpande (eds), *Anthropology in the East: Founders of Indian Sociology and Anthropology*, pp. 64–105. Chicago, IL: University of Chicago Press, 2008.

Rendal, J. 'Women and the public sphere.' *Gender & History*, 11: 3 (1999), 475–88.

Renshaw, M. 'Accommodating the Chinese: The American Hospital in China, 1880–1920.' PhD dissertation, University of Adelaide, 2003.

Reser, A. and L. McNeill. *Forces of Nature: The Women Who Changed Science*. London: Frances Lincoln, 2021.

Richards, P. S. 'The movement of scientific knowledge from and to Germany under national socialism.' *Minerva*, 28: 4 (1990), 401–25.

Richardson, S. S. *The Maternal Imprint: The Contested Science of Maternal-Fetal Effects*. Chicago, IL: University of Chicago Press, 2021.

Riley, D. *War in the Nursery: Theories of Mother and Child*. London: Virago, 1983.

Rimland, B. *Infantile Autism: The Syndrome and Its Implications for a Neural Theory of Behavior*. New York: Appleton-Century-Crofts, 1964.

Rogaski, R. *Hygienic Modernity: Meanings of Health and Disease in Treaty-Port China*. Berkeley: University of California Press, 2004.

Rose, H., 'Hand, brain, and heart: A feminist epistemology for the natural sciences.' *Signs*, 9: 1 (1983): 73–90.

Rose, H. *Love, Power and Knowledge: Toward a Feminist Transformation of the Sciences*. Cambridge: Polity Press, 1994.

Rosser, S. V. *The Science Glass Ceiling: Academic Women Scientists and the Struggle to Succeed*. London: Routledge, 2004.

Rossiter, M. W. 'The Matthew Matilda effect in science.' *Social Studies of Science*, 23: 2 (1993), 325–41.

Rossiter, M. W. *Women Scientists in America: Struggles and Strategies to 1940*. Baltimore, MD: The Johns Hopkins University Press, 1982.

Rossiter, M. W. *Women Scientists in America: Before Affirmative Action, 1940–1972*. Baltimore, MD: The Johns Hopkins University Press, 1995.

Rossiter, M. W. '"Women's work" in science, 1880–1910.' *Isis*, 71: 3 (1980), 381–98.

Røstvik, C. M. and A. Fyfe. 'Ladies, gentlemen, and scientific publication at the Royal Society, 1945–1990.' *Open Library of Humanities*, 4: 1 (2018), 1–40.

Russett, C. E. *Sexual Science: The Victorian Construction of Womanhood*. Cambridge, MA: Harvard University Press, 1989.

Scheich, E. 'Science, politics and morality: The relationship of Lise Meitner and Elisabeth Schiemann.' *Osiris, Women, Gender, and Science: New Directions*, 12 (1997), 143–68.

Schiebinger, L. 'The history and philosophy of women in science: A review essay.' *Signs: Journal of Women in Culture and Society*, 12: 2 (1987), 305–32.

Schiebinger, L. *Has Feminism Changed Science?* Cambridge, MA: Harvard University Press, 2001.

Schiebinger, L. 'Feminist history of colonial science.' *Hypatia*, 19: 1 (2004), 233–54.

Schiebinger, L., A. D. Henderson and S. K. Gilmartin. *Dual-Career Academic Couples: What Universities Need to Know.* Stanford, CA: Stanford University, 2008.

Schmalzer, S. *Red Revolution, Green Revolution: Scientific Farming in Socialist China.* Chicago, IL: University of Chicago Press, 2016.

Schott, L. K. *Reconstructing Women's Thoughts: The Woman's International League for Peace and Freedom Before World War II.* Stanford, CA: Stanford University Press, 1997.

Scott, A. F. 'On seeing and not seeing: A case of historical invisibility.' *The Journal of American History*, 71: 1 (1984), 7–21.

Sdrobiş, D. *Limitele meritocraţiei într-o societate agrară: Şomaj intelectual şi radicalizare politică a tineretului în România interbelică* [*The Limits of Meritocracy in an Agrarian Society: Intellectual Unemployment and Political Radicalization in Interwar Romania*]. Iaşi: Polirom, 2015.

Sehrawat, S. *Colonial Medical Care in North India: Gender, State and Society, c.1840–1920.* New Delhi: Oxford University Press, 2013.

Shah, S. 'Representation of female sexuality in the Ayurvedic discourse of the early medieval Period.' *Studies in History*, 22: 1 (2006), 45–58.

Shapin, S. 'The house of experiment in seventeenth-century England.' *Isis*, 79: 3 (1988), 373–404.

Shapin, S. 'The invisible technician.' *American Scientist*, 77: 6 (1989), 554–63.

Shapin, S. *The Scientific Life: A Moral History of a Late Modern Vocation.* Chicago, IL: University of Chicago Press, 2008.

Sharma, J. *Arogya Darpan.* Vol 3. Prayag, 1898.

Sharma, J. *Santan Palan.* Benares, 1933.

Shastri, C. *Brahmacharya Sadhan.* Lucknow, 1928.

Shemo, C. A. ' "Wants learn cut, finish people": American missionary medical education for Chinese women and cultural imperialism in the missionary enterprise, 1890s–1920.' *The Chinese Historical Review*, 20: 1 (2013), 54–69.

Shemo, C. A. *The Chinese Medical Ministries of Kang Cheng and Shi Meiyu, 1872–1937: On a Cross-Cultural Frontier of Gender, Race, and Nation.* Bethlehem: Lehigh University Press, 2011.

Shen, G. Y. 'Women and the transnational dynamics of science education in early twentieth century China: A quiet revolution.' *Chinese Annals of History of Science and Technology*, 3: 2 (2019), 62–93.

Shetterly, M. L. *Hidden Figures: The American Dream and the Untold Story of the Black Women Who Helped Win the Space Race.* London: William Collins, 2016.

Shi, X. *At Home in the World: Women and Charity in Late Qing and Early Republican China.* New York: Columbia University Press, 2018.

Shields, S. A. 'To pet, coddle, and "do for": Caretaking and the concept of maternal instinct'. In M. Lewin (ed.), *In the Shadow of the Past: Psychology Portrays the Sexes*, pp. 256–73. New York: Columbia University Press, 1984.

Shrivastava, J. S. *Vaidya Priya.* Lucknow: Naval Kishore Press, 1924.

Shteir, A. and B. Lightman, eds. *Figuring It Out: Science, Gender and Visual Culture.* Hanover, NH: University Press of New England, 2006.

Silverman, C. *Understanding Autism: Parents, Doctors, and the History of a Disorder.* Princeton, NJ: Princeton University Press, 2012.

Sime, R. L. *Lise Meitner: A Life in Physics.* Berkeley: University of California Press, 1996.

Simpson, R. *How the PhD Came to Britain: A Century of Struggle for Postgraduate Education*. Guilford: Society for Research into Higher Education, 1983.

Sinha, M. *Mother India: Selections from the Controversial 1927 Text, Edited and with an Introduction by Mrinalini Sinha*. Ann Arbor: Michigan University Press, 2000.

Sinha, N. and P. Kumar, eds. *Lesser Lives: Stories of Domestic Servants in India*. New Delhi: Pan Macmillan, 2021.

Sinopoli, C. M. *The Himalayan Journey of Walter Norman Koelz: The University of Michigan Himalayan Expedition, 1932–1934. Anthropological Papers*, No. 98. Ann Arbor, MI: Museum of Anthropology, 2013.

Sivaramakrishnan, K. *Old Potions, New Bottles: Recasting Indigenous Medicine in Colonial Punjab, 1850–1945*. New Delhi: Orient Longman, 2006.

Solovey, M. *Social Science for What? Battles over Public Funding for the 'Other Sciences' at the National Science Foundation*. Cambridge, MA: The MIT Press, 2020.

Stamhuis, I. H. and A. B. Vogt. 'Discipline building in Germany: Women and genetics at the Berlin Institute for Heredity Research.' *British Journal for the History of Science*, 50: 2 (June 2017), 267–95.

Stanek, Ł. *Architecture in Global Socialism: Eastern Europe, West Africa and the Middle East in the Cold War*. Princeton, NJ: Princeton University Press, 2020.

Stanley, A. *Mothers and Daughters of Invention: Notes for a Revised History of Technology*. New Brunswick, NJ: Rutgers University Press, 1995.

Steel, D. 'Carl Sagan: Practitioner, popularizer and proponent of science.' *Contemporary Physics*, 42: 4 (2001), 247–49.

Subramanian, C. V. 'Edavaleth Kakkat Janaki Ammal.' *Resonance*, 12: 6 (2007), 4–9.

Sun, Q. 'Reluctant "Maocare": The shaping and practice of the rural healthcare system in the Wenzhou region, 1949–1978.' PhD diss., the University of Hong Kong, 2017.

Sundar, N. 'In the cause of anthropology: The life and work of Irawati Karve.' In P. Uberoi, N. Sundar and S. Deshpande (eds), *Anthropology in the East: Founders of Indian Sociology and Anthropology*, pp. 360–416. Chicago, IL: University of Chicago Press, 2008.

Sur, A. *Dispersed Radiance: Caste, Gender, and Modern Science in India*. New Delhi: Navayana Publishing, 2011.

Susen, S. 'Critical notes on Habermas's theory of the public sphere.' *Sociological Analysis*, 5: 1 (2011), 37–62.

Taylor, K. *Chinese Medicine in Early Communist China, 1945–63: A Medicine of Revolution*. New York: Routledge, 2004.

Thane, P. 'The careers of female graduates of Cambridge University, 1920s–1970s.' In D. Mitch, J. Brown and M. H. D. van Leeuwen (eds), *Origins of the Modern Career*, pp. 207–24. Aldershot: Ashgate, 2004.

Thaxton, R. Jr. *Catastrophe and Contention in Rural China: Mao's Great Leap Forward Famine and the Origins of Righteous Resistance in Da Fo Village*. New York: Cambridge University Press, 2008.

Theriot, N. M. 'Women's voices in nineteenth-century medical discourse: A step toward deconstructing science.' *Signs*, 19: 1 (1993), 1–31.

Tilley, H. 'Global histories, vernacular science, and African genealogies; or, is the history of science ready for the world?' *Isis* 101: 1 (2010), 110–19.

Tonn, J. 'Extralaboratory life: Gender politics and experimental biology at Radcliffe College, 1894–1910.' *Gender & History*, 29: 2 (2017), 329–58.
Tonn, J. 'Laboratory of domesticity: Gender, race, and science at the Bermuda Biological Station for Research, 1903–30.' *History of Science*, 57: 2 (2019), 231–59.
Turda, M. *Eugenism și modernitate. Națiune, rasă și biopolitică în Europa (1870–1950)* [*Eugenics and Modernity: Nation, Race and Biopolitics in Europe (1870–1950)*]. Iași: Polirom, 2014.
Vaiou, Nt. and A. Psarra. Εισαγωγικό σημείωμα επιμελητριών [Editors' introductory note]. In Nt. Vaiou and A. Psarra (eds), *Εννοιολογήσεις και πρακτικές του φεμινισμού: Μεταπολίτευση και «μετά»* [*Meanings and Practices of Feminism: The Era of the Democratic Transition and Onwards*]. Athens: Workshop proceedings, the Hellenic Parliament Foundation for Parliamentarism and Democracy, ix–1, 2018.
Varma, N., N. Sinha and P. Jha, eds. *Servants' Pasts: Sixteenth to Eighteenth Century, South Asia*. Hyderabad: Orient Blackswan, 2019.
Văcărescu, T. *Personajele acestea de a doua mână: Din publicațiile membrelor Școlii Sociologice de la București* [*These Second-Hand Characters: From the Publications of the Members of the Bucharest Sociological School*]. Bucharest: EIKON, 2018.
Vernon, K. "A truly taxonomic revolution? Numerical taxonomy 1957–1970." *Studies in the History and Philosophy of Biological and Biomedical Sciences*, 32 (2001), 315–41.
Vicedo, M. *Intelligent Love: The Story of Clara Park, Her Autistic Daughter Jessica, and the Myth of the Refrigerator Mother*. Boston, MA: Beacon, 2021.
Vicedo, M. 'The social nature of the mother's tie to her child: John Bowlby's theory of attachment in post-war America.' *British Journal of the History of Science*, 44 (2011), 401–26.
Vicedo, M. *The Nature and Nurture of Love: From Imprinting to Attachment in Cold War America*. Chicago, IL: University of Chicago Press, 2013.
Visvanathan, S. *Carnival for Science: Essays on Science, Technology and Development*. Oxford and New York: Oxford University Press, 1997.
Vishwanath, L. S. 'Female infanticide: The colonial experience.' *Economic and Political Weekly*, 39: 22 (2004), 2313–18.
Walsh, M. R. *'Doctors Wanted, No Women Need Apply': Sexual Barriers in the Medical Profession, 1835–1975*. New Haven, CT: Yale University Press, 1977.
Wang, D. *Street Culture in Chengdu: Public Space, Urban Commoners, and Local Politics, 1870–1930*. Stanford, CA: Stanford University Press, 2013.
Warner, M. *Publics and Counterpublics*. New York: Zone Books, 2005.
Watson, J. *The Double Helix: A Personal Account of the Discovery of the Structure of DNA*. New York: Atheneum, 1968.
Watts, R. *Women in Science: A Social and Cultural History*. London and New York: Routledge, 2007.
Weatherall, M. 'Making medicine scientific: Empiricism, rationality, and quackery in mid-Victorian Britain.' *Social History of Medicine*, 9: 2 (1996), 175–94.
Webner, P. 'Political motherhood and the feminization of citizenship: Women's activism and the transformation of the public sphere.' In N. Yuval-Davis and P. Webner (eds), *Women, Citizenship and Difference (Postcolonial Encounters)*, pp. 221–45. London: Zed Books.

Wemheuer, F. *Famine Politics in Maoist China and the Soviet Union.* New Haven, CT: Yale University Press, 2014.
White, P. 'Darwin's emotions: The scientific self and the sentiment of objectivity.' *Isis*, 100 (2009), 811–26.
Wilson, E. B. *The Cell in Development and Heredity.* New York: The Macmillan Company, 1925.
Wray, K. B. 'Scientific authorship in the age of collaborative research.' *Studies in History and Philosophy of Science Part A*, 37: 3 (2006), 505–14.
Wu, Y.-L. *Reproducing Women: Medicine, Metaphor, and Childbirth in Late Imperial China.* Berkeley: University of California Press, 2010.
Xu, S. 'Minguo shiji Shanghai nü xiyi yanjiu, 1919–1937' [Study on female doctors of Western medicine in Shanghai during the Republican Period, 1919–1937]. Master's dissertation, Jinan University, 2020.
Yang, B. and S. Cao. 'Cadres, grain, and sexual abuse in Wuwei County, Mao's China.' *Journal of Women's History*, 28: 2 (2016), 33–57.
Yang, D. *Calamity and Reform in China: State, Rural Society, and Institutional Change Since the Great Leap Famine.* Stanford, CA: Stanford University Press, 1996.
Yang, J. *Tombstone: The Great Chinese Famine, 1958–1962.* Translated by S. Mosher and J. Guo. New York: Farrar, Straus and Giroux, 2012.
Yip, K.-C. *Health and National Reconstruction in Nationalist China: The Development of Modern Health Services, 1928–1937.* Ann Arbor, MI: Association for Asian Studies, 1995.
Yong, E. 'The women who contributed to science but were buried in footnotes.' *The Atlantic*, 11 February 2019.
Yoon, J.-R. 'Korean women in science and technology.' *Asian Women*, 10 (2000), 33–42.
Zamfirache, O., ed. *Ea. Perspective feministe asupra societății românești* [*She: Feminist Perspectives on Romanian Society*]. Bucharest: Curtea Veche, 2019.
Zanish-Belcher, T. 'Documenting the sometimes invisible: Working with women scientists.' *Humanities Collections*, 1 (2001), 3–17.
Zanish-Belcher, T. 'The archives of women in science and engineering and future directions for oral history: Questions for women scientists.' *Centaurus*, 54 (2012), 292–98.
Zavarache, C. '"Numărul mare de copii … așteaptă o asistență medicală serioasă și la vreme făcută." Consecințele primei conflagrații mondiale asupra serviciului medical școlar din România.' ["The great number of children … is in need of serious and timely medical care." The impact of World War I on the medical care in schools in Romania]. In C. Mihalache and N. Roman (eds), *Copilării trecute prin război: Povești de viață, politici sociale și reprezentări culturale în România anilor 1913–1923* [*Childhood in War: Life Stories, Social Policies and Cultural Representations in Romania 1913–1923*], pp. 196–97. Iași: Editura Universității Alexandru Ioan Cuza, 2020.
Zengin, B. *Women Engineers in Turkey: Gender, Technology, Education and Professional Life.* Saarbrücken: Lambert Academic Publishing, 2010.
Zhao, L. 'Qingmo Minchu Zhongguo Nü Xiyi Yanjiu' [Study of Chinese female doctors of Western medicine in late Qing dynasty and early Republic of China]. Master's dissertation, Hunan Normal University, 2013.
Zhu, H., ed. *Zeng Jingguang*. Beijing: Zhongguo Zhong yiyao chubanshe, 2018.
Zhu, Z. *Zhu zi qing san wen ji* [*Collection of Zhu Ziqing's Prose*]. Nanjing: Nanjing chuban she, 2018.

Index

Adegbohungbe, Ebun 132
Adler, Saul 113
adultery 227
affirmative action 27, 329, 330
Ajakaiye, Deborah 132
alcohol block therapy 204, 206
All India Vaidya Sammelan 219
All India Women's Conference (AIWC) 237, 240
All-China Women's Federation (ACWF) 198, 199
American College, Istanbul 105
American Orthopsychiatric Association 271
American Rose Society 83
Amour, Anna 134
Anarchist Informational Bulletin 290
Andics, Erzsébet 325
Angelescu, Constantin 150, 152, 156
anti-nepotism rules xxx, 259, 315
Anti-Rightist Campaign 194–95, 197, 200, 208
Asperger, Hans 261
Association for Women in Science (USA) 304
Association of Child Care Workers (USA) 270
Association of Greek Women Scientists 289
Association of Hungarian Women Graduates of University and College 323
Association of Scientific Workers (A.Sc.W.) (UK) 50
Associazione Italiana Donne Ingegneri e Architetti (AIDIA) 133

authoritarian regimes 9, 27
autism 24, 25, 205, 258, 260–63, 265, 266, 268–69, 271–73
autobiography 49, 246
Ayrton, Hertha 46
Ayurveda 23–24, 194, 202, 217–27, 229

Baghdadi Jews 109, 112, 246
Bama (writer) 246
Basalla, George 13
Baum, Jiří 104
Baur, Erwin 79
behavioural conditioning 266
Beit Railway Trust Rhodesian Fellowship 14, 39
Bernal, John Desmond 48, 50
Bettelheim, Bruno 261–62, 266
biographical dictionaries 5, 6, 164
birth control 26, 282, 284–85, 288, 289, 291–93
Blanchard, Frieda 88, 93, 96
Bobula, Ida 323–24, 326–28
Bowlby, John 261–62
Brahmacharya (celibacy) 222, 224–25
Brahmo Samaj 241
Braniște, Valeriu 152
British Association for the Advancement of Science 50
British Communist Party 50
British Medical Journal 217
Brooks, Harriet 42
Broom 290
Bulletin of the Association of Greek Women Scientists 291
Bulletin of the Democratic Women's Movement 291

Cambridge Scientists' Anti-War Group 15, 48–50
Cambridge Socialist Society 15, 48
Carson, Rachel 63–64
caste 3, 12, 24, 28, 88–89, 94, 219–22, 225, 246, 249
Cavanagh, Aileen 134
Cavendish Laboratory, University of Cambridge 14, 39–42, 44–45, 47, 52
Cercle d'études des femmes ingénieurs de l'Association des Françaises diplômées des universités 133
Chadwick, James 39, 42
Chand, Rai Pooran 219–20
Charles University, Prague 103
Chen, Weixi 182
Chengdu College of Chinese Medicine 203
Chengdu Superior Medical Vocational School 172–73, 184
child science 264
China's war with Japan 172
Chinese Communist Party 193–94, 196, 198–99, 202, 208–09
Cholnoky, Viktor 323–24, 328
Clancy, Kathryn B. H. 307, 308
class 3, 12, 24, 28, 50, 63, 219–22, 224–25, 236–38, 240–41, 243, 246–50, 252, 258, 329
Cold War xxxi, 2, 6, 18–19, 125–29, 135–38, 250
colonial criticism of Indian systems of healing 218, 237
contraception 26, 285–92
Convention on Elimination of All Forms of Discrimination Against Women xxxi
Cori, Gerty xxx
counterpublics 282–86, 288–89, 291–93
Covid-19 xxi, 3, 305
Cranbrook Institute of Science 304
Crick, Francis 64
Cultural Revolution 196
Cupcea Popovici, Maria 158
Curie, Marie xxviii, 5, 325
Current Scene 192–93, 195–98, 201, 203, 206, 208
Current Science 86, 96–97

cytogenetics 11, 17, 80–81, 84, 87, 97, 118
Czaplicka, Maria 10
Czechoslovak Institute for Tropical Diseases Research, Baghdad 106

Daily Mail 51
dais (midwives) 221
Darlington, Cyril Dean 81, 87
Davies, Ann 44–45, 47
Deng, Xiaoli 178, 185
dengue fever 116–17
Devi, Yashoda 23–24, 217–19, 222–27, 229
Dhanvantari 221
Di-Lotti, Vlasta Kálalová 17, 18, 23, 103–14, 116–18
Discover Magazine 57
domestic science 21, 24, 174, 236–37, 239–42, 249, 252–53
domestic servants 24, 241
domesticity 4, 23, 25, 243, 252–53
Drbohlav, Jaroslav 112–14, 118

Eldredge, Niles 65–66
endosymbiosis 57–58, 61–62, 66
engineering xxi, xxvi, xxxi, 1–4, 10–11, 14, 19, 28, 48, 125–39, 303, 305, 321
equal opportunities xxiv, xxxi, 27, 131, 328
Erlanson, Eileen Whitehead 17, 23, 79–84, 86–94, 96–98, 118
eugenics 150, 224
European Union xxi, 325–29, 331
evolutionary biology 66, 68
experimental taxonomy 87

famines 1, 20–22, 192–94, 198–99, 200–4, 207–8
Famintsyn, Andrei Sergeyevich 60
Federation of Greek Women 289
female infanticide 220
female theorists 14, 16, 57, 59, 66, 68–70
Ford Foundation 250
Franklin, Rosalind 15, 63, 64, 91
Free Women's Movement 289
Fremlin, John 15, 23, 47, 49–50

friendship 16, 17, 79, 82, 86, 91
Frisch, Rose Epstein 201

Gaia theory 69
Gayton, Anna Hardwick 81
gender gap xxxi, 52, 98, 180
gender segregation 151, 160, 183
gender studies 9
geography 4, 13, 28, 324
Ghelbruch, Gudlea 158
Gifford, Retta 176, 183
Goeppert-Mayer, Maria xxviii, xxxi
Gohara, Sawako xxvi
Gomez, Eunice 88
Gopalan, Shyamala 24, 242–43, 253
Gordon, Isabella 94
Götz, Irén 325
Gould, Stephen J. 65
grassroots science 22, 192–93, 195, 196, 198, 203, 208
Great Depression 17, 88
Great Leap Forward 1, 193–99, 200–3, 206–8
Greek junta (Junta of the Colonels) 26, 288
Greek Orthodox Church 290
Guo, Shiduan 185
gynaecology and obstetrics 26, 104, 108, 147, 172–73, 182–85, 203–6, 282

Habsburg monarchy 105
Haldane, J. B. S. 50, 87
Hardwich, Isabel 131
Hargittai, Magdolna 6
Harris, Kamala 242–43, 254
Helsinki Group on Gender in Research and Innovation (now the Standing Working Group on Gender in Research and Innovation, SWG GRI) 325, 328
Herschel, Caroline 46
Hodgkin, Dorothy Crowfoot xxviii, xxxi
home economics 24, 237–39, 250, 253
home science 24, 236–43, 246, 248–49, 250–54
Home Science Association of India 250
Hopper, Grace xxxi

Horizon 2020 325, 326
Horthy era 321
Hugonnai, Vilma 325–26, 328, 331
Human Genome Project 70
Hungarian Academy of Sciences 322, 325, 327–28
Hungarian Women's Review 321, 323
Hunter, Carla Desi-Ann 308

Illinois Distributed Museum 26, 302–3
Imperial Council of Agricultural Research 85
imperialism 6, 8, 10, 12, 118, 195
infertility 227
Institute of Hygiene and Public Health, Bucharest 153
International Conference of Women Engineers and Scientists (ICWES) xxx, xxxi, 3, 18–19, 125–39
International Council of Women 127
International Federation of Business and Professional Women 127
International Federation of University Women (IFUW) 16, 127
International Federation of University Women in the Balkans 11
International Network of Women Engineers and Scientists 127
International Women's Congress 117
intersectionality 12, 20, 28, 40, 147, 308
intimate knowledge 265–66, 273
Iron Curtain 126–27, 129, 136
Irwin, Dorothy 240
Iyengar, M. O. Parthasarathy 96
Iyer, L. K. Ananthakrishna 94

Janaki Ammal, Edavaleth Kakkat 11, 17, 23, 80, 84–89, 90–92, 96–98
Janet, Mercia 88
Japan Inter-Society Liaison Association Committee for Promoting Equal Participation of Men and Women in Science and Engineering (EPMEWSE) xxvii
Japan Medical Women's Association xxx
Jiang, Liangying 183, 184

John Innes Horticultural Institute 81–82, 84, 87–88
John, Sosa P. 88
Joint Commission on Childhood Mental Illness 268
Journal of Autism and Child Schizophrenia 270
Journal of Czech Physicians 18, 113, 117
Journal of Social Hygiene 154
Journal of the Indian Botanical Society 97
Journal of Theoretical Biology 60

Kanner, Leo 261
Kanya Sarvasva 222
Katina 290
Kausalya, C. K. 91
Keller, Evelyn Fox 67
King Faisal I of Iraq 112, 116
Koelz, Walter Norman 90
Koyama, Shizuko 238
Kozo-Polyansky, Boris 60
Kuroda, Reiko 6

Lady Irwin College, New Delhi 24, 236–37, 240–50, 253–54
Lady Science 304
Lamson, Eleanor 63
leishmaniasis 17, 18, 104, 112–14
Levi-Montalcini, Rita 130
Li, Shuyuan 184
Linnean Society 85
Liu, Yunbo 172–74, 178, 181, 184
Lonsdale, Kathleen 47
Lost Women of Science 304
Lushan Conference 200

Maasdorp, Reinet 14–15, 23, 39, 40–49, 50–52
Madách, Imre 322–24
Madak-Erdogan, Zeynep 307, 309–10
Madras University 84–85, 88, 93, 96
Maeda, Sonoko xxx
Magyar Női Szemle 321
Maharaja's College for Women, Trivandrum 87
Maharaja's College of Science, Trivandrum 84–85, 87–88

Manchester Society for Women's Suffrage 42
Manhattan Project 48
Mao, Zedong 22, 193–97, 199, 200, 203, 208
Margulis, Lynn 1, 15–16, 23, 57, 58–59, 60–63, 65–70
Martinis, Susan 308–9
Masaryk, Alice 106–7
Masaryk, Tomáš Garrigue 17–18, 104, 106, 107, 109, 110, 111–12, 114, 116–17
masculinity 238
Mason, Eleanor Dewey 86, 96–97
mass science 193, 208
Matthew Matilda effect xxviii, 7
May Fourth Movement 179
Mayo, Katherine 241
McClintock, Barbara 91
McGaw, Helen 45
medical education 103, 176, 179–81, 183
Medical Technical School, Sichuan 182, 190
Medical Women's International Association (MWIA) xxx
medicalization of starvation 200, 208
Mehta, M. M. 91
Meitner, Lise xxviii, 79
Men's League for Women's Suffrage 331
Manchester Branch 42
mentorship 16, 84, 88, 90, 223, 246, 307, 309, 312, 329
Mereschkowski, Konstantin 60
Michigan Academy of Sciences 96
Michigan Association for Emotionally Disturbed Children 271
The Michigan Daily 81, 85, 87
midwifery 173, 176, 179, 183–85, 196, 221
Mind: A Quarterly Review of Psychology and Philosophy 263
Miyagawa, Kanaeko xxvi
Modern Woman 291
motherhood 23, 257, 259, 262, 290, 291
Movement of the Liberation of Women 289
multilingual archive 4

Murray Washington, Mary 239
museums 2, 26, 87, 89, 90, 96, 301, 302–4, 315, 326
Musil, Alois 104–7
Muthunayagom, Daisy 88
Muwanga, Miriam 132

Naba Bibhakar Sadharani 237
Naidu, Sarojini 248
NASA 5, 12, 129
National Association of Hungarian University and College Students 323
National Council of Polish Women 133
National Institute of Health, Prague 18, 104, 114
National Midwifery School, Peking 173
National Science Foundation 130
National Society for Autistic Children (NSAC) 268–69, 271
Nationalist Government of the Republic of China 172
Nature 49, 50, 67
Nazi Germany 2, 49, 132
Nehruvian science 241
neo-Darwinism 57, 68
Nobel Prize xxiii, xxvii, xxviii, xxx, xxxi, 41, 81, 182
Normanton, Helena xxiv
Nova 65
numerus clausus 325
nursing 179
Nylon Curtain 136

Obeng, Leticia 132
Ōhashi, Hiro xxvi, 23
Old Kingdom of Romania 149, 153
Oliphant, Mark 45, 47, 49
oral history 4, 236, 239, 243, 246
Oriental Institute, Prague 105, 107
Ormos, Maria 327
Ouwerkerk, Louise 88

Panhellenic Socialist Movement 288
Park, Clara 24, 25, 205, 258–63, 265–73
Park, Jessica 25, 258–67, 269–71
Partition of India 243
Paulová, Milada 103
peace movements 49, 52, 125
Peking Union Medical College 173, 177, 184–85

Peng, Dehuai 200
Penson, Lillian M. xxiv
Petřivalský, Julius 105, 106
PhD training 326
physical anthropology 87–88, 92, 97–98
Polányi, Laura 325
Polish Women's Council 136
population genetics 7, 68, 69
Presidency College, Madras 85, 90
Project Phaedra, Harvard University 48
psychoanalysis 25, 260–62, 266, 270
Pu, Fuzhou 199
public memory 21, 41
public spheres 26, 64, 179, 219, 281–86, 291–93
purdah 219, 228

Qing dynasty 174, 176

racism 41, 307, 308
Radcliffe College, Cambridge, Massachusetts 258
Ralef, Xenia 157
Ranchi Women's College 245
Ray, Satyajit 252
Red Cross 106, 117, 178
Refling Hagen, Ingeborg 117
refrigerator mothers 25, 261, 266
reproduction 159, 164, 194, 199, 202, 206, 286, 291
Rimland, Bernard 25, 262
Rischowski, Ira 19, 127, 132
Rockefeller Foundation 177, 181
Rossiter, Margaret xxiv, xxvii, xxviii, 7, 59, 80
Royal Society 15, 39, 41, 46–48, 52
Rutherford, Ernest 5, 14, 42–44, 46

Sagan, Carl 23, 59, 65–66
Savinescu, Olga 163
Schafer, Ruth 125, 129, 131, 134, 136
Schiemann, Elisabeth 79
Schopler, Eric 270
Science 83
science and archives 2, 3, 4, 6, 12–13, 21, 22, 25–26, 28, 147, 149, 161, 179, 185, 301–2, 304, 314–15
science and audiences 25,–27, 64, 150, 264, 281–82

science and emotions 91, 258, 264, 272
science and mass media 26, 285
science and mobility xxvii, 11, 14, 16, 22, 98, 172
science and objectivity 12, 25, 59, 258, 263, 265, 273
science and political activism 4, 15, 48, 52, 283
science and statistics 1, 4, 9, 28, 52, 111, 137, 160, 195, 309, 313, 329
science communication 4, 25–26, 281–82, 284, 292–93
scientific careers and marriage xxvi, 17, 23, 43, 51, 80, 91–92, 98, 238, 240,–42, 245, 248, 254, 259
scientific collecting 81, 84–85, 87, 89, 90, 94, 96, 98, 113, 264, 314, 315
scientific couples xxvii, xxx, 23, 79, 257
scientific rebels 14–16, 58–59, 62, 65–66, 68–70
second-wave feminism 65, 130
secondary schools 1, 20, 148–49, 150–56, 158–61
Sen, Hannah 240, 247
seroanthropology 92, 98
Sex Disqualification (Removal) Act, 1919 (UK) xxiv
sexism 16, 41, 47, 62, 64, 307
sexuality 3, 12, 28, 159, 220, 241, 284, 286, 289–91
Shanghai Dongnan Medical College 185
Shastri, Chatursen 222
Šikl, Heřman 113
Silliman, Flower 24, 243, 246–49, 254, 334
Simpson, Evelyn xxiv
Smedley, Ida xxiv
Smith, Rebecca Lee 307
smoke fumigation (medical treatment) 205, 206
social hygiene 149, 154
socialism 13, 49, 50, 125, 128, 131, 136, 138, 195, 199, 241, 325–26, 331
Society of Czech Physicians 117

Society of Japanese Women Scientists (Nihon josei kagakusha no kai, SJWS) xxxi, 133
Society of Women Engineers (SWE) 126–29, 130–34
Soviet Women's Committee 136–37
space race 18, 126, 129, 134, 138
Spanish Civil War 49
Sparshott, Marie 45
Spitz, René 261–62
Sportime 282
Sputnik (satellite) 128
Stambul Seriyati 105
Stănescu, Virginia 159
Stanier, Roger Y. 62, 65
State Institute of Health, Prague 18, 112–13
state-led economic development 194
Ștefănescu, Eliza 162–63
STEMM xxi, xxiii, xxx, xxxi, 1, 4, 5, 7–9, 11, 14, 19, 28, 41, 47, 52, 127–28, 130–31, 302, 306, 312–14, 321, 326
Stephenson, Marjorie 47
Stri Chikitsak 217, 222
Stridharma Shikshak 222
Sullivan, Ruth 268–69
Superior School of Midwifery, Nanchang 184
Superior School of Midwifery, Tianjin 184

Tange, Ume xxvi
Tauer, Felix 104
Tauerová, Marie 104
Techert, Margit 321
technicians 63, 179
technoscience and peace 126–27, 136–38
Tereshkova, Valentina 128
The Times 42–43
Thomson, J. J. 42
Thurlow, Sylva xxiv
Tian, Yingzhao 184
Tissot, Samuel-Auguste 225
Tobolářová, Ruth 109–10, 113
Tohoku Imperial University xxvi
Tokyo Women's Medical School 183
Traditional Chinese Medicine (TCM) 22, 193, 195, 197, 203–6, 208

Trancu Rainer, Marta 147
transnational science 17, 19, 22, 79, 125–27, 133, 139, 181
Tsuda, Umeko 10

Unani medicine 218, 219
Ungureanu, Alexandrina 155
Union of Greek Women 289
University of Illinois Archives 26, 302–5, 314–15
Urata, Tada xxvi, 23
uterine prolapse 22, 192–95, 198–99, 200–8

Varshaneya, Govind Prasad 221
Vasiliu, Victoria 163–64
venereal diseases 149, 152
Victoria University of Manchester 42

Watson, James D. 58, 62, 64, 70
West China Union University (Huaxi xiehe daxue, WCUU) 22, 176, 177, 179, 183
Western medicine 21, 172–75, 177, 179, 182, 184–85, 192, 195, 203, 218–19, 221
White Jews of Cochin (Paradesi Jews) 92
Whiteley, Martha Annie xxiv
Woman's International League for Peace and Freedom 127
Women Chemical Engineers of the Philippines 133
Women in Science Lecture Series, University of Illinois 26, 302–4, 313, 316

Women's Christian College, Madras 86, 91
Women's Engineering Society (WES) 127, 132–33, 138
Women's Group of Self-Examination 289
women's history 9, 48
women's illnesses (*funü bing*) 20, 22, 147, 192–95, 198, 200–4, 206–8
Women's Liberation Movement 130
Women's Medical College of, Pennsylvania xxiv
Women's Missionary Society 22, 176
World Conference of Women, Mexico 135
world wars xxiii, xxvi, xxx, xxxi, 18, 20, 47–48, 51, 64, 117, 125, 130, 147–49, 150, 159, 164, 175, 247, 250, 259, 264, 324
World Women Congress, Berlin 135
Wu, Chien-Shiung 6
Wuhan Medical College 192, 197

Yang, Chongrui 173
Yasui, Kono xxvi
Yuasa, Toshiko xxvi
Yue, Yicheng 179

Zeng, Jingguang 203–6, 208
Zhang, Fengtong 184
Zhang, Jingfen 184
Zhejiang School of Midwifery 184
Zheng, Qiyin 183–84
Zhou, Jixian 184
Zhu, Ziqing 172, 174

EU authorised representative for GPSR:
Easy Access System Europe, Mustamäe tee 50,
10621 Tallinn, Estonia
gpsr.requests@easproject.com

www.ingramcontent.com/pod-product-compliance
Ingram Content Group UK Ltd.
Pitfield, Milton Keynes, MK11 3LW, UK
UKHW021826140426
5217IPUK00004B/115